The rise of early modern science

This is a study of the long-standing question of why modern science arose
only in the West and not in the civilizations of Islam and China, despite the
fact that medieval Islam and China were more scientifically advanced. To
find an explanation, the author examines the differences in religious, philosophical, and legal institutions of the three civilizations, focusing on the
legal concept of *corporation*, which is unique to the West and gave rise to
neutral space and free inquiry, concepts integral to modern science.

D0886030

The rise of
early modern science

Islam, China, and the West

Toby E. Huff

University of Massachusetts Dartmouth

CAMBRIDGE
UNIVERSITY PRESS

Published by the Press Syndicate of the University of Cambridge
The Pitt Building, Trumpington Street, Cambridge CB2 1RP
40 West 20th Street, New York, NY 10011-4211, USA
10 Stamford Road, Oakleigh, Melbourne 3166, Australia

© Cambridge University Press 1993

First published 1993
First paperback edition 1995

Printed in the United States of America

Library of Congress Cataloging-in-Publication Data is available.

A catalogue record for this book is available from the British Library.

ISBN 0-521-43496-3 hardback
ISBN 0-521-49833-3 paperback

For Judylein, Erik, and Niki,
three of life's companions who have
made it brighter

Contents

Illustrations

Preface

This book is about the rise of modern science and how the world got to be the way it is. The twentieth century has witnessed extraordinary collisions of societies, cultures, and civilizations. As a by-product of the newly intensified global economy, the last quarter of this century has experienced unprecedented fusions of cultures. What has not been sufficiently recognized, however, is the degree to which the cultural and legal forms forged in the twelfth and thirteenth centuries in the West laid the foundations for the present world order. Among these early modern cultural forms are those that created forums of free and open discourse that have led to universal forms of participation – in the world of thought, in government, and in commerce. Modern science is one striking example of a universalizing form of social discourse and participation. The continuing globalization of the practice of modern science represents a prime test of the proposition that universal forms of dialogue and participation exist and that they appeal to peoples of diverse cultures of origin. The possible shift of the center of modern science from the West to the East further dramatizes the universality of this mode of dialogue.

Nevertheless, alongside these universalizing forms of discourse and participation are equally strong forces asserting the priority of ethnic and local particularities. There are also those who fear more sinister uses of the fruits of scientific understanding. Likewise, the battle over the ascendancies of the various forms of reason and rationality will continue unabated. The present moment is filled with anticipation and apprehension as to whether the forces of equality and inclusiveness will prevail, or whether the forces of ethnic exclusivity and indigenous identities will further divide the communities of the world.

Acknowledgments

This book has been a long time in the making. Consequently I owe a debt of gratitude to many individuals and organizations. The National Endowment for the Humanities granted me a year of study at the University of California, Berkeley in 1976–7 (Grant F76–240), where I attended a seminar, "Tradition and Interpretation," directed by Robert Bellah. That fellowship gave me the first opportunity to write down my thoughts on the problem of Arabic science.

In 1978–9 the Institute for Advanced Study in Princeton, New Jersey, sponsored a year of study during which I was supposed to work on the present study. Instead, the year was devoted to Benjamin Nelson's book, *On the Roads to Modernity*, due to his sudden death. That period in Princeton, however, was invaluable in many ways for the present work.

During the fall of 1980 I was granted a sabbatical leave by my own university, and I spent it as a visiting scholar in the history of science department at Harvard University. It was during that fall semester that I first presented the outline of the thesis of this book to the History of Science Seminar at Harvard. I am very grateful to Professor A. I. Sabra of Harvard for his support of my project and for his many comments over the year. I twice partially audited his course on the history of Arabic science and gained many invaluable insights from his discussions. It should be understood, however, that Professor Sabra and I hold different points of view.

Another sabbatical leave from my university in the fall of 1987 allowed me the leisure to explore a variety of questions in the comparative history of law, and without that opportunity, the thesis of this book would be weaker and differently stated. I am most grateful for that leave.

I trust that it will be evident to my readers that this study could not have been carried out without access to a formidable array of library resources and that I have benefited from libraries from Maine to California. The computer-based OCLC (Ohio College Library Consortium) system through the library at the University of Massachusetts Dartmouth gave me access to many materials that I otherwise would not have been able to consult. The Consortium deserves a special note of thanks and recognition. I also owe a special debt to the new Thomas P. O'Neill, Jr., Library at Boston College, where large portions of Chapters 4 through 8 of this book were written. The O'Neill Library's exceedingly pleasant surroundings, highly efficient information retrieval system, and well-arranged open stacks made progress on this book in its advanced stages much easier and more rapid than would otherwise have been possible. That is a benefit I gratefully acknowledge. Most of the dates of historical figures referred to in this study have been standardized according to *Webster's New Biographical Dictionary;* otherwise, I followed the *Dictionary of Scientific Biography.*

Finally, I must acknowledge that this study would not have been undertaken at all but for the example and encouragement of my New School mentor, Dr. Benjamin Nelson. Although he died shortly after reading what was a mere sketch of the present study, which I had written at Berkeley in 1977, by the early seventies he had already published the essays I needed to guide this study. I can only hope that this book evokes the spirit of his wide-ranging knowledge, the generosity of his person, and the prescience of his many insights.

Introduction

For the past five hundred years in the West, the pursuit of science has been more or less unfettered. If, in the light of more recent assessments of the freedom of thought and inquiry that existed in the universities of the twelfth and thirteenth centuries, we add another three hundred years, then we may say that the pursuit of science in the West has been carried on undiminished for nearly nine hundred years. This flight of the imagination, if you will, was both sponsored by and motivated by the idea that the natural world is a rational and ordered universe and that man is a rational creature who is able to understand and accurately describe that universe. Whether or not men and women can solve the riddles of existence, so this view goes, they are able to advance human understanding mightily by applying reason and the instruments of rationality to the world we inhabit.

The breakthrough that allowed freedom of scientific inquiry is undoubtedly one of the most powerful intellectual (and social) revolutions in the history of humankind. As the paradigmatic form of free inquiry, science has been given a roving commission to set all the domains of thought aright. Science is thus the natural enemy of all vested interests – social, political, and religious, including those of the scientific establishment itself. For the scientific mind refuses to let things stand as they are. The organized skepticism of the scientific ethos is ever present and always doubtful of the latest (and even the long-standing) intellectual consensus.

Given this intellectual commission to investigate all forms and manner of existence, science is especially the natural enemy of authoritarian regimes. Indeed, such regimes can exist only if they repress or otherwise subvert those forms of scientific inquiry that reveal the true nature of the social – economic, political, and medical – consequences of their rule. We must be careful here not to confuse journalism and the press with science. The free

1

press is unquestionably an indispensable institution for maintaining democracy, but we should not confuse press reports, or even investigative journalism, with science. Moreover, it is clear that journalists look to science and scientists for guidance in their inquiries. In the final analysis, good investigative journalism must meet the tests of scientific inquiry, which in the social realm demands adequacy of sampling, appropriate instruments of data gathering, and sound techniques of inference and analysis as understood by the prevailing standards of social science. In the natural realm, likewise, the so-called canons of science must be observed and the function of the press generally is to explain the findings of science to laymen, not to undertake scientific inquiry per se. Only in rare and dramatic cases do journalists undertake to determine whether a particular set of research findings was produced according to acceptable scientific standards. And in such cases, we should note, the task is the affirmation of the canons of science by pointing out their possible breach.

Can we now say that men and women in all civilizations have equally shared (or do share) the view that science is and ought to be free to state its views on all matters of inquiry? Can we say that the other civilizations of the world equally held a fully rationalist conception of the orderliness of the cosmos and equally valued the rational capacities of man to the extent that they institutionalized the means by which men could fully apply their reason in the interests of advancing the most consistent and theoretically powerful explanatory systems? The fact that modern science arose only in the West – despite the fact that Arabic-Islamic science was more advanced up until the twelfth and thirteenth centuries – suggests a negative answer to those questions.

Should the world be thought to be fully rational, understandable, and explainable by mere mortals? If we answer in the affirmative, should we continue to support and extend the neutral zones of free inquiry even further so that researchers may continue to develop their scientific systems of thought in all domains, which will raise ethical issues and quite possibly bring some harm (through misuse as well as continuing scientific ignorance) but a great deal of human benefit? And if we go that far, how shall we design such institutions for the future? What are the sociological foundations – the philosophical, metaphysical, and institutional assumptions – that enable us to carry on this enterprise of freely pursuing inquiry wherever it leads? Can they be put in place in all civilizations without seriously unbalancing those societies and civilizations, and disturbing the vested interests? Or is modern science just a Western "disease"?

As we enter the global world as never before, these are surely questions of fundamental importance. To create a truly global order there must be a set of fundamental shared principles – legal, philosophic, humanitarian – which enable us to communicate freely and resolve conflicts peacefully. Perhaps the conditions that allowed the development of modern science can tell us something about how societies (and civilizations) can and should be ordered so that men and women are fully enabled to participate in the construction and design of their social orders. We have much to learn about such questions, and the study of the sociological foundations of modern science may have much to tell us about the ingredients that go into the making of the "open society," freedom of expression, and the peaceful resolution of conflict.

To understand the evolution of modern science, we must consider several different levels of social and cultural process. These may be conceptualized as in Figure 1. The relationships between the domains shown in this diagram are far more complex and interactive than I have indicated, with various interactions and feedback loops between domains.

I have given law priority of place in this scheme because law (sacred law) has been the penultimate directive structure of classical Arabic-Islamic civilization. Law has been equally important in the West, but the coloring of the legal domain there has been so different (and progressive) that law seems less important than in Arabic-Islamic civilizations. Ultimately, as we shall see, revolutions in the structure of Western law have been of overwhelming importance in shaping Western social, political, and intellectual experience. In the case of China, legal concepts and codes, so it is said, have played a much smaller role. Nevertheless, it is imperative to understand the role of law in China as well.

In the present context, I have been particularly struck by the manner in which legal thought has structured conceptions of rational deliberation as well as action. Legal thought in both Islam and the West has established canons of rational inquiry and placed limits on the forms of legitimate inquiry. Moreover, legal systems create the operative canons of rationality for settling disputes that arise within their domain. Viewed from another perspective, legal systems institutionalize a whole range of social and cultural forms by mandating the forms of human relationships and the means for dispute resolution. Accordingly, the study of legal systems becomes a useful window through which we may grasp the underlying structural properties of a society and civilization. Both the forms of rationality and the institutional apparatus created by the legal system are of paramount interest.

Because scientists take it as their mission to explore and define the elementary structures and processes of nature, it is equally important to consider the images of order, chaos, and process that appear in systems of religious and theological thought. It is obvious to the modern reader that the tradi-

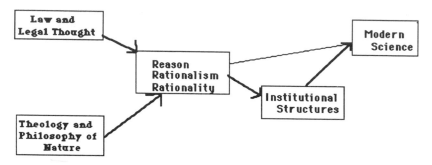

Figure 1. The domains of social process.

tional images of man, nature, and the universe ensconced in religious thought have been profoundly shaken by the rise of modern science. It has not been widely appreciated, on the other hand, that some religious and theological systems have contained images of order, regularity, and even system-processes that have been conducive to the development of science. Theological systems have shaped conceptions of reason and rationality as attributes of man and nature. These metaphysical presuppositions have been particularly fertile for encouraging scientific thought. The absence of theology in a strict sense in China is a matter of some importance, both for the history of thought in China and in terms of contemporary understandings of that civilization.

As we shall see, indigenous systems of legal thought, as well as religious and theological ones, function to create images of man's rational capacities as well as images of the rationality of nature. From the point of view of the evolution of modern science and intellectual life, these early intellectual systems have been of paramount significance. It is true also, of course, that one must recognize the independent influence of philosophical systems of thought, which, due to the Greek tradition in the West, have been of very great significance for the development of modern science. In short, a study of the sociological foundations of modern science takes one into the metaphysical and philosophical foundations of science, and this takes one further into the anthropology of man and the philosophy of nature.

Finally, a major factor in this developmental perspective that must be scrutinized is the nature of the institutional structures which are the repositories and intellectual laboratories within which conceptions of reason and rationality are set to work. Sociologists have not attended enough to the deep structures of social institutions and the ways in which they are products of the legal concepts within which the institutions are situated. Once social institutions are embodied in a normative order, and particularly under the aegis of the legal formalisms of the day, a new level of social and cultural process is put into effect. Such institutions may function in a conservative fashion, giving the society and culture a perdurable cast; or the social institutions may embody progressive and even revolutionary thematics and thus, over the course of time, may function to powerfully reshape and transform the social, political, and economic orders. Such is what transpired in the West and we are well advised to understand the dynamic nature of those institutional arrangements. The study of the rise of modern science from this point of view is the study of institution building. From the perspective of late modernity it may be the paradigmatic story of the creation of modern social institutions.

In Chapter 1 I attempt to locate the present study in the literature of the comparative and historical sociology of science. Aside from a monumental study by Joseph Needham and the seminal responses to that work by the late Benjamin Nelson, little has been done to establish a framework within which comparative studies can be fruitfully undertaken between the scientific enterprises of the East and the West. Joseph Ben-David's inquiry into the *role* of the scientist provides a useful point of departure, but his complete omission of any discussion of either Arabic or Chinese science necessarily creates a tunnel vision that prevents one's grasping the import of the religious, legal, and philosophical contexts within which science must always be carried on. In Chapter 2 I set out the problem of Arabic science, and in Chapter 3 I analyze the differing philosophies of man and nature found in Arabic-Islamic civilization and the West. In Chapters 4 and 5 I attempt to elucidate the philosophical and legal foundations of institution building in the two civilizations. In Chapter 6 I summarize the major elements of the great social and intellectual transformation of the West in the twelfth and thirteenth centuries as they bear on questions of the *ethos of science*. I have done so with a view to spelling out the cultural and institutional impediments that prevented the emergence of modern science and its ethos in the Arabic-Islamic world. In Chapters 7 and 8 I extend the framework of analysis to the case of China.

The great intellectual struggles that went into the fashioning of the institutional foundations of modern science are the very ones that have shaped the structures of modernity more generally. Accordingly, readers familiar with the writings of Max Weber will be aware that the problem of the rise of modern science is exactly parallel to the problem of the rise of capitalism in the West – and only there. Throughout this undertaking I have been aware of this parallelism and Weber's suggestion that exploring this question was "the next step."[1] I agree with this suggestion and with the judgment of Benjamin Nelson and Joseph Needham that this problem of the unique rise of modern science in the West is more fundamental for sociological inquiry than the rise of capitalism. I record these thoughts here only to alert the reader to the fact that this has been uppermost in my thinking about the problem at hand, but that I deliberately chose not to burden the reader with constant references to Weber's far-ranging thoughts on all these questions, for that would entail another volume altogether.

I must add one last comment which I hope will prevent the mistaken interpretation that I have overly identified reason and rationality with modern science. As I have tried to make plain, the European medievals had a great faith in reason that gave birth to a variety of new forms of rational discourse. Before the rise of modern science, there was the science of faith (theology) and the science of law (jurisprudence). As any reader knows who has read Max Weber's introduction to his *Collected Essays on the Sociology of Religion*, which was inserted as the author's introduction to the English translation of *The Protestant Ethic and the Spirit of Capitalism*, the West is distinguished from the Middle East and Asia not just in the successful birthing of modern science, but in its rationalizing pursuit of all forms of thought and action, art and music included. Indeed, Weber's 1911 essay, *The Rational and Social Foundations of Music*, is a brilliant reminder of the fact that in the West music too was subjected to all the rationalizing impulses of the Western spirit, with the result that contrapuntal polyphonic music, the full development of equal temperament in music, and the symphonic orchestra are all unique Western creations. It is to this work of Weber's that one should look, therefore, in attempting to formulate the meaning of the idea of *rationality* and the processes of rationalization. Beyond that we must remember that modern science is but one domain in which we should look for the embodiment of rationality. When Joseph Needham suggests that the rise of modern science is some kind of package deal, he ought to insist, therefore, not on the linkage between science and capitalism, but on the linkage between the rise of a great faith in reason and the application of this real or imaginary agency to the study of the natural world and to all the other domains of cultural existence.

As we envision the world of the future and ask whether or not there will be a pacific renaissance, the central issues are the same: will the developing countries of the world allow their citizens full participation in the realms of the mind – scientific, political, and literary – or will they continue to erect

[1] Max Weber, *The Protestant Ethic and the Spirit of Capitalism* (New York: Scribners, 1958), p. 182.

barriers to freedom of thought, expression, and action in the interests of primordial religious and ethnic identities? One can only say that significant portions of Asia, especially Southeast Asia, appear poised to go forward. At the same time, there are many forces arrayed against such openings. The struggle to achieve this global awakening is being waged, but it is somewhat premature to speak of the postmodern era: the conditions of modernity have yet to be achieved among the greater part of the peoples of the world.

1

◁══════════════════════════════════════▷

The comparative study of science

The modernity of science

In the present world, science and its offshoots appear to be the epitome of modernity. The scientific method of treating every conceivable natural, human, or social malady is everywhere in evidence. If the scientific approach has not been applied to the problem at hand, the treatment and analysis are thought to be either defective or suspect. This state of affairs is not bereft of moral critics who think that science itself has too much power or that the strictly scientific point of view, especially in medicine, claims too much, is overly confident, arrogant, and even capable of reaching false diagnoses. In the Western world there are those who think that science itself is a "social problem."[1] To them the technological products of science – excessive levels of radiation released into the atmosphere of local communities, the inadequately monitored use of pesticides, the general degradation of the natural environment caused by the dumping of toxic substances, and even global warming – are all to be laid at the door of modern science and technology. Nevertheless, alternative forms of knowledge – those derived from religion, mysticism, or occult sciences such as astrology – must offer their own defenses against the prevailing scientific posture. If they are to be accredited, these alternatives must be shown to produce their results and achieve their effects in ways that are consistent with either scientific ignorance ("about this we have no knowledge") or scientific wisdom ("this outcome is perfectly conceivable within expanded parameters of our present scientific knowledge").[2]

moi ?!

[1] Sal Restivo, "Modern Science as a Social Problem," _Social Problems_ 35 (1988): 206–25.
[2] For discussions of the growing dominance of the scientific worldview, see the essays in _The Knowledge Society: Its Growing Impact on Scientific Knowledge_, ed. Gernot Böhme et al. (Dordrecht: Reidel, 1986).

8

Indeed, it is common to refer to the privileged status of scientific knowledge in the modern world. This phrase means several things. First, it implies that the knowledge claims of scientific experts are given pride of place in public discussion and, above all, in matters of health, public and private. Second, expert witnesses, who are reputed to be scientific experts, are permitted to testify in courts of law regarding arcane and abstruse topics that laymen are hard pressed to understand. In such circumstances these experts are permitted to use their scientific knowledge to establish possible facts as well as the probable causes of events. Readers of mystery novels know of forensic experts who, through laboratory techniques, match fragmentary samples of fibers or hair to clothing and possible suspects and thereby link individuals to the scene of a crime. Such scientific knowledge is not based on firsthand observation of the events but is after-the-fact knowledge gleaned through the techniques of scientific analysis and inference. In short, the very idea of scientific knowledge has dramatically altered what is considered legitimate evidence and testimony in courts of law.[3]

Third, we may speak of scientific research as privileged in the sense that the legitimating authority of science grants permission to researchers to observe and even publicly describe those areas of life that are generally hidden from public view out of a sense of privacy or that are ruled off-limits by moral or religious scruples. For example, physicians are permitted to physically examine disrobed bodies in the most intimate of fashions. This is done in the name of science.

Similarly, social scientists as well as press reporters are often permitted to gain inside information on all aspects of public and private life. For sociologists, political scientists, and social anthropologists, the justification of such unrestricted observation is based on the desire to advance social scientific knowledge about how the social and political worlds work. Thus, a sociologist studying the police of a local community will attempt to observe every aspect of the daily routine of police officers. He will not only observe the apprehension and arrest of suspects but also listen in on phone conversations (as a surrogate detective) received by the vice squad from potential informants at police headquarters. This also is done "in the name of science."[4]

But there is another level of privilege that should be noted. This is the privilege granted to researchers to gather evidence and information unhindered and to possess this information free from seizure by political authorities. This is the limited protection granted social scientists and journalists to

---

[3] See, for example, Hans Zeisel, "Statistics as Legal Evidence," *International Encyclopedia of Statistics*, vol. 2 (New York: Macmillan, 1978), pp. 1118–22.

[4] See Jerome Skolnick's defense of all these practices in *Justice Without Trial*, 2d ed. (New York: Wiley, 1976), chap. 2.

freely gather and dispose of information as they see fit without any obligation to publish it or reveal its sources to political authorities. Likewise, they are free to criticize the social order or segments of it based on their inquiries. But the privilege I would like to emphasize is that which is granted by the courts of law and which establishes the inviolability of the researcher's right to withhold his knowledge from public scrutiny.[5] These, then, are some of the ways in which we may speak of the privileged status of scientific knowledge and inquiry in the Western world. They help to give substance to the view that the institutions of science are among the most important in modern society.[6]

While we in the West take the scientific point of view as the standard by which all others are to be judged, however, it often escapes our attention that the scientific point of view (which I shall deliberately leave undefined for the present) had to fight its way to success through many long battles. Beyond that, modern science, as we know it, failed to materialize in the other civilizations of the world (in India, China, and Islam), despite the fact that some of them had great cultural and scientific advantages over the West up until the thirteenth and fourteenth centuries. That realization ought to encourage us to consider the possibility that the arrival of modern science at its destination in the West was in fact the outcome of a unique combination of cultural and institutional factors that are, in essence, nonscientific. In other words, the riddle of the success of modern science in the West – and its failure in non-Western civilizations – is to be solved by studying the nonscientific domains of culture, that is, law, religion, philosophy, theology, and the like.

---

[5] I refer here to the case of a sociology graduate student in the early 1980s, Mr. Mario Brajuha at the State University of New York, Stony Brook, who was engaged in a field study of a restaurant on Long Island that subsequently burned down. The circumstances surrounding the fire suggested to police investigators that arson was a strong possibility. Upon discovering that a graduate student had been studying this establishment and recording extensive research notes, an attorney in the case attempted to subpoena the sociologist's notes in the hope that they might contain clues regarding the fire. After a prolonged court battle, including testimony from a variety of social scientists regarding the importance of protecting such research from undue interference, the court ruled that the sociologist's notes were in fact protected from unreasonable seizure and that it was not in the interest of society at large to coerce the revelation of the research notes because of the chilling effect such an action would have on future research, which the court believed enriches our knowledge of how social systems operate. See American Sociological Association, *Footnotes*, Aug. (1984), p. 11, and the *New York Times*, Apr. 5–6, 1984.

[6] The view that science is the most significant modern institution has been suggested recently, albeit with agnostic affirmation, by Harriet Zuckerman, "The Sociology of Science," in *The Handbook of Sociology*, ed. Neil J. Smelser (Beverly Hills, Calif.: Sage, 1988), pp. 511–74. At p. 511 Zuckerman also cites Derek L. De Solla Price who put the claim more boldly: "It [science] has transformed the life and destinies of more of the world's peoples than any . . . religious and political event," above all by controlling economic and military forces as well as the quality of life of the peoples of the world.

From such a point of view, the rise of modern science is the result of the development of a civilizationally based culture that was uniquely humanistic in the sense that it tolerated, indeed, protected and promoted those heretical and innovative ideas that ran counter to the grain of accepted religious and theological teaching. Conversely, one might say that critical elements of the scientific worldview were surreptitiously encoded in the religious and legal presuppositions of the European West.

It seems paradoxical to suggest that modern science emerged because of the uniquely humanistic dimensions of Western culture only because we have not considered "those commitments without which no man would be a scientist"[7] in their religious, philosophical, and legal guises. Put differently, it might be suggested that the foundations of modern science, both cultural and institutional, are to be found precisely in those areas outside of science where men speculate about the nature of the cosmos, in its deepest and most mystical sense, and where the human imagination forges the institutions that allow individuals to perpetually enjoy *neutral spaces* free from the incursions of political and religious censors. It is the task of this book to explore the philosophical, legal, and institutional origins of such neutral zones.

Science as a civilizational institution

Many social scientists have difficulty working with *civilizational* frames of reference, believing as they do that this abstraction is too global for scientific analysis. Only a moment's reflection, however, is needed to arrive at the observation that modern science – as an ongoing enterprise of self-correcting investigation – is, above all else, an enterprise that simultaneously engages the attention and participation of groups and individuals scattered all over the globe. For the past five hundred years in any particular field, individuals living in diverse societies (though mainly in Europe, Europe-overseas, and later, the Americas) have made seminal contributions to the advancement of the various sciences. Indeed, there have been fairly constant rivalries between nationals of these countries – Italians, Englishmen, Frenchmen, Germans, Americans, and others – for the honors and prizes modern science bestows on those who display scientific originality.[8]

---

[7] Thomas Kuhn, *The Structure of Scientific Revolutions*, enlarged ed. (Chicago: University of Chicago Press, 1970), p. 42.

[8] Here one can consult the large literature on the awarding of Nobel prizes. See Harriet Zuckerman, *Scientific Elite: Nobel Laureates in the United States* (New York: The Free Press, 1974); Robert K. Merton, "Singletons and Multiples in Science," in *The Sociology of Science: Theoretical and Empirical Investigations*, ed. Norman Storer (Chicago: University of Chicago Press, 1973), pp. 343–70; "Priorities in Scientific Discovery," in ibid., chap. 16; and

From this point of view, science is and has always been a transnational activity and product that continues its advance despite, and perhaps because of, linguistic differences and national rivalries. It is, therefore, a preeminently civilizational activity and can be understood in its fullest sociological sense only in a civilizational context. It is a cultural activity carried on by individuals and groups living in $2 + n$ different societies[9] across time and space – societies that share certain fundamental metaphysical assumptions, canons of evidence and proof, as well as rules of etiquette and reciprocity. There is not only an ethos of science (more on which below), but a far larger set of metaphysical assumptions, as Thomas Kuhn put it, assumptions "without which no man is a scientist." It is precisely this underlying civilizationally based institutional apparatus that makes it possible for the enterprise of science to succeed at all.

One may note, moreover, that modern science has not required either political unity – in the sense of a bureaucratically unified world government – or linguistic unity. The final breakthrough to modern science and its spread in Europe in the sixteenth and seventeenth centuries, paradoxically, occurred virtually simultaneously with the breakdown of linguistic unity (created by the medieval use of Latin for official communication), along with the rise of nationalism based on indigenous languages and local literary symbols. Both in England and Italy, scientists deliberately published their major works – or translated classic works – into the vernacular so that laymen and disinterested others could be brought into the circle of scientific discourse.[10] Despite this apparent nationalizing of science, there was a movement toward the *universalization* of scientific discourse, a deliberate turn toward breaking down the barriers between the elite cognoscenti (the initiated) and Everyman (the layman, the uninitiated, one of the masses). This was a dramatically different thrust than was to be found in either Islamic or Judaic culture, where the law and the secrets of God were carefully guarded.[11]

"Institutional Patterns of Evaluation in Science," in ibid., chap. 21. For a recent overview of the sociology of science, see Zuckerman, "The Sociology of Science," in *Handbook of Sociology*, pp. 511–74.

9 This definition of civilizational phenomena was worked out by Benjamin Nelson; see *On the Roads to Modernity: Conscience, Science, and Civilizations. Selected Writings by Benjamin Nelson*, ed. Toby E. Huff (Totowa, N.J.: Rowman and Littlefield, 1981), chaps. 5 and 13.

10 For the English case, see Christopher Hill, *The Intellectual Origins of the English Revolution* (Oxford: At the Clarendon Press, 1965), chap. 2 and passim. In Italy, see the biographical essays in *Galileo: Man of Science*, ed. Ernan McMullin (New York: Basic Books, 1967); Stillman Drake, ed. and trans., *Discoveries and Opinions of Galileo* (New York: Doubleday, 1957); and De Santillana's discussions in *The Crime of Galileo* (Cambridge, Mass.: MIT Press, 1955).

11 This is a repeated theme in both medieval Islamic and Judaic thought, as seen in the writings of Averroes and Maimonides. See below, Chap. 2 as well as Chap. 6.

Given the undeniable civilizational dimensions of science as an ongoing social activity, it is neither ethnocentric nor orientalist to speak of the directive structures and institutions that served as the guiding moral, religious, and legal frameworks for intellectuals in medieval Islamic civilization, in China, or in the European West. I am referring, of course, to a level of symbolic and intellectual discourse that was relatively institutionalized and shared to a great extent (though by no means perfectly or uniformly) by informed individuals living in widely scattered places across all these civilizations. Modern science is, therefore, not only a civilizational but an inter-civilizational outcome.

In the first instance, the contributions that Arabic-Islamic civilization made to the development of modern science – its contributions to the fund of knowledge, logical, mathematical, and methodological – prior to its demise after the thirteenth and fourteenth centuries, were significant. As we shall see, the eventual transmission to the West of scientific and philosophical knowledge built up and stored in Arabic-Islamic civilization through the great translation effort of the medieval Europeans had a powerful fructifying effect on the course of Western intellectual development. Thus, modern science is the product of intercivilizational encounters, including, but not limited to, the interaction between Arabs, Muslims, and Christians, but also other "dialogues between the living and the dead" involving Greeks, Arabs, and Europeans. Indeed, some would say that it was the Greek heritage of intellectual thought, above all its commitment to rational dialogue and decision making through logic and argument, that set the course for intellectual development in the West ever after.[12] One does not have to subscribe to such a view to recognize the great importance of the Greek tradition to Western science. The larger point is, however, that modern science is the end product of several such sustained intercivilizational encounters over the centuries.[13]

Secondly, the modern science that emerged in the West became increasingly a universal science in that it was available to all peoples of the world. It

---

[12] A. C. Crombie, "Designed in the Mind: Western Visions of Science, Nature, and Humankind," *History of Science* 26 (1988): 1–12.

[13] For the present I am setting aside the question of the possible contributions of Chinese science to modern science. Joseph Needham, in *Science and Civilisation in China* and elsewhere, has said much about this. The problems connected with an evaluation of the contributions of Chinese science to modern science are many. One of the most significant difficulties hinges on the distinction between science and technology, which Needham has intentionally resisted. Since, as we shall see later, the putative technological superiority of Chinese science to Western technology from the second century to the sixteenth (in Needham's account) did not give rise to modern science in China, the connection between the two is rendered problematic. My reasons for not discussing the possible contributions of Chinese science to modern science will become evident in Chaps. 2, 7, and 8.

became, in Joseph Needham's phrase, "ecumenical science," and as such it was applied to appropriate conditions and bodies of knowledge throughout the world. Although Arabic-Islamic civilization resisted an indigenous development of modern science long after its development in the West, there is today little doubt that great numbers of people living in Muslim lands (like others) desperately want access to the knowledge and the benefits that modern science offers. In short, whatever defects modern science brings in its train, its benefits in terms of modern standards of living, especially health (both mental and physical), are universally acclaimed and universally claimed as the birthright of all peoples whether or not their tribe, country, or community ever made a contribution to that wealth of ecumenical knowledge.

Elements of the sociological perspective

In 1904 Max Weber wrote, "The belief in the value of scientific truth is not derived from nature but is a product of definite cultures."[14] Some thirty-four years later, Robert K. Merton added the following emendation:

This belief [in scientific truth] is readily transmitted into doubt and disbelief. The persistent development of science occurs in societies of a certain order, subject to a peculiar complex of tacit presuppositions and institutional constraints. What is for us a phenomenon which demands no explanation and secures many self-evident cultural values, has been in other times and still is in many places abnormal and infrequent. The continuity of science requires the active participation of interested and capable persons in scientific pursuits. But this support of science is assured only by [the existence of] appropriate cultural conditions.[15]

The suggestion is that the pursuit of science requires the presence of certain cultural and institutional supports if it is to steadily advance. As the philosopher Karl Popper insists, if theoretical, knowledge-producing activities are to merit the title of science, they must be progressive, that is, constantly in search of innovation and the elimination of error.

With these perspectives in mind, we may say that the rise of modern science in the West – and the fact that it did not develop in China, Arabic-Islamic civilization, and elsewhere – parallels the problem of the rise of modern capitalism (and the fact that it did not develop in the Orient). In 1920 when Weber wrote the introduction to his *Collected Essays on the Sociology of Religion*, he saw his subject as one centered on the history and development of rationality and rationalism. He wrote, "It is . . . our first

---

[14] Max Weber, *The Methodology of the Social Sciences* (New York: Free Press, 1949), p. 110.
[15] Merton, "The Normative Structure of Science," in *The Sociology of Science*, p. 254.

concern to work out and to explain genetically the special peculiarity of Occidental rationalism, and within this field that of the modern Occidental form."[16]

Given the civilizational frames of reference noted above, there are four remaining strands in the sociology of science that must be brought together to yield a workable comparative and historical sociology of science which is adequate to the task Weber set before us. The first of these is the idea of the role of the scientist, which has been discussed by Joseph Ben-David in *The Scientist's Role in Society*.[17] The second concerns the social norms of science. These were set out by the young Robert Merton as the ethos of science. The third strand focuses on scientific communities and asks what such communities have in common that makes them work. It was the work of Thomas Kuhn in *The Structure of Scientific Revolutions* that first attempted to answer this question and resulted in the idea of scientific *paradigms*.

The fourth tradition in the sociology of science is the comparative, historical, and civilizational study of science. Although the comparative, historical perspective is the oldest in the sociology of science – stemming from Robert Merton's classic dissertation, *Science, Technology, and Society in Seventeenth-Century England*[18] – this tradition has been the least developed by sociologists. This is due to the fact that most of Merton's students opted to study the reward system in science,[19] rather than following the mainstream of Merton's classic. Likewise, the seventies and eighties have witnessed the growth of an interactional approach to the study of science, and consequently it has become even more ahistorical and further removed from the comparative study of sciences and the cultural and institutional conditions that enable their growth.[20]

---

16 Weber, "Author's Introduction," *The Protestant Ethic and the Spirit of Capitalism* (New York: Scribners, 1958), p. 26.

17 Joseph Ben-David, *The Scientist's Role in Society* (Englewood Cliffs, N.J.: Prentice-Hall, 1971).

18 Robert Merton, *Science, Technology, and Society in Seventeenth-Century England* (New York: Harper and Row, 1970), first published in 1938 in *Osiris*. A useful collection of articles debating this thesis is now available. See I. B. Cohen (with the assistance of R. E. Duffin and Stuart Strickland), *Puritanism and the Rise of Modern Science* (New Brunswick, N.J.: Rutgers University Press, 1990).

19 There is now a sizable literature with this focus; see Cole and Cole, *Social Stratification in Science* (Chicago: University of Chicago Press, 1973); Norman Storer, *The Social System of Science* (New York: Holt, Rinehart and Winston, 1966); Jerry Gaston, *The Reward System in British and American Science* (New York: Wiley, 1978); H. Zuckerman and R. K. Merton, "Age, Aging, and Age Structure in Science," in *The Sociology of Science*, pp. 497–559; and Merton's classic discussion, "The Matthew Effect in Science," reprinted in *The Sociology of Science*, pp. 439–59.

20 For this trend see the essays in *Science Observed*, ed. Karen Knorr-Cetina and Michael Mulkay (Beverly Hills, Calif.: Sage, 1983), but especially Karen Knorr-Cetina, *The Manu-*

With the appearance of Joseph Needham's monumental study, *Science and Civilisation in China,* however, and the publication of Needham's own sociological thoughts about the reasons why Chinese science failed to give birth to modern science,[21] a new chapter in this tradition was written. The late Benjamin Nelson referred to this new development as "Needham's Challenge,"[22] which was precisely the task of going beyond Max Weber and other pioneers in the comparative sociology of sociocultural process and institutions to tackle Needham's riddle regarding the uniqueness of the West as an incubator for modern science. Needham clearly placed his emphasis on the social and cultural conditions that may either speed up or retard the development of science.

These four strands in the sociology of science have hitherto developed exclusively without benefit from each other. In the following discussion I will highlight some of the strengths and weaknesses of these perspectives so that they can be recast as a workable comparative and historical sociology of science which would aspire to throw new light on the development and fate of science in the modern world.[23]

The role of the scientist

One might justifiably argue that the focal point for studying the rise of modern science ought to be the evolution and development of the *role* of the scientist. This is the perspective the late Joseph Ben-David adopted in his well-known study.[24] The central insight of that perspective is that

the persistence of a social activity over long periods of time, regardless of changes in the actors, depends on the emergence of roles to carry on the activity and on the understanding and positive evaluation ("legitimation") of these roles by some social

facture of Knowledge: Toward a Constructivist and Contextual Theory of Science (Oxford: Pergamon, 1981), and Bruno Latour and Steve Woolgar, *Laboratory Life: The Social Construction of Scientific Facts* (Beverly Hills, Calif.: Sage, 1979). For a recent overview of this literature and its trends, see Jan Golinski, "The Theory of Practice and the Practice of Theory: Sociological Approaches in the History of Science," *Isis* 81 (1990): 492–505, as well as Zuckerman, "The Sociology of Science," in *The Handbook of Sociology,* especially pp. 546–58.

21 See Joseph Needham, *The Grand Titration* (London: Allen and Unwin, 1969); hereafter cited as *GT.*

22 See *On the Roads to Modernity,* chaps. 6 and 10.

23 Although I do not attempt to discuss Karl Popper's ideas about the sources and uses of reason in the open society, it is apparent that Popper's thoughts on this subject represent a philosophical counterpart to the sociological thrust of the present study. Popper is aware that in a deep sense the Western belief in reason and rationality is a leap of faith, even an irrational commitment. See Karl Popper, *The Open Society and Its Enemies* (New York: Harper and Row, 1945), vol. 2, p. 231, but also 226–7.

24 Ben-David, *The Scientist's Role.*

group. . . . In the absence of such a publicly recognized role, there is little chance for the transmission and diffusion of the knowledge, skills, and motivation pertaining to a particular activity and for the crystallization of all this into a distinct tradition.[25]

As a first approximation to the problem at hand, this articulation of the importance of the role of the scientist is very suggestive, but a more careful scrutiny of it reveals many defects. The first of these is Ben-David's lack of attention to well-known distinctions in the sociological theory of roles stated so powerfully years ago by Robert Merton. As Merton put it, individuals in society are not, by virtue of occupying a single status, called upon to enact a single role, but to participate in a *role-set*. That is, "We must note that a particular social status involves, not a single associated role, but an *array of associated roles.*"[26] Social actors are involved in a role-set which is composed of that "*complement of role relationships which persons have by virtue of occupying a particular social status.*"[27] In that single position the social actor is called upon to interact with multiple others who are part of his role-set. For example, the public school teacher must teach students, work with other teachers, deal with the parents of students, respond to the school principal, and even relate to local school boards and committees. In the interaction with each of these complementary others, a different repertoire of behavioral responses and vocabulary of motives is required, thereby making the status one of multiple dimensions, skills, and attitudes.

We should be very careful, therefore, to separate the idea of a role-set from the vague idea of "multiple roles." It should be plain, Merton wrote, "that the role-set differs from the pattern which has long been identified by sociologists as that of 'multiple roles.'"[28] To use Merton's classic example, the role-set of the medical student involves regularized interaction with other medical students, with physicians, with nurses, with medical technicians, and with social workers, and these role expectations all derive from a single social status, that of medical student. On the other hand, medical students may also be husbands (or wives), fathers (or mothers), brothers (or sisters), and members of political parties and religious beliefs. Those affiliations, however, point to another dimension of social structures.

The role-set of the scientist is most typically comprised of a college or university professor, a teacher of students, a member of a disciplinary department, a researcher, a writer and author, and, quite possibly, a gatekeeper who referees knowledge claims produced by other scientists. Nor should we

---

[25] Ibid., p. 17.
[26] *Social Theory and Social Structure*, enlarged ed. (New York: Free Press, 1968), pp. 423 and 42.
[27] Ibid.
[28] Ibid., p. 423.

ignore the role of the scientist as expositor to the public of authoritative knowledge, above all, when these knowledge claims are published. In that form they purport to carry the imprimatur of the scientific community at large to which the scientist belongs.

In sum, it is an extremely abbreviated view which neglects to observe that every social role entails a complex of associated roles attached to a single status. As a participant in a role-set, one is always engaged in interaction with multiple others who each may have their own definition of the proper role of the other. For our purposes the implication is that scientists – ancient and modern – are not isolated practitioners sequestered in laboratories, but cultural actors whose very existence depends upon multiple others who (1) provide essential institutional support in the form of teaching and research opportunities, (2) provide vehicles for the publication of scientific results, and (3) provide tacit support both for the role of the scientist and the values and worldview of the scientific enterprise. Without these cultural and institutional formations (as has been suggested) there can be no scientific role. The role of the scientist is in fact a construct composed of a complementary array of role performances that are essential and indispensable to the status of the scientist.

The formulation of this broader conception of the role of the scientist as a role-set should serve to suggest that the various generic elements of the scientist's position are embedded in an institutional history, and that these elements evolved over time and at different rates. One must also note that there are a great many specialized practitioners – astronomers, astrologers, mathematicians, physicists, chemists, opticians, biologists, physicians, and so forth – who each may claim the title of scientist. Furthermore, each of these scientific specialties emerged and achieved its scientific status at a different point in time. Thus, mechanics and astronomy had reached high levels of precision and theoretical development long before the Middle Ages. Even today, Kuhn reminds us, the ancient works of Archimedes and Ptolemy, that is, *Floating Bodies* and the *Almagest,* "can be read only by those with developed technical expertise."[29]

The formulation of the problem in this manner helps to shift our attention away from the purely *internal* aspects of scientific inquiry, that is, on the methods, theories, paradigms, and instrumentation of science, and onto those *external* cultural and institutional structures that give scientific inquiry a secure place in the intellectual life of a society and civilization. What is more, one is drawn to the historical study of the evolutionary and incremen-

---

[29] Thomas Kuhn, "Mathematical vs. Experimental Traditions in the Development of Physical Science," *Journal of Interdisciplinary History* 7, no. 1 (1976): 5.

tal steps by which each of the components of the role-set of the scientist came into existence, including the points of friction (very largely philosophical and ideological) that the scientific worldview had to overcome on its way to full institutionalization. Furthermore, it may turn out that many aspects of the scientist's role are in fact generic modes of thought, inquiry, and knowledge production that those who are scientists merely appropriated (quite legitimately) for their specialized endeavor.

As this shift of focus brings into view the limitations on the strictly internal history of science, it throws into sharp relief the limitations of the view that the scientific role first occurred in England in the seventeenth century.[30] To say this is not, however, to challenge the historiographic truth which maintains that modern science began in the seventeenth century with the fusion of the mathematical movement of the Continent with the empirical and experimental tradition of England – though that is challenged by Kuhn.[31] While Ben-David is unequivocal about this fusion point in the seventeenth century, Thomas Kuhn seems to place the fusion in the mid-nineteenth century. Kuhn writes that the antimathematical but highly empirical/experimental wing of the scientific movement that centered in England "had little effect on scientific theory or conceptual structure" until the mid-eighteenth century,[32] and in a later discussion, he pushes the fusion point into the late nineteenth century.[33] Furthermore, Kuhn suggests that Newton is actually a British anomaly, that his methods, "sources, colleagues, and rivals . . . were all Continentals."[34] My point is, however, that while this fusion of intellectual traditions was an essential event in the full emergence of modern science, the essentially external or social foundations of modern science – constituted by deeper philosophical as well as institutional underpinnings – were established much earlier. Considered from the point of view of the revolutionary intellectual and social transformation of the twelfth and thirteenth centuries – which I will discuss in Chapter 4 – the breakthrough laying the legal and social foundations of modern science took place much earlier than is generally realized.

If the social role of the scientist is to be the focus of comparative and historical inquiry, to reiterate the point, it must be remembered that the apparent singularity is in fact a multiplicity, and it is very likely that these

---

[30] Ben-David, *The Scientist's Role*, p. 17; and ibid., "The Scientific Role: The Conditions of Its Establishment in Europe," *Minerva* 4, no. 1 (1965): 15–54 at p. 15.

[31] Thomas Kuhn, "Scientific Growth: Reflections on Ben-David's 'Scientist's Role'," *Minerva* 10, no. 1 (1972): 166–78.

[32] Ibid., p. 174.

[33] "Mathematical vs. Experimental Traditions," pp. 19–27.

[34] "Scientific Growth," p. 173.

different aspects of the role-set of the scientist emerged and became institutionalized at different points in time. In addition, the legitimating cultural values may come from very different cultural spheres, not from a singular source of scientific values. In other words, scientific values and the ethos of science are constructs that emerged over time and evolved out of nonscientific contexts. Furthermore, many elements of scholarly research that are thought to be generically associated with scientific research were well established and widespread before the term *scientist* came into use in the nineteenth century. The word *scientist* did not come into use until the first half of the nineteenth century when the Cambridge philosopher of science William Whewell coined the term. Whewell had become aware that the English language had no term to refer collectively to chemists, mathematicians, physicists, electrochemists, and those who studied the natural world. When he took it upon himself to invent such a term, that is, the word *scientist*, his brainchild was at first rejected and then later treated as a barbarous innovation.[35] Writing in 1834, Whewell lamented the fact that the English language had no term "to designate the student of the knowledge of the material world collectively." Whewell says that this fact

was very oppressively felt by the members of the British Association for the Advancement of Science at their meetings at York, Oxford, and Cambridge, in the last three summers. There was no general term by which these gentlemen could describe themselves with reference to their pursuits. *Philosopher* was felt to be too wide and too lofty a term . . . *savans* was rather assuming.[36]

At that point, he says, "some ingenious gentleman [who was Whewell himself][37] proposed that by analogy with *artist*, they might form *scientist*, but

---

35 The story of the multiple invention of this term has now been lucidly told by two writers; see Sydney Ross, "Scientist: The Story of a Word," *Annals of Science* 18, no. 2 (1962): 65–85 (published in 1964); and Robert Merton, "Le molteplici origini e il carattere epiceno del termine inglese *Scientist*. Une episodio dell'interazione tra scienza, linguaggio e soceità" ("The Multiple Origins and Epicene Character of the Word *Scientist*: An Episode in the Interaction of Science, Language, and Society"), in *Scientia: L'immagine e il mondo* (Milano), pp. 279–93. Merton's account lays heavier stress on the multiple and independent invention of the term, as well as intended and unintended sociological effects.

36 Whewell in *The Quarterly Review* 51 (1834):58–61, as cited in Ross, "Scientist," p. 72. The resistance to introducing the term on philological grounds are discussed in both Ross, ibid., pp. 75ff, and Merton, "Le molteplici origini," pp. 281ff. Such objections continued to be heard to the very end of the nineteenth century.

37 Evidence identifying Whewell as this "ingenious gentleman" is in Ross, "Scientist," p. 71, n9, and Merton, "Le molteplici origini," p. 291, n6, and pp. 279–83. Still, at least three more individuals apparently coined the term independently during the nineteenth century. Whewell's use of the term *scientist* has generally been located in the 1840 edition of his *Philosophy of the Inductive Sciences,* but the accounts of Ross and Merton clearly place his actual invention of the term in 1834 in his review of the book by Mary Somerville, *The Connexion of the Sciences.*

this was not generally palatable."[38] This suggests that as late as the nineteenth century, the role of scientist as a unitary identity was still in doubt and that it would be anachronistic at best to expect individuals prior to that time to have a completely formed self-image.

Still another defect of Ben-David's account of the rise of the role of the scientist was pointed out by Thomas Kuhn when he noted that "unless science and the scientist's role are identifiable by criteria immune to social influence and [are] the same throughout all time and place, fundamental elements of Professor Ben-David's method are in jeopardy."[39] Insofar as the content of the scientific role is concerned, Ben-David only suggests that the emergence of this "new social role" was an outcome in which "the acceptance of the search for truth through logic and experiment" became the paramount criteria.[40] This clearly gives the impression that Ben-David remained an "unregenerate positivist, a man who believes that scientific ideas are, if only by definition, responses to logic and experiment alone."[41] As a result, his position "is impotent to deal with the development of technical specialities, later recognized as having been science, during the many centuries when nothing quite like the concept of science, much less the scientific role, can be found."[42] This outcome is all the more puzzling for a sociologist, since the sociological point of view has always stressed the importance of the nexus between social collectives, especially institutional formations, and the behavior of those affected by them. This was powerfully articulated by Robert Merton in his new preface to *Science, Technology, and Society in Seventeenth-Century England* and is a logical extension of the fact, noted above, that individuals are always required to play multiple roles. Merton pointed to a seminal aspect of comparative and historical sociology, which is

that the socially patterned interests, motivations and behavior established in one institutional sphere – say, that of religion or economy – are interdependent with the socially patterned interests, motivations, and behavior obtaining in other institutional spheres – say, that of science. There are various kinds of such interdependence, but we need touch upon only one of these [for example, between religion and science]. . . . The same individuals have multiple social statuses and roles: scientific and religious and economic and political. This fundamental linkage in social structure in itself makes for some interplay between otherwise distinct institutional spheres even when they are segregated into seemingly autonomous departments of life. Beyond

---

[38] As cited in Ross, "Scientist," p. 72, and *The Oxford English Dictionary*, 2d ed. (1989), vol. 14, 652.

[39] Thomas Kuhn, "Scientific Growth," p. 169.

[40] Ben-David, *The Scientist's Role*, p. 17.

[41] Ibid., p. 168.

[42] Ibid., p. 169.

that, the social, intellectual and value consequences of what is done in one institutional domain ramify into other institutions. . . . Separate institutional spheres are only partially autonomous, not completely so.[43]

The inability of Ben-David's conception of the scientific role to explain why modern science failed to rise earlier in history or in other civilizations is due to its circular reasoning: modern science did not arise because modern scientists did not emerge; this (according to Ben-David) occurred only in England in the seventeenth century. In Ben-David's words, ancient science failed to give birth to modern science "because those who did scientific work did not see themselves . . . as scientists."[44] Accordingly, Ben-David suggested that the question is "what made certain men in seventeenth-century Europe and nowhere before, view themselves as scientists."[45] But as we have seen, the term scientist did not exist in the English language until its invention in the nineteenth century by William Whewell. In short, without giving the role of scientist and the self-image of the scientist some specific content, Ben-David's argument lapses into tautology and anachronism.

Given the fact that all the conventional accounts of the scientific revolution would place it in the sixteenth and seventeenth centuries, we need to take a broader look at the nature and sources of those intellectual commitments (prior to the seventeenth century) that made the production and pursuit of scientific knowledge a matter of honor as well as urgency.

The ethos of science

To remedy this latter defect of Ben-David's excessively narrow and substantively vague conception of the scientific role, we need to borrow another element from the work of Robert Merton, that is, his description of the *ethos of science*. According to Merton,

the ethos of science is that affectively toned complex of values and norms which is held to be binding on the man of science. The norms are expressed in the form of prescriptions, proscriptions, preferences, and permissions. They are legitimatized in terms of institutional values. These imperatives, transmitted by precept and example, and reinforced by sanctions are in varying degrees internalized by the scientist, thus fashioning his scientific conscience or, if one prefers, his "superego."[46]

Although there are defects and unresolved tensions in Merton's formulation of the social norms of science, his formulation remains the most influ-

[43] *Science, Technology, and Society in Seventeenth-Century England,* pp. ix–x.
[44] Ben-David, "The Scientific Role," p. 15.
[45] Ibid.
[46] Merton, "The Normative Structure of Science," in *The Sociology of Science,* pp. 267–80 at p. 268f.

ential and most promising starting point for the analysis of the ethos of science in comparative perspective. Following Merton's original formulation, there are four sets of "institutional imperatives"[47] associated with scientific activity: universalism, communalism,[48] disinterestedness, and organized skepticism. Merton later added the norm of originality, while other commentators on the ethos stressed rationality as well as individualism, intending to stress the importance for science of the individual's freedom and autonomy in choosing his own problems. While these norms were intended to identify the social norms of science, Merton also recognized that there are methodological canons which are "both technical expedients and moral compulsives."[49] They too can be strong directives regarding scientific behavior, and later discussions raised the question of whether or not these methodological canons and technical rules might not be more important than purely social norms as directives of scientific activity. Nevertheless, Merton (in the early 1940s) thought it both appropriate and possible, "in only a limited introduction to a large problem, [that is,] the comparative study of the institutionalized structure of science,"[50] to focus on "the mores with which [the methods of science] are hedged about."[51] "For the mores of science possess a methodologic rationale but they are binding, not only because they are procedurally efficient, but because they are believed to be right and good. They are moral as well as technical prescriptions."[52] In short, one might consider these norms and mores with which the practice of science is hedged about to be essential components of the role of the scientist, components that Ben-David largely ignored. As I review these elements of the ethos of science, I invite the reader to keep in mind comparative, historical, and civilizational perspectives in which the Mertonian norms might not be so readily assented to.

1. *Universalism:* This norm suggests two imperatives: first, that knowledge claims should be judged impersonally according to standard criteria and without regard to the personal characteristics of the researcher; and second, that all persons, regardless of ethnic or kinship ties, or religious knowledge, should be freely admitted into the universe of scientific discourse.[53]

---

[47] Ibid., p. 270.
[48] Merton originally called this the norm of communism, but as this term implies a political and economic theory, it seems best to use the term *communalism*. Bernard Barber, in *Science and the Social Order* (New York: Free Press, 1952), p. 130, proposes the term *communality*.
[49] Merton, in *The Sociology of Science*, p. 268.
[50] Ibid., p. 269.
[51] Ibid., p. 268.
[52] Ibid., p. 270.
[53] Ibid.

2. *Communalism:* According to this imperative, the actual findings of research belong to the community at large and are not to be secreted or appropriated solely by the researcher. One is enjoined to make results available through publication as soon as normal cautions regarding error and precision are taken.

3. *Disinterestedness:* According to this norm, the scientist is expected to display a dispassionate pursuit of the truth through publicly available means and to forgo all forms of personal gain and aggrandizement.[54]

4. *Organized skepticism:* This institutional imperative enjoins "temporary suspension of judgment and the detached scrutiny of beliefs in terms of empirical and logical criteria,"[55] and the application of this attitude toward all knowledge claims, even those issuing from other well-regarded institutions.[56] One may remark that this norm is particularly volatile and that traditional (or late-developing) societies are especially sensitive to criticism and questioning directed at their central and sacred values. This attitude prevails today among Muslims who are reluctant to allow any form of public skepticism, fictional or scientific, regarding the Prophet Muhammad or his teachings (on which more in Chapter 4).

With the publication of his seminal paper, "Priorities in Scientific Discovery,"[57] Robert Merton raised his earlier discussion of the competition for recognition in science through the pursuit of originality to a fifth normative element of science. Here the imperative is the clear injunction to seek all rewards in science through displays of originality, with the highest reward being the eponymous naming of a scientific discovery after the researcher.

I shall leave it for later discussion to decide whether or not this outline of the ethos of science fully or adequately articulates the unique value system of science. Several critics have suggested that organized skepticism and in fact, all the norms together, "may well be characteristic of the Western academic community in general."[58] From my point of view, that critique is well taken, and I will discuss it later. Nevertheless, in 1942 when Merton first published this piece on the ethos of science, he asserted that "the institutional imperatives" of science, which enjoin "the extension of certified knowledge," derive from "the goal and the methods" of science.[59] But this appears to be a

---

[54] Ibid., p. 276.
[55] Ibid., p. 271.
[56] Ibid., p. 277.
[57] Merton, in *The Sociology of Science*, pp. 286–324, originally published in 1957.
[58] Michael Mulkay, "Some Aspects of Cultural Growth in the Natural Sciences," *Social Research* 36 (1969): 22–52, at p. 27. S. B. Barnes and R. G. A. Dolby say that they "are nonspecific to science," in "The Scientific Ethos: A Deviant Viewpoint," *European Journal of Sociology* 11 (1970): 3–25 at p. 14.
[59] Merton, in *The Sociology of Science*, p. 270.

tautology. So I would argue that insofar as we can speak of a specific institution of science, its normative imperatives are derived from a far more general cultural ambience and, above all, rely upon religious and legal presuppositions that long antedate the rise of modern science in the seventeenth century.

Paradigms and scientific communities

The third strand of theory and research in the comparative sociology of science that should be considered is that initiated by Thomas Kuhn's book, *The Structure of Scientific Revolutions*, probably the most influential book regarding the sociology of science in the second half of the twentieth century.

It was the appearance of this work that led many critics of Merton's scientific ethos to suggest that it is "the body of established knowledge,"[60] "the technical norms of paradigms," not social norms, which generate the "cohesion, solidarity, and commitment" of scientists and their communities.[61] It is a paradoxical turn of events, therefore, that sociologists took up so enthusiastically the Kuhnian position which strongly suggests that it is the internal (technical and intellectual) history of science that provides the key to understanding revolutions in science.[62] This was largely because Kuhn attempted to locate his internalist discussion within "the sociology of the scientific community."[63] But we should remind ourselves that Kuhn's book is really an answer to the following question: if communities of scientific practitioners can be identified, what do they have in common that allows them to maintain such intense and relatively full communication about their search? His answer to that question was paradigms: "those universally recognized scientific achievements that for a time provide model problems and solutions to a community of practitioners." At first glance this appears to be a very powerful thesis that establishes the basis for writing a truly internal history of science. That is, if we accept Kuhn's thesis that normal science really begins with the acquisition and development of a paradigm, then students of the history of science would be well advised to study the history of the specialized sciences precisely from the point of view

---

[60] Mulkay, "Some Aspects of Cultural Growth," p. 22.
[61] Barnes and Dolby, "The Scientific Ethos," p. 23.
[62] See Barry Barnes, *Scientific Knowledge and Sociology Theory* (London: Routledge and Kegan Paul, 1974), chap. 5, for an overview of the internal/external debate. A selection of papers discussing these issues can be found in George Basalla, ed., *The Rise of Modern Science: Internal or External Factors?* (Lexington, Mass.: D. C. Heath, 1968).
[63] Kuhn, *The Structure of Scientific Revolutions*, p. vii.

of the development and overthrow of paradigms. Such a view clearly suggests that the main story in the history of science is just such an internalist story, which focuses on the technical, theoretical, and instrumental applications of a particular paradigm. It is such features that establish a universally agreed upon solution to a long-standing set of scientific problems, which thereby gives a new and coherent focus to scientific inquiry in that field.

It should be kept in mind, however, that Kuhn himself did not reject external factors as influences on science. While he alerted readers of *The Structure of Scientific Revolutions* to the fact that he said nothing "about the role of technological advance or of external social, economic, and intellectual conditions in the development of the sciences,"[64] he was quite prepared to acknowledge such factors, since "one need . . . look no further than Copernicus and the calendar to discover that external conditions may help to transform a mere anomaly into a source of acute crisis." The analysis of external factors "would surely add an analytic dimension of first-rate importance for the understanding of scientific advance."[65]

Kuhn's subsequent use of the term paradigm in the book often expanded to include all the "accepted rules" of a scientific community.[66] For example, he writes that close "historical investigation of a given specialty at a give time, discloses a set of recurrent and quasi-standard illustrations of various theories in their conceptual, observational, and instrumental applications. These are the community's paradigms, revealed in its textbooks, lectures, and laboratory exercises."[67] At the same time Kuhn admitted that identifying a shared paradigm for a scientific community does not result in identifying all the shared rules. In fact, he deliberately expanded the concept of *rule* to encompass many other items beyond the paradigm. For example, there are rules that take the form of "explicit statements of scientific law and about scientific concepts and theories."[68] In addition, "At a level lower or more concrete than that of laws and theories, there is . . . a multitude of commitments to preferred types of instrumentation and to the ways in which accepted instruments may legitimately be employed."[69]

Thirdly, there are "the higher level, quasi-metaphysical commitments" of scientists and these are "both metaphysical and methodological."[70] For ex-

[64] Ibid., p. x.
[65] Ibid.
[66] For the many different meanings of the term in Kuhn's study, see Margaret Masterman, "The Nature of a Paradigm," in *Criticism and the Growth of Knowledge*, ed. Imre Lakatos and Alan Musgrave (Cambridge: Cambridge University Press, 1970), pp. 59–89.
[67] *The Structure of Scientific Revolutions*, pp. 43, 54.
[68] Ibid., p. 40.
[69] Ibid.
[70] Ibid., p. 41.

ample, in the seventeenth century, "most physical scientists assumed that the universe was composed of microscopic corpuscles and that all natural phenomena can be explained in terms of corpuscular shape, size, motion and interaction." These assumptions, as metaphysical commitments, "told scientists what sort of entities the universe did and did not contain." As a methodological injunction, this set of commitments "told [scientists] what ultimate laws and fundamental explanations must be like: laws must specify corpuscular motion and interaction, and explanations must reduce any given natural phenomenon to corpuscular action under these laws."[71]

Finally, Kuhn identified a still higher set of commitments, those "without which no man is a scientist."[72] By this means Kuhn elaborated a picture of constantly evolving, but also shifting and reconstituted, *communities* of scientific practitioners who are held together by the "existence of this strong network of commitments – conceptual, theoretical, instrumental, and methodological." But this network of commitments also contains a large set of metaphysical commitments. In effect, this litany of scientific commitments far exceeds the bounds of technical and instrumental considerations and surely stretches the usual connotation of the term *scientific*. The richness of Kuhn's description of science and its practice clouded the distinctions between paradigms, rules, and other commitments, some of which would be accounted external and nonscientific, since they are in the nature of philosophical speculations. But having mentioned that domain of commitments "without which no man is a scientist," Kuhn remained silent on the issue throughout his work. No doubt much of the difficulty in seeking to specify those commitments stems from the fact that during the course of history many sorts of commitments – religious, philosophical, metaphysical, and political – have been held by those who have made lasting contributions to the history of science. It is imperative, nevertheless, to consider the nature of the metaphysical commitments that were operative during the rise of modern science. Here again, perspective on this problem can only be gained by adopting comparative and civilizational frames of reference, as will become apparent in Chapter 2 when I consider the case of Arabic science.

In the meantime, it should be noted that when Kuhn wrote the postscript to the second edition of *The Structure of Scientific Revolutions*, he took pains to clarify the conceptual muddle that had overtaken some of his earlier discussions. He did this by acknowledging that at least two main senses of the term paradigm are operative in his book. In one sense, the idea of a paradigm "stands for the entire constellation of beliefs, values, techniques,

---

[71] Ibid.
[72] Ibid., p. 42.

and so on shared by the members of a given [scientific] community."[73] This formulation Kuhn calls the sociological sense. In contrast, there is the sense that "denotes one sort of element in that [sociological] constellation, the concrete puzzle-solutions which, employed as models or examples, can replace explicit rules as a basis for the solution of the remaining puzzles of normal science."[74]

Although Kuhn believes that this second meaning of paradigm is philosophically deeper, I would suggest that, at least for sociological purposes, the reverse is the case, since the larger constellation of commitments clearly entails those philosophical and metaphysical commitments that always remain vague and capable of infinite permutation and transformation, while the paradigmatic exemplars, despite their great practicality, can be overthrown and, in the event, are relegated to the category of "once respectable, now forgotten errors" of scientific history. Likewise, it might be suggested that among the elements of this constellation of commitments constituted by the sociological sense of paradigm are those commitments referred to above as the ethos of science. In reformulating the sociological sense of paradigm, however, Kuhn introduced the term *disciplinary matrix* and suggested that this is the broad rubric under which we should consider those commitments which a community of practitioners have in common. In the first instance, this includes symbolic generalizations, which are those lawlike statements such as $f = ma$.[75] A second component of this matrix consists of metaphysical paradigms or the metaphysical part of paradigms. These, according to Kuhn, serve to "supply the group with preferred or permissible analogies and metaphors." When these elements are also referred to as models, we see once again that this array of symbolic commitments within the scientific community ranges from technical rules to quite abstract and rule-of-thumb approximations whose actual moorings are in philosophical and metaphysical commitments. Nevertheless, Kuhn sought to separate values from this previous set of disciplinary factors.[76] In this revised formulation, Kuhn uses the term *values* to refer to preferences regarding the nature of predictions (whether they are quantitative or qualitative), as well as criteria used in judging the virtue of theories under various test conditions.[77] Such criteria would include standards such as consistency, simplicity, and plausibility. When new explanations and experimental results emerge they will be judged in terms of their internal consistency as well as their plausibility vis-à-vis

[73] Ibid., p. 175.
[74] Ibid.
[75] Ibid., p. 184.
[76] Ibid.
[77] Ibid., p. 185.

existing fact and theory. And those theoretical formulations that are more parsimonious will be preferred over those that are less so. While value standards such as these may be differently interpreted and applied, Kuhn suggests that during periods of scientific crisis, when the prevailing paradigms are overwhelmed by anomalies, researchers do "resort to shared values rather than to shared rules" and these provide a stable basis for deciding research outcomes.

It is evident, then, that in this reformulation of the way values, norms, rules, and paradigms affect normal science, Kuhn attempted to abstract a domain of values that would be inherently scientific and speak to the issue of when to reject, when to accept, and when to remain neutral regarding theoretical and empirical claims that stand as candidates for canonization as received scientific wisdom. To some degree such notions as consistency, simplicity, and plausibility are offshoots of the rules of logic and mathematics, but only in a vastly extended sense of those terms. Conversely, such standards doubtless apply to the social sciences, to the science of linguistics, possibly even to literary criticism, as well as law. Such values are not, however, of the same order as those norms of science that Robert Merton sought to identify in the ethos of science. Kuhn's values seem much more technical and seem to occupy some intermediate zone between social norms, on the one hand, and methodological rules on the other.

Finally, Kuhn attempted to narrow the concept of a paradigm to that of an exemplar, a "concrete problem solution" that "students encounter from the start of their scientific education, whether in laboratories, on examinations," or in textbooks.[78] These, Kuhn claims, "provide the fine-structure of science." By studying these exemplars or, more precisely, working through similar problems using the exemplar as a model solution, the aspiring scientist learns "to see a variety of situations as like each other."[79] In effect, this is a means of learning "tacit knowledge," which, Kuhn seems to suggest, is not the same as learning rules.[80] A very suggestive example of this is one that Kuhn originally used in the first edition of the book. It concerns the use of a set of legal rulings in law as a paradigm. A paradigm, once learned, serves as a model for future cases and teaches students how to see new cases as similar to old ones. Hence by studying the paradigmatic cases, "like an accepted judicial decision in the common law,"[81] one learns how to solve future cases. This example of the role of exemplar in the law will be a useful parallel for later discussion.

[78] Ibid., p. 187.
[79] Ibid., p. 189.
[80] Ibid., p. 191ff.
[81] Ibid., p. 23.

With this I will draw my review of Thomas Kuhn's contribution to the sociology of science to a close. Because Kuhn persisted in treating scientists as working members of communities (within which some set of sociological principles ought to operate), his work has attracted a great deal of attention among sociologists. Nevertheless, Kuhn's analysis, despite its reference to values and metaphysical commitments, has focused on those scientifically indigenous elements that provide the fine structure of scientific practice and solidify group commitments among scientists. Sociologists of science notwithstanding, the thrust of Kuhn's work drifted toward an internalist account of science. That is to say, unless and until students of the history of science can show that particular items of the disciplinary matrix or explicit parts of paradigms were derived from extrascientific – that is, religious, legal, economic, political, or philosophical – contexts, Kuhn's method, rightly, focuses attention on the internal dialogue and crises of science, albeit in the competition between schools of scientific practice within the same field.[82] Even if the thrust of Professor Kuhn's work inevitably stressed the so-called internal aspects of science, it must be said that Kuhn's discussion of the nature and role of paradigms in the history of science vastly expands our understanding of the whole enterprise. By accentuating the notion of metaphysical commitments and pointing out their role in the theory and practice of science (for example, in seventeenth-century corpuscular theory), Kuhn has highlighted the centrality of philosophies of nature and their importance for the history of science. In this regard, there are affinities on this level between the ideas of Kuhn and those of Joseph Needham regarding the nature and role of philosophies of nature in Chinese science.

In short, there is no reason to suspend pursuit of the sources and functions of those values and metaphysical commitments "without which no man would be a scientist" along with a description of the philosophies of nature that can provide explanatory models for the sciences at various points in their development. In pursuing that strategy, however, the most powerful context is that of comparative, historical, and civilizational analysis. It is by examining one or more cases in which modern science did not emerge that we can arrive at those values, commitments, and institutional arrangements that enable science as we know it to flourish. If Kuhn's work does not provide an exemplary model for that endeavor, it does at least draw our attention to the kinds of elements one would be likely to encounter in protoscientific communities. Such an orientation raises the question of

---

[82] The classic pre-Kuhnian effort to explain the rise of Newton's theories by economic factors is found in Bernard Hessen, "The Social and Economic Roots of Newton's Principia," in *Science at the Cross-Roads*, ed. J. D. Bernal (London: Kniga Ltd., 1931).

whether or not the community dynamics of paradigm-emergence (and over-throw) can be seen in non-Western settings, and this is so whether or not we can settle the question of the unequivocal identification of the role of the scientist.

There is, however, another contribution to the history of science that Kuhn has made and which we should consider before turning our attention to a sketch of Joseph Needham's outline of a universal history of science. That contribution is to be found in Kuhn's probing of the empirical versus the mathematical tradition in the Western scientific revolution, which I mentioned earlier. Kuhn's suggestion is that these two types of science – the one practical and experimental, the other more mathematical and abstract – remained separate enterprises well into the nineteenth century, and in some cases into the early twentieth. Thus, if we distinguish between Baconian and classical traditions and ask how the two traditions interacted, Kuhn says, "not a great deal and often with considerable difficulty. . . . Into the nineteenth century the two clusters, classical and Baconian, remained distinct."[83] "Excepting chemistry, which had found a variegated institutional base by the end of the seventeenth century," he writes, "the Baconian and classical sciences flourished in different national settings from at least 1700."[84] While practitioners of both can be found on the Continent and in England, England was home to the Baconian sciences, while the Continent, especially France, was the home of the mathematical sciences. Furthermore, Kuhn points out that the French Academy of Sciences did not have a section on experimental science (*physique experimentale*) until 1785, "and it was grouped in the mathematical division (with geometry, astronomy, and mechanics)."[85] In fact, there were few experimentalists among the Academy members. Considering "the 18th century as a whole, the contributions of academicians to the Baconian physical sciences were minor compared with those of doctors, pharmacists, industrialists, instrument makers, itinerant lecturers, and men of independent means." In England the situation was the reverse, which is to say that the Royal Society was composed chiefly of amateurs, "men whose careers were first and foremost in science."[86] In addition, Newton's apparent participation in both traditions (in the classical via the *Principia*, and in the experimental via his *Optiks*) was unique. Kuhn suggests that the readers of the *Optiks* found a "non-Baconian use of experiment," which was "a product of Newton's deep and simultaneous immer-

---

[83] Kuhn, "Mathematical vs. Experimental Traditions," p. 16.
[84] Ibid., p. 25.
[85] Ibid., p. 20.
[86] Ibid., p. 21.

sion in the classical scholastic tradition."[87] In short, Newton's points of reference were mostly Continental and his work seems much more at home there.[88] This division was made sharper by the fact that the classic sciences had been made part of "the standard curriculum of the medieval universities," whereas for the experimental sciences "the universities had no place [for them] before the last half of the nineteenth century."[89] Thus, it is not so surprising that it was not until the 1840s that a common term, scientist, was found acceptable to describe all those engaged in the study of the natural world.

In sum, Kuhn's suggestion is that the experimental and mathematical traditions fused much later than the putative time of the rise of modern science in the sixteenth and seventeenth centuries. This fusion was achieved in the mid-nineteenth century, and for some sciences, not until the twentieth century. Indeed, Kuhn further suggests that it was the unique ability of the German universities to modify the institutional arrangements of their classic and medieval foundations that gave them an edge in the early twentieth-century development of modern physics.[90] This account suggests that something other than experimentalism was the new driving force of modern science and that, whatever it was, it was something triggered prior to Galileo. It therefore suggests that the origins of the classical tradition that culminated in Copernicus, Kepler, Galileo, and Newton had sources much deeper and earlier than the seventeenth century.[91]

Comparative civilizational sociology of science: Joseph Needham

Without doubt Joseph Needham's monumental study, *Science and Civilisation in China*,[92] did more than any other work in the twentieth century to draw attention to the need for a comparative, historical, and sociological study of the rise of modern science. The need for such a study had been

---

[87] Ibid., p. 18.
[88] Kuhn, "Scientific Growth," p. 173f.
[89] Ibid., p. 19.
[90] Ibid., p. 31.
[91] Speaking of the early successes of the mathematical wing of the scientific movement, Kuhn says: "This movement, which centers on the Continent, is responsible for virtually all the early-modern achievements in the mathematical and physical sciences, including analytic geometry and calculus, heliocentric astronomy, the new optics and mechanics. Newton was perhaps its only first-rank British representative, and his sources, colleagues and rivals (excepting the idiosyncratic Boyle) were all Continentals," in "Scientific Growth," p. 174.
[92] *Science and Civilisation in China* (New York: Cambridge University Press, 1954–), 7 vols., in progress; hereafter cited as *SCC*.

broached by Weber in 1920 when he wrote the introduction to his *Collected Essays on the Sociology of Religion.*[93] That introduction was later translated by Talcott Parsons and published as the "Author's Introduction" to *The Protestant Ethic and the Spirit of Capitalism.*[94] However, in his study of the religion of China (published in 1916), Weber spoke of the failure of "systematic and naturalistic thought" to mature in China,[95] though it is clear that Weber's sources of information were deficient, probably even for that period of time.[96]

Although Needham does not refer to Max Weber's comparative and historical writings, it is clear from everything he has written about the social aspects of Chinese science that Weber's concerns were never distant from the issues Needham felt compelled to explore. As Benjamin Nelson pointed out in his long commentary on Needham and Weber, Needham has in many respects gone beyond Weber in probing the social, cultural, and ontological groundings of Chinese science and civilization,[97] most noticeably, perhaps, in Needham's protracted discussion of the idea of law and laws of nature in Chinese civilization.[98] But there are many other areas of Chinese thought and philosophy in which Needham's command of the original sources reaps

---

[93] *Gesammelte Aufsätze zur Religionssoziologie* (Tübingen: J. C. Mohr, 1920–1), 3 vols.

[94] Weber, *The Protestant Ethic and the Spirit of Capitalism,* first translated in 1930.

[95] Weber, *The Religion of China,* trans. Hans Gerth (New York: Free Press, 1951), p. 150.

[96] This is a point made by Nathan Sivin in "Max Weber, Joseph Needham, Benjamin Nelson: The Question of Chinese Science," in *Civilizations East and West: A Memorial Volume for Benjamin Nelson,* ed. E. V. Walter et al. (Atlantic Highlands, N.J.: Humanities Press, 1985), pp. 37–49, at p. 46. Nevertheless, Weber's suggestions and insightfulness on many scores remain. The fact that Weber says "there was no rational science," in *The Religion of China,* p. 151, however mistaken it appears to be, does in fact point to an actual difference between the science of China and that of the West until the seventeenth century. This may be illustrated by Nathan Sivin's argument regarding biology. Needham restates the argument "that for medieval and traditional China 'biology' was not a separated and defined science. One gets its ideas from philosophical writings, books on pharmaceutical natural history, treatises on agriculture and horticulture, monographs on groups of natural objects, miscellaneous memoranda and so on" (*SCC* 5/2:xxii). Similarly, Sivin says: "The sciences were not integrated under the dominion of philosophy, as schools and universities integrated them in Europe and Islam. [The] Chinese had sciences, but no science, no single conception or word for the overarching sum of all of them. Words for the level of generalization above that of the individual science were too broad. They referred to everything that people could learn through study, whether of Nature or human affairs" (Sivin, "Why the Scientific Revolution Did Not Take Place in China – or Didn't It?" in *Transformation and Tradition in the Sciences,* ed. Everett Mendelsohn [New York: Cambridge University Press, 1984], pp. 531–54 at p. 533). In other words, the specifically theoretical sciences, as we shall see, were not as highly developed in China as in Arabic-Islamic civilization.

[97] Nelson, "Sciences and Civilizations, 'East' and 'West,'" in *On the Roads to Modernity,* pp. 152–200.

[98] This was originally discussed in *SCC* 2: 518–83 and later in "Human Law and the Laws of Nature," in *GT,* pp. 239–330.

rich new descriptions of Chinese intellectual, philosophical, and religious life that greatly surpass Weber's.[99]

While Needham is deeply appreciative of Chinese science and technology and their achievements, he is also acutely aware of the weaknesses of Chinese science as a theoretical endeavor. Indeed, Needham's sensitivities to these very weaknesses seem to have contributed to his puzzlement regarding the failure of Chinese science to give birth to modern science. That puzzlement led him to articulate the central question, that is, given China's scientific traditions and its apparent technological superiority over Western Europe up until the seventeenth century,[100] how is it that it "failed to give rise to distinctively *modern* science," despite its "having been in many ways ahead of Europe for some fourteen previous centuries."[101] It should be noted that Needham equates modern science with the apparent mathematization of nature associated with the work of Galileo, though on other occasions he places more stress on the experimental thrust of modern science. In Needham's view, until science "had been universalized by its fusion with mathematics, natural science could not become the common property of all mankind":[102]

When we say that modern science developed only in Western Europe at the time of Galileo in the late Renaissance, we mean surely that there and then alone there developed the fundamental bases of the structure of the natural sciences as we have them today, namely the application of mathematical hypotheses to Nature, the full understanding of the experimental method, the distinction between primary and secondary qualities, the geometrization of space, and the acceptance of the mechanical model of reality.[103]

Throughout his writings Needham refers to this episode as the birth of the "new, or experimental, philosophy,"[104] and while he does not wholly dis-

---

[99] Compare Weber's abbreviated discussion of law in China in *The Religions of China*, pp. 147–50, with Needham, *GT,* chap. 8.

[100] Needham's turning point for modern science has consistently been +1600; but China's technological superiority, if it was that, disappeared by the mid-1400s. Needham's last assessment of the relative achievements of Western and Chinese technology is his "Provisional Balance Sheet" in *SCC* 4/2: 222–25. As we shall see, there are many reasons for rejecting Needham's chronology.

[101] *SCC* 5/2: xxii; and *GT,* p. 16. I must mention at this juncture the fact that the distinguished historian of Chinese science Nathan Sivin takes a rather different view than Needham regarding the utility of raising the questions that Needham raises, and he challenges the assumption of the universality of modern science ("Why the Scientific Revolution Did Not Take Place in China," p. 537). I shall leave further consideration of Sivin's perspective for later.

[102] *GT,* p. 15.

[103] Ibid.

[104] For example, *SCC* 3:156; and "The Evolution of Oecumenical Science: The Roles of Europe and China," *Interdisciplinary Science Reviews* 1, no. 3 (1976): 202–14, at p. 202.

miss the significant continuity between the science of Galileo and that of his medieval predecessors, Needham tends to place an acute stress on the experimental elements of the new philosophy and to neglect the larger intellectual, philosophical, and metaphysical contexts of European thought in which the new science was embedded. It must be said, however, that in the course of Needham's writings, especially his lectures and occasional essays, he has dealt with virtually every aspect – social, cultural, linguistic, and technological – that might conceivably have a bearing on the question of the rise of modern science. The problem is that Needham's results and sociological speculations are so many and varied that they require a separate study to weigh their consistency and validity.

Insofar as the European context is concerned, there is much recent research on the history of medieval science, as well as Arabic science, that would require the reformulation of many of Needham's assumptions. As we shall see in Chapter 4 and elsewhere, the pursuit of natural science in medieval Europe was far more advanced and sophisticated than Needham's account generally concedes. A great deal of research on Chinese science has also been done since the publication of Needham's early volumes in the 1950s.

Second, Needham's case for the superiority of Chinese science before the Galilean revolution rests on the highly contested claim that no meaningful distinction can or should be made between science and technology in history. Conversely, many historians of technology would argue the reverse, claiming that it was only in the late nineteenth and early twentieth centuries that science and technology were intimately connected.[105] In the past, knowledge of the principles of the natural world generally lagged far behind technology, whereas today knowledge of the principles of mechanics, motion, hydraulics, thermodynamics, chemistry, genetics, microparticle forces and behavior, and so forth, is frequently the source of technological innovations. The idea that technology is applied science could only exist where there was a firsthand knowledge of scientific principles that could be applied to the manipulation of the natural world. Not to separate the two in this case may radically blur our inquiries since our attention must be directed toward the development of the systems of ideas and the groups of practitioners who were uniquely identified with the cultivation and development of these new symbol systems. On the other hand, the fact that Chinese technology re-

---

[105] Melvin Kranzburg and Carroll Pursell, eds., *Technology in Western Civilization* (New York: Oxford, 1967), 2 vols.; Sivin, "Why the Scientific Revolution Did Not Take Place in China," p. 532, also doubts the integration of science and technology in earlier history. Likewise, Nelson rejects Needham's claims regarding the affinity between science and technology; see Nelson, *On the Roads to Modernity*, chap. 10.

mained superior to that of the West from the second to the mid-fifteenth century poses its own problem – namely whether or not technology per se has any intrinsic relationship to science and, if it does, why that superior technology (in China) did not spawn the growth of modern science but began itself to stagnate after the sixteenth century.

As noted, during the course of his inquiries, and in his voluminous reflective essays, Joseph Needham seemingly explored every factor and combination of factors – both internal and external – that could be imagined to be a significant impediment to the development of modern science in China. These factors include the inherent structural properties of the Chinese language, the geographic isolation of China, the need for large irrigation networks, philosophies of nature, time, and the cosmos (especially the unique influences of Taoism, Buddhism, Confucianism, and the Mohists), the presence or absence of mathematical ideas and symbolisms, the use of the experimental method, the absence of the idea of a creator God and the idea of laws of nature, and the overwhelming dominance of the Chinese bureaucracy. In several places Needham treats these factors as "facilitating" and "inhibiting" factors.[106] In other places he groups them as four sets of factors: geographical, hydraulic, social, and economic.[107] Insofar as social factors are concerned, Needham's analysis is very eclectic and lacking in focus, despite its brilliance. For example, he never clearly defines the nature of social factors, though he is convinced that social and economic forces, had they been present in China, would have overcome whatever defects existed in Chinese science, thus allowing it to gestate modern science.[108]

For example, Needham wants to dismiss altogether the role and influence of Confucianism in Chinese civilization. He writes: "All explanations in terms of the dominance of Confucian philosophy . . . may be ruled out at the start, for they only invite the further question, why was Chinese civilization such that Confucian philosophy did dominate."[109] Yet, conversely, no one would dismiss explanations regarding Christianity and Christian philosophical ideas as irrelevant in the West and propose to raise the further question of why Christian ideas and doctrines (rather than Judaic, Islamic, and so forth) dominated (and still dominate) the West. In brief, Needham's unabashed commitment to the principle "that the vast historical differences

---

106 A good outline of these factors is provided by Sal P. Restivo, "Joseph Needham and the Comparative Sociology of Chinese and Modern Science," in *Research in Sociology of Knowledge, Science, and Art*, ed. Robert A. Jones (Greenwich, Conn: JAI Press, 1979), 2: 25–51.
107 *GT*, p. 150.
108 *SCC* 3: 167–8.
109 *GT*, p. 150.

between the cultures can be explained by sociological studies"[110] is biased toward a Marxist and materialist account, such that cultural and ethnic differences which are evident in attitudes and patterns of thought, and which otherwise would surely be social factors, are immediately suspect. This leads Needham to undervalue cultural factors and local symbolic idioms.

On the other hand, Needham has much to say in scattered places about Chinese bureaucracy and its impact on Chinese life and thought. Surely Confucianism had much to do with the shaping of this social factor and both Confucianism and bureaucracy ought to be central subjects of inquiry in this connection. The point here, however, is that Needham affirms that the bureaucracy "absolutely prevented the rise of the merchants and the coming of capitalism,"[111] but he does not systematically follow up the implications of this conclusion for the development of the role of the scientist and the rise of modern science. Although he may later expand these considerations, his analysis neglects the study of the institutional contexts within which science was taught and pursued. He is clearly aware that in China, "so long as 'bureaucratic feudalism' remained unchanged, mathematics could not come together with empirical nature-observation and experience to produce something fundamentally new."[112] In the same vein, he elsewhere argues that "there cannot be much doubt . . . that the failure of the rise of the merchant class to power in the state lies at the base of the inhibiting of the rise of modern science in Chinese society,"[113] and it was the Chinese bureaucracy that "absolutely prevented the rise of the merchants" and the coming into being of capitalism. Such an analysis seems to undercut Needham's early Marxist leanings, which lead him to argue: "Whoever would explain the failure of Chinese society to develop modern science had better begin by explaining the failure of Chinese society to develop mercantile and then industrial capitalism."[114] In other words, whether or not one considers the direct effect of the bureaucracy on the development of science (above all, its near monopoly of many domains of scientific practice), or whether one considers the indirect effects of the bureaucracy (through its alleged influence on the merchant class as a carrier of modern science), in either case the bureaucracy must be treated as an independent variable, and in this case, according to Needham's own analysis, it was of overwhelming importance as an inhibiting factor vis-à-vis the rise of modern science. But this is not the conclusion that Needham chooses to emphasize.

[110] Ibid., p. 191.
[111] Ibid., p. 152.
[112] Ibid., p. 212.
[113] Ibid., p. 186.
[114] *GT*, p. 40.

Given the all-pervasiveness of the Chinese bureaucracy, it would seem that Needham has in fact put together (in separate places) a social explanation of the arresting of modern scientific development in China that has little to do with capitalism and a great deal to do with the bureaucratic nature of Chinese education and scientific practice. Such insights ought logically to induce a study of the motives (and processes) whereby the bureaucratic elite prevented the rise of autonomous social collectivities, that is, communities, cities, guilds, colleges, and universities. Such a study might also suggest that the Chinese bureaucracy created a reward system which, by rewarding classical, ethical, and literary scholarship, systematically deflected scholarly pursuits away from natural philosophy and scientific inquiry.

Despite these lacunae in his work, Needham has prepared the way for a comparative and historically grounded sociological analysis of the rise of modern science in civilizational perspective. He has done this, in the first instance, by undertaking a monumental history of the natural sciences and technology in China – albeit a project that still continues. More importantly, his work has shifted the focus from internal factors to external factors that encompass ideas about nature, time, cosmology, natural laws, and cultural ontologies, as well as patterns of conduct and institutional structures. He has done so with the abiding commitment that there is "only one unitary science of nature" and the faith that someday "we shall have an historical account which will allow us to trace an absolute continuity between the first beginnings of astronomy and medicine in Ancient Babylonia, through the advancing natural knowledge of medieval China, India, Islam, and the Classical Western world, to the breakthrough of late Renaissance Europe, when, it has been said, the most effective method of discovery was itself discovered."[115]

I must touch on one additional weakness of Needham's sociological account. Although Needham was among the first in the history of science to stress the potential importance of such metaphysical and extrascientific factors as cultural ontologies, images of law, and natural process, Needham did not focus attention on the corresponding images of man that all societies and civilizations contain. Is man, for example, in a particular time and civilization, thought to be completely rational and hence fully capable of discovering, decoding, and explaining the mysteries of nature? Or is man thought to be too weak in his intellectual powers to divine the secret and unknown processes and mechanisms of Nature that the naked eye can rarely see? Is he permitted to speak openly and possibly critically about the wisdom of the ages or about the official and public accounts of nature and its processes? In

[115] *SCC* 5/5: xxvi.

what forums may these dissenting thoughts be expressed, and can they be freely expressed, discussed, and publicly passed on to wider audiences? These are matters central to an anthropology of man and deserve the highest consideration in the context of the rise of modern science. They are also central to any examination of the institutional location of scientific practice. This angle of vision is generally submerged even by those who are otherwise sympathetic to the idea of exploring indigenous philosophies of nature. Thus Nathan Sivin comments that although one may find probing theoretical discussions in Chinese science that affirm the usefulness of such inquiries, "their authors did not believe that empirical investigations integrated by theory could completely explain physical phenomena. . . . [The] texture of reality is too fine and too subtle to be completely apprehended."[116] This account of Chinese thought about nature is exceedingly valuable, but is it not possible to explore this philosophy of nature from the other direction, that is, from its implicit assumptions about the capacities of man? Is it really the case that Chinese thought – Confucian, Taoist, Mohist, and so forth – had no integral speculation about the intellectual powers and limits of man? And how did these conceptions of man's reason and rationality contribute to (or inhibit) the conduct of scientific research and discourse?

Needham's writings display a great reluctance to deal with these themes, and one feels that this is due to both Needham's Marxist commitments and his great fear that consideration of such culturally located differences would lead to racist characterizations. However that may be, no account of the rise of modern science can be complete without a corresponding analysis of the theories of man, of mind, soul, psyche, and conscience that have animated and authorized men over the centuries to speak freely their deepest thoughts about the world and its ontologies. Inevitably such discussions must focus on the social institutions that legitimate and standardize such viewpoints and on how such official (and frequently legal) structures shape social and intellectual discourse. It must be remembered that in the original projection for the seven volumes of *Science and Civilisation in China* Needham intentionally set aside a section for a final analysis of all the social and economic factors bearing on the "poverties and triumphs" of Chinese science. But as the project grew ever larger, the hope of realizing that goal receded.[117]

---

[116] Sivin, "Max Weber, Joseph Needham, Benjamin Nelson," p. 46.

[117] In his comments on Needham and his project, the late Derek J. de Solla Price hinted that Needham was beginning to feel that the effort to deal with this summing up of the social and economic background might take another lifetime, and thus it seemed equally far off. See "Joseph Needham and the Science of China," in *Chinese Science: Explorations of an Ancient Tradition*, ed. S. Nakayama and N. Sivin (Cambridge, Mass.: MIT Press, 1973), p. 16.

Benjamin Nelson: universalization and wider
spheres of discourse

We may gain further insights into the importance of Needham's work by considering the contributions to the comparative sociology of science made by Benjamin Nelson. During the last decade of his life, Nelson became increasingly concerned with the problem of the origins and development of modern science. In part this was a product of his training as a medieval historian, and in part it was an expression of his abiding interest in philosophy and its interaction with theology, law, and science. These interests were further solidified by his ongoing friendships with such philosophers of science as N. R. Hanson, Karl Popper, Imre Lakatos, Stephen Toulmin, and others.

But Nelson also knew that during the High Middle Ages, the seminal thinkers of the West were well versed in all these fields such that great university teachers were experts in natural philosophy, metaphysics, theology, and even canon law. Furthermore, from his knowledge of the great debates of that era, Nelson knew that the debates regarding the permissibility of arguing questions in science and natural philosophy, which paved the way for the scientific revolution associated with the names of Copernicus and Galileo, had precious little to do with the central issues at stake in the Reformation – except, of course, that both movements represented strong challenges to the authority of the Catholic church. In other words, the mathematical-cosmological aspect of the scientific revolution was born in Catholic culture areas and obviously preceded the Reformation.[118] Nelson was keen to rectify the impression that Max Weber's thesis regarding religion and the rise of capitalism (the Protestant ethic thesis) ought to be extended (as Weber seemed to hint in his closing pages) to explain the scientific revolution, which clearly had cultural roots antedating the Reformation, and the scientific movement that became so strong in England in the seventeenth century.

Given Nelson's historical training, it was natural for him to have a broader view of the issues and debates surrounding the Continental scientific revolution than those who took a narrow and positivistic view of the matter, under the assumption that the rise of modern science was the more or less direct process of rejecting all metaphysics, religiously grounded conceptions included, and the replacement of it with the unequivocal method of experiment.[119] In contrast to such a view, Nelson argued that the road to modern

---

[118] See *On the Roads to Modernity*, chaps. 7, 8, and 9.
[119] This is, of course, a long-standing view among nineteenth- and early twentieth-century

science was paved with stepping-stones fashioned in the argot of Christian theology and Western philosophy (a mix of Aristotle, Plato, and even Averroes), as well as by uniquely Western legal conceptions. The idea, for example, that the world is a rational and coherent order, that the world is a machine, that a divine being created the world according to "number, weight and measure," are all medieval themes enunciated by Christian clerics cum natural philosophers, theologians, and even canonists.[120] Indeed, the idea of laws of nature had Judeo-Christian groundings far stronger than any purely scientific arguments available at the time. In addition, Nelson was taken with his own suggestion that the idea of conscience was central to the empowerment of men as rational actors (above all in ethical matters) and in the grounding of subjective certitude about natural phenomena. From his point of view, there was little doubt that Copernicus and Galileo were committed to a realist interpretation of the world, and that this commitment was founded on the theological conceptions that men have reason and conscience, which empower them to arrive at *subjective certitude* beyond *objective* demonstration, and that this is acceptable in the sight of God as well as man.[121] In brief, Nelson's entrée to the sociology of science and the special question of the rise of modern science requires that one take all the symbolisms – theological, natural, and mathematical – equally seriously in unraveling the unique success of modern science in the West. The deficiencies of other civilizations with regard to the development of science, Nelson believed, were not matters of scientific technique in the narrow sense, but rather deficiencies in the symbolic technologies of the sociocultural domain. They were deficiencies in the structures and institutions, which either opened up wider spheres of public discourse and participation or placed severe limitations on such openings.

It is at this point that Nelson's concern for the fashioning of new universal structures – social, intellectual, and political – comes into view. From his earlier study, *The Idea of Usury*,[122] Nelson had argued that a very important dynamic of Western civilization was the pursuit of universal communities of

historians and historians of science; see, for example, W. E. H. Lecky, *History of the Rise and Influence of the Spirit of Rationalism in Europe,* rev. ed. (New York: Appleton, 1871), and even George Sarton, *Introduction to the History of Science* (Baltimore, Md., Williams and Wilkins, 1927–48), 3 vols. in 5 parts. Due to the work of historians of science during the last twenty years or so, this view is largely outmoded, though it would be too strong to say that the sociological view has triumphed.

[120] Nelson, "Certitude and the Books of Scripture, Nature, and Conscience" in *On the Roads to Modernity,* especially pp. 158–9.

[121] See ibid., as well as Chapter 3 below.

[122] Benjamin Nelson, *The Idea of Usury: From Tribal Brotherhood to Universal Otherhood,* 2d enlarged ed. (Chicago: University of Chicago Press, 1969).

discourse and participation. His study of the fate of the idea of usury revealed in the Old Testament – whereby usury was forbidden among Jews but allowed between Jews and others – demonstrated that there had been a progression "from tribal brotherhood to universal otherhood" whereby each person became equally an "other" rather than a "tribal brother." Nelson saw this new ethic worked out most clearly in the nexus between Christian theology and law, above all in the writings of John Calvin. Calvin's argument in favor of the permissibility of usury was that now, in the Christian era, all men were equally brothers in Christ, and therefore usury was permissible provided that one always remembered to temper the practice of lending with Christian charity. Hence, a new level of universalism was achieved by transposing a religious commitment into a legal principle, a principle whereby all are equally "others" and the rule applies to all regardless of one's denomination.

The study of movements toward such universalisms captured Nelson for the rest of his life. When he turned to the question of the rise of modern science, he therefore saw it as a paradigmatic setting within which we might see yet another passage from an elitist and exclusive discourse to a freely open and public discourse. Modern science could then be seen as the triumph of a universalistic mode of discourse that is in the service of a completely open and unending intellectual quest for new knowledge. It was such a point of view that led him to assert that in the comparative and historical sociology of science an "indispensable reference point . . . will be found in the study of the factors working to promote and those working to retard the forging of new types of universalities and universalizations necessary for the institutionalization of innovation in the 'advancement of science.' "[123] Joseph Needham's exceedingly acute stresses on the idea of ecumenical science represented a profound meeting of two minds, for this was exactly the formulation that Nelson was seeking in his inquiries regarding the development of modern science. If we take that ecumenical view as the reference point, then a consideration of the breakthrough to modern sciences (and the realization of "the highest levels of universalization") ought to focus on three interrelated sets of issues. The first concerns the processes by which the "bars to freedom of entry and exit from the communities of learners and participants in the communities of discourse" can be overcome and "inherited invidious dualisms" transcended; second, attention must be given to the means and mechanisms by which "incentives to produce and distribute warranted knowledge, including new [theoretical] mappings and innovative procedures" are produced; and, third, attention ought to be given to the

---

[123] *On the Roads to Modernity*, p. 11.

processes by which "blocks to the achievement of ever-higher levels of generality in the language structures – written and spoken" have been, and can be, surmounted.[124] This last entails the creation of new, more abstract and universal symbolisms that allow the resolution of long-standing technical and conceptual puzzles blocking social and intellectual advancement. Nelson was fully persuaded that the high road to cultural and civilizational advancement depended upon the continuous creation of new translocal and transnational symbolisms, which open up new freedoms of discourse and participation. From a sociological point of view, the study of legal structures and formalisms would also be indispensable, for legal formalisms serve to institutionalize patterns of behavior. Therefore, the study of the rise and development of modern science ought to be viewed from this point of view, in addition to the more narrowly technical and mathematical. For, from a civilizational point of view, Nelson wrote,

it is not nearly so important whether in any given science a given people did or did not actually make an advance upon the Greeks in respect to one or another discipline – for example, chemistry, optics, and mathematics. The fundamental issue is whether there did occur a comprehensive breakthrough in the moralities of thought and in the logics of decision which open out the possibility of creative advance in the direction of wider universalities of discourse and participation in the confirmation of improved *rationales*.[125]

The advancement of science, from this point of view, is not just a technical question of new mathematical solutions, greatly refined experiment and observation, or new theoretical formulations. It is the result of intellectual breakthroughs that allow thinkers to apply the new symbolisms and new conceptions which break the bounds of traditional wisdom and association, as well as the inherited logics of decision. Such breakthroughs allow the fashioning of new and expanded *neutral spaces* wherein people are free to express their individual and collective wills, to freely exchange ideas with others, and to openly argue for new scientific, legal, ethical, theological, and social conceptions. Only by such means can public discourse work toward the uncoerced realization of individual and collective aspirations. And only by such means can the highest levels of creativity be realized. One can find, as we shall see later, impressive expressions of scientific daring and innovation among all the cultures and peoples of the world (above all, among the Arabs), but for these innovations to mature and yield modern science, new social and institutional forums of discourse and participation must be opened up.

---

[124] Ibid., p. 111f.
[125] Ibid., pp. 98–9.

In Nelson's view, the early modern revolution in science and philosophy of the sixteenth and seventeenth centuries was precisely this kind of struggle: "The founders of modern science and philosophy," he argued, "were anything but skeptics. They were instead committed spokesmen of the new truths clearly proclaimed by the Book of Nature which, they supposed, revealed secrets to all who earnestly applied themselves in good faith and deciphered the signs so lavishly made by the Author of Nature."[126]

In the case of Galileo and the Copernican theory, these new truths "challenged the dominant fictionalism [of the Church] [and] evidently raised questions about the more or less accredited interpretations of scriptural passages by the theologians."[127] Here, one might say, was a classic confrontation between the accepted logics and official interpretations that reoccur in history and have often come out badly for the people involved. In this case, Galileo was only politely confined to his villa. But what was at stake was an important question. "The fundamental issue at stake in the struggle over the Copernican hypothesis was not whether the particular theory had or had not been established but whether in the last analysis the decision regarding truth and certitude could be claimed by anyone who was not an officially authorized interpreter of revelation."[128] Although this was indeed a dramatic confrontation, a showdown between official authority and the freedom of the individual with many ramifications, it was in fact the last gasp of a restrictive ideology, which no longer had the power to regulate such questions. The earlier architects of Christian theology, canon law, and the universities had created the dynamic structures and processes that virtually guaranteed the continuing expansion of the realms of the mind and open discourse. In addition, given the significance of the Reformation for political and intellectual life, no central authority in Western Christendom could any longer adjudicate the fundamental issues of intellectual life as practiced in science. But this is to get ahead of our story.

Benjamin Nelson's pioneering exploration of these questions – building on the insights of Weber, Needham, Henry Sumner Maine, Durkheim, and many others, especially medieval legal and ecclesiastical historians – yielded a powerful new alternative means for exploring questions in the comparative historical sociology of science – above all, in civilizational perspective. It is on the basis of these insights that I shall consider the development of science in Arabic-Islamic as well as Chinese civilization.

---

[126] Nelson, "The Early Modern Revolution," in *On the Roads to Modernity*, p. 132.
[127] Ibid., p. 133.
[128] Ibid.

Conclusion: the issues at hand

It may now be said that, with major but limited exceptions, the comparative historical sociology of science has been neglected.[129] At the same time, a few rich and suggestive works have been written, though these achievements have been realized in almost complete isolation from each other. The masterly work of Robert K. Merton, *Science, Technology, and Society in Seventeenth-Century England,* first published in 1938, did not result in others following his path.[130] Instead, his students turned to studies of the reward system in science.[131]

On the other hand, the one person who undertook to study the origins and development of the role of the scientist, Joseph Ben-David, almost wholly neglected Merton's work on the ethos of science, as well as Merton's theoretical discussion of roles and role-sets. He also neglected religious and legal history, as well as the available materials describing the rich tradition of medieval science located, as we shall see, in the universities of the West. And, although Ben-David went back to the Greeks, he completely omitted any reference (must less discussion of) the nature of scientific practice and its institutional location among the Arabs during the intervening thousand years. Accordingly, his analysis provides no clues as to why science suddenly took off in the West when scientific theory and practice among the Arabs, Christians, and Jews of Arabic-Islamic civilization, especially in astronomy and mathematics, was vastly superior to the West up until the thirteenth and fourteenth centuries.

By bringing together the insights of Merton, Ben-David, Kuhn, Needham, and Nelson, we can see the need for a broader and yet more integrated approach to the origins and development of modern science. Even in the writings of the most internalist of these figures, Thomas Kuhn, we discover a whole array of values and metaphysical commitments that are part and parcel of scientific paradigms. One cannot adequately understand the rise of modern science as an idea system, nor the rise of the role of the scientist, without consideration of such factors.

[129] In her impressive review of the current state of the sociology of science, Harriet Zuckerman does not even attempt an overview of the comparative and historical sociology of science.
[130] Some thoughts about this can be found in my review of Merton's *The Sociology of Science* in *The Journal for the Scientific Study of Religion* 14, no. 1 (1975): 70–2, and in Nelson's review of the reissue of *Science, Technology, and Society in Seventeenth-Century England* (in 1970) in the *American Journal of Sociology* 78, no. 1 (1972): 233–31; reprinted in *Varieties of Political Expression in Sociology,* ed. Howard Becker (Chicago: University of Chicago Press, 1973).
[131] See n. 19 above.

Similarly, those scholars, such as Joseph Needham and Benjamin Nelson, who have cast an eye to other civilizations with the hope of gaining further insight into the making of modern science and the uniqueness of the West, have found it necessary to consider such matters as philosophies of nature, conceptions of law and natural law, as well as images of man and his rationality. The deeper one digs into such questions and specialized histories, the more one becomes impressed with the strength, originality, and vitality of indigenous scientific conceptions in earlier periods. Conversely, the more one marvels at the intellectual and theoretical successes of the diverse peoples of the world, the more one becomes struck by the social, institutional, and indeed legal impediments that have blocked access to the open society and prevented the opening of broader spheres of discourse and the free flow of information. To focus on such questions puts one's attention squarely on the evolution of the institutional arrangements in the civilizations of the world that have either facilitated or inhibited the breakthrough to wider spheres of discourse. At the same time it provides a reminder that institutions are ideas that have been given concrete realization through the use of forensic devices. It is astonishing in retrospect that the Chinese invented movable type printing (as well as paper) four hundred years before the West, yet neither in China where this invention first appeared nor in Arabic-Islamic civilization (which had the most direct access to the new print technology) was there anything like the social and intellectual revolution that occurred in the West in the twelfth and thirteenth centuries. Indeed, Arabic-Islamic civilization placed a ban on the use of printing until the early nineteenth century (with a brief interlude of toleration in the early eighteenth century), and in both civilizations (Islamic and Chinese), the new and more advanced Western print technology of the fifteenth century was introduced in the nineteenth century as if printing had never existed before.[132] The path to modern science is the path to free and open discourse and that is a major puzzle for the sociological imagination, as well as the inquiry at hand.

---

[132] For the invention and use of printing in China, see *SCC* 5/1; *Paper and Printing* by Tsien Tsuen-hsuin; as well as T. F. Carter, *The Invention of Printing in China and Its Spread Westward*, rev. ed. by L. C. Goodrich (New York: Ronald Press, 1955), especially chap. 15, "Islam as a Barrier to Printing." On printing in Islam, see Johannes Pedersen, *The Arabic Book* (Princeton, N.J.: Princeton University Press, 1984); and for the Western case, Elizabeth Eisenstein, *The Printing Press as an Agent of Change: Communications and Cultural Transformation in Early Modern Europe*, 2 vols. (New York: Cambridge University Press, 1979).

2

◁══▷

Arabic science and the Islamic world

The problem of Arabic science

The problem of Arabic science has at least two dimensions. One concerns the failure of Arabic science to give birth to modern science; the other concerns the apparent decline and retrogression of scientific thought and practice in Arabic-Islamic civilization after the thirteenth century. Although the question of why intellectual thought retrogressed after the golden era[1] is a matter of considerable interest to the inhabitants of the contemporary Middle East, it is a problem that lies outside the bounds of the present inquiry.[2]

[1] In this I follow Marshall Hodgson's view that in the five centuries after 945 A.D., "the former society of the caliphate was replaced by a constantly expanding, linguistically and culturally international society ruled by numerous governments," and that "it was this international Islamicate society [that] was certainly the most widely spread and influential on the globe." Marshall G. S. Hodgson, *The Venture of Islam*, 3 vols. (Chicago: University of Chicago Press, 1974), 2: 3. But I would draw the line in terms of significant cultural and scientific growth at the end of the thirteenth century. There were significant scientific events after that, but they were minor in comparison to what was taking place in Europe. Also see Ira Lapidus, *A History of Islamic Societies* (New York: Cambridge University Press, 1988). For an overview of Arabic-Islamic civilization, including Islamic art, science, and architecture, see Bernard Lewis, ed., *Islam and the Arab World* (New York: Knopf, 1976).

[2] A likely explanation of it, however, is to be found in the pattern of conversion to Islam over the centuries. For the first several centuries of the Islamic empire, the percentage of subjects who were Muslims in many areas of the empire remained less than a majority. It was not until about the tenth century that the preexisting communal structures of the non-Islamic peoples were weakened so that widespread conversion to Islam took place. Accordingly, the tenth century marks a turning point when rates of conversion soared, and with this new wave of conversion to Islam, the percentage of freethinkers who were not fearful of the corroding effects of the foreign sciences also dramatically declined, and this dynamic probably had negative consequences for the pursuit of the natural sciences and intellectual life in general. Although there will be those who will challenge this hypothesis, it has, I believe, the virtue of

47

Our concern is with the fact that from the eighth century to the end of the fourteenth, Arabic science was probably the most advanced science in the world, greatly surpassing the West and China. In virtually every field of endeavor – in astronomy, alchemy, mathematics, medicine, optics, and so forth – Arabic scientists (that is, Middle Eastern individuals primarily using the Arabic language but including Arabs, Iranians, Christians, Jews, and others) were in the forefront of scientific advance. The facts, theories, and scientific speculations contained in their treatises were the most advanced to be had anywhere in the world, including China.[3] This is so, it seems to me, for two reasons.

The first is that while the Greek scientific heritage was lost to the Western world for the centuries between the collapse of the Roman Empire in the fifth century and the great translation movement of the twelfth and thirteenth centuries, the Arabs[4] had virtually full access to that heritage from the eighth century onward. This occurred because of a momentous translation effort whereby the great works of Greece and other cultures were translated into Arabic. While the transmission of these ancient sciences into Arabic-Islamic civilization was selective, it was thoroughly representative of Greek

being consistent, not only with the facts noted above, but with the additional fact that even today there is no Islamic equivalent to a Hong Kong, a Singapore, a Taiwan, or, much less, a Japan, among the Islamic countries of the world, despite the fact that at least six of them have enormous oil wealth, including the per capita richest, Brunei, which could be directed toward this goal were it considered desirable. (More on this subject at the end of Chap. 6 and in the Epilogue.

[3] The superiority of Arabic science during the Middle Ages was first documented in great detail by George Sarton in his magisterial *Introduction to the History of Science*, 3 vols. in 5 parts (Baltimore: Williams and Wilkins, 1927–48). Since then a number of synoptic overviews of science in classical Islamic civilization have been written. These include, among others, Max Meyerhof, "Science and Medicine," in *The Legacy of Islam*, 1st ed., ed. T. Arnold and A. Guillaume (London: Oxford University Press, 1931), pp. 311–56; G. Anawati, "Science," in *The Cambridge History of Islam*, edited by P. M. Holt (New York: Cambridge University Press, 1970), 2:741–80; Martin Plessner, "Science," in *The Legacy of Islam*, 2d ed., ed. J. Schacht and C. E. Bosworth (Oxford: The Clarendon Press, 1974), pp. 425–60; A. I. Sabra, "The Scientific Enterprise," in *Islam and the Arab World*, ed. Bernard Lewis, pp. 181–92; and a great variety of essays in specialized journals as well as the *Dictionary of Scientific Biography* (hereafter cited as *DSB*). To my knowledge, no one has attempted to compare the achievements of Arabic science with those of Chinese science during the same period as a result of Joseph Needham's monumental study, *Science and Civilisation in China* (New York: Cambridge University Press, 1954–), 7 vols., in progress (hereafter cited as *SCC*). However, there is strong evidence, as we shall see, that Arabic science contributed the most intellectual capital to the breakthrough to modern science in the West in the twelfth and thirteenth centuries and thereafter.

[4] When I use the term *Arab* in this context, I use it in the broad sense of those individuals using the Arabic tongue for purposes of communication. This often included ethnically diverse peoples, such as Iranians, Syrians, Egyptians, and Turks, as well as Jews, Spaniards, and Christians.

scientific and philosophic thought as a whole.[5] Moreover, the Arabic borrowing of the Hindu numeral system must be accorded high recognition.

On the other hand, present scholarship regarding Chinese science still maintains that China developed independently along its own lines in most scientific areas. There were interchanges between the Chinese and the Indians, as well as between the Chinese and the Arabs of the Middle East, but these did not result in major Chinese alterations. The important fact is that the Chinese knew virtually nothing of Aristotle, Euclid, Ptolemy, or Galen, whereas their works, especially in modified and amplified Arabic forms, became major points of departure in the development of modern Western science.

In the case of mathematics, for example, the Chinese path of development was clearly independent of that of the Arabic Middle East so that its early achievements, especially in algebra, occurred in a form that could not easily (if at all) be translated into Arabic and Western idioms. In the famous Chinese work called *The Nine Chapters on the Mathematical Procedures* (from about the first century A.D.), there are discussions of arithmetic fractions, the statement of formulas for the computation of areas and volumes, the solution of systems of simultaneous equations, and procedures of square and cube root extraction.[6] Moreover, during the Sung dynasty (ca. 960–1279), Chinese mathematics underwent a vigorous new growth spurt in algebraic computation.[7] Nevertheless, the Chinese system of representation and positional notation as well as its techniques of computation (with counting rods), were cumbersome and not nearly as generalizable and easy to use as the Arabic-Hindu numeral system. And this system, which was clearly located in a decimal place-value system, had been available in al-Khwarizmi's work since about 825.[8] Consequently, the development of mathematics in China required a move from computation with counting rods to the use of the abacus (generally in about the sixteenth century) and the incorporation and use of the zero (in the thirteenth and fourteenth centuries). Only in the

---

5 See F. E. Peters, *Aristotle and the Arabs* (New York: New York University Press, 1968); R. Walzer, *Greek into Arabic* (Columbia, S.C.: University of South Carolina Press, 1962); as well as Max Meyerhof, "Von Alexandria nach Baghdad," in *Sitzungsberichte der Prüssischen Akademie der Wissenschaften* no. 23 (1930): 389–429.

6 "Mathematics in China and Japan," *Encyclopedia Britannica* 23 (1991): 633b–e. A parallel description of the status and achievements of the medieval Arabs is in ibid., "Mathematics in Medieval Islam," pp. 613–15.

7 Ibid., and see Lî Yan and Dù Shíràn, *Chinese Mathematics: A Concise History* (Oxford: Clarendon Press, 1987), pp. 109ff, as well as Needham *SCC* 3: 38ff.

8 E. S. Kennedy, "The Exact Sciences [The Period of the Arab Invasion to the Suljugs]," *The Cambridge History of Iran* 4 (1975): 380. And see Michael Mahoney, "Mathematics," in *Science in the Middle Ages*, ed. David C. Lindberg (Chicago: University of Chicago Press, 1978), pp. 151f.

seventeenth century was the use of paper and pen calculation introduced for arithmetic operations.[9] In contrast, the West had begun using the abacus in the eleventh and twelfth centuries and this, along with the Arabic-Hindu numeral system, induced the Europeans to begin the transition away from the use of the abacus to paper and pen calculations.[10]

In brief, the defect of Chinese mathematical and scientific thought was that it lacked the logic of proof, Euclid's *Elements* of geometry, and Ptolemy's planetary models contained in the *Almagest* and his *Planetary Hypotheses*. Likewise, it lacked the Hindu-Arabic numerals, and the zero until about the thirteenth century.[11] Whether or not it is of great significance, Joseph Needham found it odd that for "a people who carried algebra so far, the equation form remained implicit, and there was no indigenous development of the equality sign (=)."[12] But perhaps most important of all, trigonometry – an essential part of mathematics for astronomy – was invented by the Arabs and not developed at all by the Chinese.[13] To compensate for this, the Chinese employed Arab astronomers in the Chinese Astronomical Bureau in Peking from the thirteenth century onward.[14] Yet the transition to a geometrical astronomy (as opposed to a "mathematical point estimation" model)[15] did not occur in China until the seventeenth century when the Europeans arrived in the form of Jesuit missionaries.[16]

---

[9] Lî Yan and Dù Shíràn, *Chinese Mathematics,* p. 191.

[10] Mahoney, "Mathematics," p. 151.

[11] Needham, *SCC* 3: 10, 43.

[12] Ibid.

[13] According to Baron Carra de Vaux, the Arabs "were indisputably the founders of plane and spherical trigonometry, which properly speaking, did not exist among the Greeks," in "Astronomy and Mathematics," in *The Legacy of Islam,* 1st ed., p. 276. Likewise, E. S. Kennedy agrees that trigonometry, the study of the plane and spherical triangle, "was essentially a creation of Arabic-writing scientists." "The Arabic Heritage in the Exact Sciences," *Al-Abhath* 23 (1970): 337. Also see Kennedy, "The History of Trigonometry: An Overview," in *Studies in the Islamic Exact Sciences,* ed. E. S. Kennedy et al. (Beirut: American University Beirut Press, 1983), pp. 3–29.

[14] Nathan Sivin, "Wang Hsi-Shan," *DSB* 14: 159–68, at p. 159. Also see Sivin, "Why the Scientific Revolution Did Not Take Place in China – Or Didn't It?" in *Transformation and Tradition in the Sciences,* edited by E. Mendelsohn (New York: Cambridge University Press, 1984), pp. 531–54.

[15] Nathan Sivin, "Copernicus in China," *Studia Copernicana* 6 (1973): 63–122, and "Wang Hsi-Shan," *DSB* 14: 159–68.

[16] This episode centers on the Jesuit Matteo Ricci, and his mission in China has been made into a colorful story by Jonathan Spence in *The Memory Palace of Matteo Ricci* (New York: Viking, 1984); and see P. D'Elia, *Galileo in China: Relations Through the Roman College Between Galileo and the Jesuit Scientist-Missionaries (1610–1640)* (Cambridge, Mass.: Harvard University Press, 1960); and Sivin, "Wang Hsi-Shan," for the rethinking of Chinese astronomy in the seventeenth century. For a more detailed account, see John Henderson, *The Development and Decline of Chinese Cosmology* (New York: Columbia University Press, 1984).

The Arab astronomers and mathematicians working in the Marâgha observatory in western Iran, and especially the Damascene timekeeper Ibn al-Shatir (d. 1375), had improved the Ptolemaic system so that it was mathematically equivalent[17] to the Copernican system (though still geocentric). Or more accurately stated, the planetary models of Copernicus, appearing 150 years after the time of Ibn al-Shatir, are actually duplicates of the models developed by the Marâgha astronomers (on which more below). The Chinese, having less hostile relations with Islam, and possibly through the intermediaries of the Mongols who sponsored the Marâgha observatory, could have gained access to the most advanced astronomy in the world two centuries before the West did. In Joseph Needham's view, the Chinese astronomers had every chance to learn about it.[18] Of course Euclid's geometry was indispensable to this, but this work was also widely discussed and available in many editions. Even in the fourteenth century, after Euclid had been translated into Latin, there were far more Arabic recensions of and commentaries on Euclid than there were in Latin.[19] Furthermore, there is evidence that a copy of Euclid's *Elements*, translated into Chinese, was in the imperial library in the thirteenth century.[20] The Chinese could have found these works and adopted them if they had been so inclined.[21]

In optics, which in early science probably played something like the role of physics in modern science, the Chinese, in Needham's words, "never equalled the highest level attained by the Islamic students of light such as Ibn al-Haytham."[22] Among other reasons, this was a reflection of the fact that the Chinese were "greatly hampered by the lack of the Greek deductive geometry" that the Arabs had inherited.[23] Finally, though we think of physics as the fundamental natural science, Joseph Needham concluded that there was very little systematic physical thought among the Chinese.[24] While one can find Chinese physical thought, "one can hardly speak of a developed science of physics,"[25] and it lacked powerful systematic thinkers, who would correspond to the so-called precursors of Galileo, represented in

[17] The precise meaning of this is explained below.
[18] Needham, *SCC* 3: 50.
[19] John Murdoch, "Euclid: Transmission of the Elements," in *DSB* 4: 443–65.
[20] Needham, *SCC* 3: 105.
[21] Of course this is not the whole story regarding the strengths and weaknesses of Chinese science. My point here is simply that Arabic science appears to have had every advantage over the West as well as China in the twelfth and thirteenth centuries. More on Chinese science in Chapter 8.
[22] Needham, *SCC* 4/1: 78.
[23] Ibid., 4: xxiii.
[24] Ibid., 4/1, sec. 26, p. 1.
[25] Ibid.

the West by such names as Philoponus, Buridan, Bradwardine, and Nichole d'Oresme.[26]

Considered altogether, in mathematics, astronomy, optics, physics, and medicine, Arabic science was the most advanced in the world. In different fields it lost the lead at different points in time, but it can be said that up until the Copernican revolution of the sixteenth century, its astronomical models were the most advanced in the world. Consequently, the problem at hand asks why Arabic science, given its technical and scientific superiority built up over five centuries or so, did not give rise to modern science.

An important starting point for such an inquiry is the realization that the sciences we call the natural sciences were called the foreign sciences by the Muslims. In contrast, the so-called Islamic sciences were those devoted to the study of the Quran, the traditions of the Prophet (*hadith*), legal knowledge (*fiqh*), theology (*kalam*), poetry, and the Arabic language. For the purpose of dividing inheritances, however, arithmetic was also an important subject of study. Moreover, due to ritual and religious prescriptions, time-keepers (*muwaqqits*) found it necessary to use geometry and eventually to invent trigonometry in order to arrive at the requisite calculations to determine the direction to Mecca (the *qibla*) for prayer. In short, driven by both curiosity and religious motives, the Arab-Muslim world from the eighth to the fourteenth century achieved significant heights of scientific advance, but thereafter (and perhaps as early as the twelfth century) went into decline and even retrogression. This did not happen uniformly across all fields, but in general scientific advance was on the wane. For example, speaking of the pursuit and development of mathematics in Timurid Iran (ca. 1350–1550), E. S. Kennedy reports that in numerical analysis brilliant work was done: "Jamshid [Kashi's] computational algorisms exhibited a feel for elegance, precision, and control which had never been seen before, and which was not to be surpassed for a long time to come. . . . All things considered, Iran's scientific output, though weakening, may have maintained her in a leading position through the fifteenth century. Thereafter the lead passed to the West."[27] A similar pattern appears to have existed in medicine where great advances were also made up until the thirteenth century.[28] Thus,

[26] Ibid., 4:2.
[27] E. S. Kennedy, "The Exact Sciences in Timurid Iran," in *The Cambridge History of Iran* (1986), 6: 580.
[28] For overviews, see Michael Dols, Introduction to *Medieval Islamic Medicine: Ibn Ridwan's Treatise "On the Prevention of Bodily Ills in Egypt"* (Berkeley and Los Angeles: University of California Press, 1984), pp. 3–73; E. G. Browne, *Arabian Medicine* (New York: Cam-

in some fields advances continued to be made, yet they did not culminate in a scientific revolution.

This situation is a deep puzzle about which many have speculated for at least the last 150 years. The factors identified as responsible for the failure of Arabic science to give birth to modern science range from racial factors, the dominance of religious orthodoxy and political tyranny, and matters of general psychology to economic factors and the failure of Arab natural philosophers to fully develop and use the experimental method.[29] A common formulation of the negative influence of religious forces on scientific advance suggests that the twelfth and thirteenth centuries witnessed the rise of mysticism as a social movement. This in turn spawned religious intolerance, especially for the natural sciences and the substitution of the pursuit of the occult sciences in place of the study of the Greek and rational sciences.[30]

bridge University Press, 1962, reprinted); J. Christoph Bürgel, "Secular and Religious Features of Medieval Arabic Medicine," in *Asian Medical Systems: A Comparative Study,* ed. Charles Leslie (Berkeley and Los Angeles: University of California Press), pp. 44–62; Joseph Graziani, "The Contribution of Arabic Medicine to the Health Professions During the Eleventh Century," *Episteme* 10 (1976): 126–43; Cyril Elgood, *Medical History of Persia and the Eastern Caliphate* (Cambridge: Cambridge University Press, 1951); Sami Hamarneh, "Medical Education and Practice in Medieval Islam," in *The History of Medical Education,* ed. C. D. O'Malley (Berkeley and Los Angeles: University of California Press, 1970), pp. 39–71; Hamarneh, "Arabic Medicine and Its Impact on Teaching and Practice of the Healing Arts in the West," *Oriente e Occidente* 13 (1971): 395–426; Max Meyerhof, "Science and Medicine," pp. 311–56; and Manfred Ullmann, *Islamic Medicine* (Edinburgh: Edinburgh University Press, 1978).

29 Ernest Renan, "Islam et La Science," in *Discours et Conferences* (Paris: Calmann-Levy, 1919, 6th ed.), pp. 375–402; Sarton, *Introduction to the History of Science,* 3 vols. in 5 parts (Baltimore, Md.: Williams and Wilkins, 1927–48); Carra de Vaux, "Astronomy and Mathematics," pp. 376–97. A general approach to Islamic decadence is to be found in J. J. Saunders, "The Problem of Islamic Decadence," *Journal of World History* 7 (1963): 701–20. A broad overview of the decadence problem is also found in *Classicisme et déclin culturel dans l'histoire de l'Islam,* ed. R. Brunschwig and G. E. Von Grunebaum (Paris: G.-P. Maisonneuve et Larose, 1957), especially Willy Hartner, "Quand et comment s'est arrêté l'essor de la culture scientifique dans l'Islam?" pp. 319–37. Renan suggested the influence of racial factors, but he placed greater emphasis on the intolerance Islam had for reason. In his early writings, Sarton pointed his finger at the failure of Arabic science to develop and apply the experimental method (Sarton, *Introduction to the History of Science* 1: 29). As it turns out, there are at least three different sources of experimental method in Arabic science: in medicine, optics, and astronomy (on which more later). Carra de Vaux's only explanation had something to do with "very obscure problems of general psychology," in "Astronomy and Mathematics," p. 397.

30 This is the theme of Armand Abel, "La place des sciences occultes dans la décadence," in *Classicisme et déclin culturel dans l'histoire de l'Islam,* pp. 291–311. Others, however, argue that Abel overstates his case; see John W. Livingston, "Ibn Qayyim al-Jawziyyah: A Fourteenth-Century Defense Against Astrological Divination and Alchemical Transmutation," *Journal of the American Oriental Society* 91 (1971): 96–103. The problem is that

During the past three decades very significant advances have been made in our understanding of Arabic science, but these have not shed any light on the puzzle of why Arabic science went into decline after the thirteenth century, and why it failed to give birth to modern science. In fact, the portrait we now have only heightens the mystery. Historians of Arabic science have directed their efforts toward discovering the originality of Arabic science.[31] This is most noticeable in the history of astronomy, where researchers have impressively shown the various steps taken in astronomical thought that led to the development in the thirteenth and fourteenth centuries of a planetary system which was mathematically equivalent to that of Copernicus.[32] That is to say, (1) Copernicus uses the Tusi couple (see Figures 2 and 3) as the Marâgha astronomers did, (2) his planetary models for longitude in the *Commentariolus* are based on those of Ibn al-Shatir, while (3) those for the superior planets in *De revolutionibus* use Marâgha models, and (4) the lunar models of Copernicus and the Marâgha school are identical.[33] It is this essential equivalence of models that prompted Noel Swerdlow to ask, "not whether, but when, where, and in what form" Copernicus learned of the Marâgha theory.[34]

The achievements of Arabic astronomy

Since the modern scientific revolution of Europe in the sixteenth and seventeenth centuries is generally centered on developments in astronomy and the philosophical implications of the work of Copernicus (especially as taken up

Livingston's essay is but a single counterexample that provides no general assessment of the general pattern of change.

[31] This concern is reflected in Shlomo Pines, "What Was Original in Arabic Science," in *Scientific Change*, ed. A. C. Crombie (New York: Basic Books, 1963), pp. 181–205; as well as Willy Hartner and Matthias Schramm, "Al-Biruni and the Theory of the Solar Apogee: An Example of Originality in Arabic Science," in *Scientific Change*, pp. 206–18.

[32] For an analysis of these astronomical systems, see E. S. Kennedy and Victor Roberts, "The Planetary Theory of Ibn al-Shâtir," *Isis* 50 (1959):227–35; Kennedy, "Late Medieval Planetary Theory," *Isis* 57: 365–78; Noel Swerdlow, "The Derivation and First Draft of Copernicus's Planetary Theory," *Proceedings of the American Philosophical Society* 117 (1973):423–512; and George Saliba, "Arabic Astronomy and Copernicus," *Zeitschrift für Geschichte der Arabisch-Islamischen Wissenschaften* Band 1, 73–87; and Saliba, "The Role of Marâgha in the Development of Islamic Astronomy: A Scientific Revolution Before the Renaissance," *Revue de Synthèse* 4, no. 3/4 (1987): 361–73.

[33] Noel Swerdlow and Otto Neugebauer, *Mathematical Astronomy in Copernicus's "De revolutionibus"* (New York: Springer Verlag, 1984), p. 46. I am grateful to an anonymous reader for the concise statement of these relationships as set out in the text.

[34] Ibid. It should be pointed out, however, that in the three decades since the realization of this striking similarity between the models of Ibn Shatir and those of Copernicus, no documentary evidence has been found which would support the view that Copernicus actually had knowledge of the Arab advances.

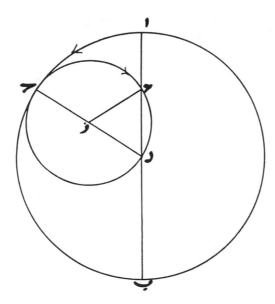

Figure 2. The Tusi couple is a modeling device to represent the motion of planetary bodies. It is composed of two nested circles, the smaller having a diameter one-half the larger. In this re-creation of Tusi's illustration, the Arabic lettering from top to bottom of the diameter of the large circle is *A, H, D,* and *B.* The smaller circle rotates in a right-hand motion while the larger circle rotates in the opposite direction and at half the speed of the smaller circle. As a result, point *H* viewed at a distance always remains between points *A* and *B* and appears to move in a straight line. This illustration appears in al-Tusi's thirteenth-century book, *al-Tadhkira* (1261). A translation of it has been made by Faiz Jamil Ragep, "Cosmology in the *Tadhkira* of Nasir al-Din al-Tusi" (2 vols. Ph.D. diss., Harvard University, 1982). Tusi's explanation of this rolling device appears in vol. 2, sec. 5, chap. 11, pp. 95–6.

by Galileo), it is useful to consider the history of astronomical thinking in medieval Islam. For astronomical work in Islam during this period was both intense and far in advance of equivalent thought in Europe. The new image we have of Islamic astronomical activity in both Eastern and Western Islam during the twelfth and thirteenth centuries suggests that Arab astronomers were hard at work reforming the Ptolemaic planetary system – otherwise known as the geocentric model – through a complex process involving mathematical models and astronomical reasoning to account for the discrepancies between theory and observation.[35] Moreover, collaboration occurred over centuries and over thousands of miles.

[35] Discussions of explicit uses of observation to test theory in Islamic astronomy are to be found in Bernard Goldstein, "Theory and Observation in Medieval Astronomy," *Isis* 63

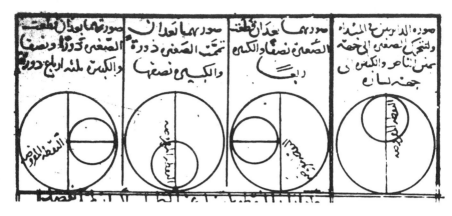

Figure 3. To illustrate the effect of his rolling device, al-Tusi drew a sequence of four diagrams to demonstrate the path of motion of a point embedded in the Tusi couple. The starting point for this illustration (reading from right to left) is the far-right diagram and the given point at the top. The larger circle turns to the left and the smaller one turns to the right. In the last diagram on the left, the small circle has traveled one and a half revolutions while the larger has traversed only three-quarters of a revolution. This sequence is meant to show that when such a point is viewed at a great distance, it will always appear to fall on a path that is very close to the diameter of the larger circle, which is represented by the vertical line. Such a path of motion would be virtually a straight line with small librations. (This illustration is a photographic reproduction of a drawing in Tusi's *Tadhkira fi 'ilm al-haya* [Ms Lalalei 2116, fol. 38b], procured by the late Professor Willy Hartner and published in *Regiomontanus-Studien*, edited by Günther Hamann [Vienna: Osterreichische Akademie der Wissenschaften], vol. 34, 1980, pl. xi, with permission of the publisher.)

For example, A. I. Sabra points out that the Cairene mathematician Ibn al-Haytham (d. ca. 1040) played an important role in stimulating Arab astronomers to surpass the work of Ptolemy and the Greeks. He did this by articulating his misgivings about Ptolemaic planetary theory in his commentary on the *Almagest*. As Professor Sabra points out, Ibn al-Haytham had the courage to draw out and boldly state "that the arrangements proposed for planetary motions in the *Almagest* were 'false' [his own word] and that the true arrangements were yet to be discovered."[36] Consequently, we find in Andalusia (Spain), a century after Ibn al-Haytham, that Arab thinkers independently led what has been called a "revolt against Ptolemaic astronomy."[37] This intellectual revolt culminated in al-Bitruji's book, *The Princi-*

(1972): 39–47; and George Saliba, "Theory and Observation in Islamic Astronomy: The Work of Ibn al-Shâtir," *Journal for the History of Astronomy* 18 (1987): 35–43.

[36] Sabra, "The Andalusian Revolt Against Ptolemaic Astronomy," in *Transformation and Tradition in the Sciences,* ed. Everett Mendelsohn (New York: Cambridge University Press, 1984), p. 134.

[37] Ibid., pp. 134ff.

ples of Astronomy – an effort to reform the Ptolemaic system by actually developing new mathematical models, though they were, in effect, a scientific failure.[38]

On the other hand, the Marâgha school in western Iran, which included such figures as al-'Urdi (d. 1266), al-Tusi (d. 1274), Qutb al-Din al-Shirazi (d. 1311), and Ibn al-Shatir (d. 1375), succeeded in producing the non-Ptolemaic planetary models that were duplicated in the work of Copernicus. Consequently, this continuous tradition, begun with Ibn al-Haytham in the eleventh century, represents a true "scientific program" – a "new Islamic research program in astronomy"[39] – including a shared set of scientific objections to existing theory along with new standards of success for scientific theory. "It can be said that both al-Bitruji and the Marâgha astronomers [though separated by a century's work] were driven by the same sort of theoretical concerns," namely, the desire to reform Ptolemaic planetary theory.[40] While the Andalusian effort (which can be extended to include Ibn Bajja [d. ca. 1138], Ibn Tufayl [d. 1185], Averroes [d. 1198], and Maimonides [d. 1204]) resulted in theoretical failure, the Marâghan astronomers succeeded. What is more, except for the heliocentricity of Copernicus's model, the similarity between the planetary models of the Marâgha school (as perfected by Ibn al-Shatir) and the Copernican models (see Figures 4 and 5) is so great that some would say that "in a very real sense, Copernicus can be looked upon as, if not the last, surely the most noted follower of the Marâgha School."[41]

In short, such historians of Arab science and astronomy as E. S. Kennedy, Goldstein, Hartner, King, Sabra, Saliba, and Swerdlow, among others, have only accentuated the puzzle by painting a portrait that almost fully assimilates the scientific activity in Arab astronomy of the twelfth, thirteenth, and fourteenth centuries with the model of modern science which arises from reconstructions of the activities of such modern scientists as Copernicus, Galileo, Tycho Brahe, and Kepler. Yet neither Ibn al-Shatir nor his successors – of whom there were many – made the big leap to the heliocentric

---

[38] Ibid., p. 137.
[39] George Saliba, "The Development of Astronomy in Medieval Islamic Society," *Arab Studies Quarterly* 4, no. 3 (1982): 223.
[40] Sabra, "The Andalusian Revolt," p. 138.
[41] Swerdlow and Neugebauer, *Mathematical Astronomy in Copernicus's "De Revolutionibus,"* p. 295, and reaffirmed by Saliba in "The Role of Marâgha in the Development of Islamic Astronomy," p. 371. However, given the present absence of any documentary evidence connecting Ibn Shatir and the work of the Marâgha school of astronomers to Copernicus, it is entirely possible that the "Tusi couple" and the other innovations found in Copernicus's work were independently discovered by Copernicus. This would then be another illustration of the sociological thesis of the simultaneous, independent and multiple discovery of scientific innovations (more on which in Chapter 5).

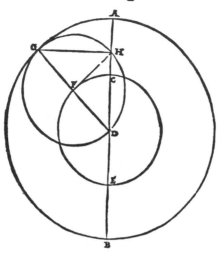

NICOLAI   COPERNICI

uel è conuerſo. H igitur in lineam A B reclinabitur: alioqui accide
ret partem eſſe. maiorē ſuo
toto, quod facile puto intel
ligi. Receſsit autem à prio-
ri loco ſecundum longitudi
nem A H retractàm per infra
ctam lineam D F H, æqualem
ipſi A D, eo interuallo quo di
metiens D F G excedit ſubten
ſam D H. Et hoc modo per-
ducetur H ad D centrum, q̃d
erit in contingente D H G cir
culo, A B rectam lineam, dū
uidelicet G D ad rectos angu
los ipſi AB ſteterit, ac deinde
in B alterum limitem perue-
niet, à quo rurſus ſimili rati

Motus in rectam   one reuertetur. Patet igitur è duobus motibus circularibus, &
lineam   hoc modo ſibi inuicem occurrentibus in rectam lineam motū

• • •

Figure 4. This illustration of a planetary model is from a first edition of Copernicus's *On the
Revolutions of the Heavenly Spheres* of 1543. The drawing clearly contains the Tusi couple —
circle *GHD* nested inside *ABG* — and appears in Book 3, Chapter 4, of Copernicus's great
work. This illustration and Copernicus's discussion of it have led many historians of science to
believe that Copernicus must have seen an Arabic manuscript containing a diagram of the Tusi
couple. The late historian of science Willy Hartner noticed the same lettering (from top to
bottom, *A, H, D, B*) in Copernicus's diagram and in the Lalalei copy of al-Tusi's *Tadhkira* (as
labeled above in Figure 2), which is in the Süleymaniye Mosque Library in Istanbul. No
researcher has yet shown a direct connection between Copernicus and such a manuscript,
however, thereby leaving open the possibility of a multiple and independent discovery of this
astronomical device. (Courtesy of the Astronomer Royal Scotland, the Crawford Library,
Royal Observatory, Edinburgh. Reproduced from *Isis* 66, no. 232 [1973], with permission of
the University of Chicago Press.)

worldview, to what might be called the great metaphysical core of the modern European scientific revolution of the sixteenth and seventeenth centuries. Instead, Arabic science stagnated and went into decline. Despite the
fact that many Arabic and Persian manuscripts relating to the history of

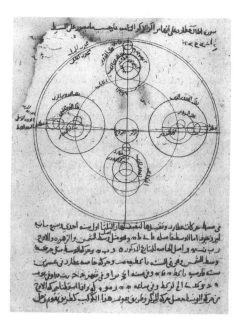

Figure 5. In these two diagrams, Ibn al-Shatir succeeded for the first time in representing the motions of the planet Mercury solely in terms of uniform circular motion, as required by Aristotle's physics. Shatir's models were still based on the earth-centered universe, but they represent the zenith of Arab astronomy. (Reproduced from Ibn al-Shatir's *Nihâya al sûl*, Ms. Marsh 139 fols. 29r–29v. Photographs courtesy of the Bodleian Library, Oxford.)

Arabic science are still unread,[42] there is little doubt, as a leading figure in the history of Arab science recently affirmed, "that the phenomenon [of scientific decline] did in fact occur" if one compares the "levels of scientific productions in, say, the fifteenth and eleventh centuries."[43]

Nevertheless, the Arab achievement is so impressive that we must ask why the Arabs did not go "the last mile" to the modern scientific revolution, which in this case required nothing more by way of mathematical modeling. The planetary models of Ibn al-Shatir and those of Copernicus are virtually identical, with only minor differences in some parameters. But the metaphysical transition would have, of course, forced an intellectual break with

[42] See Emilie Savage-Smith, "Gleanings from an Arabist's Workshop: Current Trends in the Study of Medieval Islamic Science and Medicine," *Isis* 79 (1988): 246–72, for an overview of the many manuscripts that have only recently been systematically catalogued.

[43] A. I. Sabra, "The Appropriation and Subsequent Naturalization of Greek Science," p. 238. Likewise, George Saliba says that it "incontestably took place." "The Development of Astronomy," p. 225.

traditional Islamic cosmology as understood by the religious scholars, the *'ulama'*. For the Copernican breakthrough entailed attributing three motions to the earth and placing the sun at the center,[44] an intellectual transition that likewise caused a great deal of friction in religious and intellectual circles in the West. The Arabs were perched on the forward edge of one of the greatest intellectual revolutions ever made, but they declined to make the grand transition "from the closed world to the infinite universe," to use Koyré's famous phrase.[45] Having failed to make this great transition during the early modern period, Islamic countries of the world today still cling to calendars based on lunar cycles.

It should be stated at this juncture that it is neither trivial nor partisan to raise the question of why Arabic science – which up to the fourteenth century was generally far more advanced than the science of the West – failed to give birth to modern science. Modern scientific knowledge as *episteme* is knowledge about how the world works, and it is such without being the absolute truth. Such knowledge is not inherently confined to the local community, the dominant ethnic group, the nation-state. Because of its universalistic design, it has the capacity to transcend all such boundaries wherever people are allowed to think freely. Joseph Needham has been eloquent in this regard:

> It is vital today that the world should recognize that 17th century Europe did not give rise to essentially "European" or "Western" science, but to universally valid world science, that is to say, "modern" science as opposed to the ancient and medieval sciences. Now these last bore indelibly an ethnic image and superscription. Their theories, more or less primitive in type, were culture-rooted, and could find no common medium of expression. But when once the basic technique of discovery itself had been understood, the sciences assumed the absolute universality of mathematics, and in their modern form are at home under any meridian, the common light of every race of people.[46]

---

[44] Kuhn, *The Copernican Revolution* (New York: Vintage, 1957), pp. 155ff and pp. 164–5. These three motions are (1) the earth's diurnal rotation on its axis, (2) the earth's rotation around the sun, and (3) the oscillation of the earth's axis such that the north pole shifts during the course of the earth's revolution around the sun so that the $23^{1}/_{2}°$ tilt away from the perpendicular is maintained.

[45] Alexandre Koyré, *From the Closed World to the Infinite Universe* (Baltimore, Md.: Johns Hopkins University Press, 1957). Although Koyré chose to accentuate the fact that Copernicus's universe is bounded by the "sphere of the fixed stars," he readily admits that Copernicus himself suggested that the heavens give "the impression of infinite size" and that many interpreters of Copernicus also adopted this view (pp. 31–1). Consequently, I use this phrase only to suggest that Copernicus did indeed take this first step toward cosmological infinitude.

[46] Needham, *SCC* 3: 448.

Needham's stress here on the putative ethnic character of premodern science is probably overstated, especially with regard to mathematics and astronomy and, above all, in the case of Ptolemy's *Almagest,* which, though a product of Greek culture, contained no cultural impediments that prevented its assimilation in either Middle Eastern Islamic culture or medieval European Christian culture. Likewise, the advent of modern science is not most usefully described as the discovery of "the technique of discovery itself." The new science of which Needham speaks, however, is universal or ecumenical science, knowledge that can be used by any people on the globe. Furthermore, it remains the case that Arabic science did contribute an enormous amount of original mathematical, methodological, and scientific knowledge to the development of what we today may call universal modern science, the system of knowledge production shared by, improved upon, and used by people all over the globe.[47] Accordingly, it is fair (and important) to ask why, from a sociological point of view, Arabic-Islamic civilization failed to continue its march toward the development of this universal institution of modernity.

To these aspects of the problem of Arabic science we might add another. It concerns the fact that until quite recently little attention was given to the existence of Arabic science and to the fact that it lies in a direct line leading to modern science. This omission, of course, does not apply to George Sarton's magisterial *Introduction to the History of Science,* but it does apply to many specialized works that purport to speak about the evolution of a particular scientific specialty, such as medicine. In the context of the present inquiry, we note that Ben-David in *The Scientist's Role in Society* said nothing about the role or contributions (positive or negative) of Arabic science in the evolution and development of the scientist's role in society, when it has long been recognized that the great influx of knowledge through translations of Arabic works (in the twelfth and thirteenth centuries) was a major galvanizing force triggering the unparalleled study of science in medieval European universities.[48] Similarly, Vern Bullough's otherwise exemplary study, *The*

[47] See, for example, A. C. Crombie, "The Significance of Medieval Discussions of Scientific Method for the Scientific Revolution," in *Critical Problems in the History of Science,* ed. Marshall Clagett (Madison, Wis.: University of Wisconsin Press, 1959), pp. 79–102; and Crombie, "Avicenna's Influence on the Medieval Scientific Tradition," in *Avicenna,* ed. G. Wickens (London: Luzac, 1952), pp. 84–107.

[48] I say unparalleled because the European universities, as we shall see later, were unique to world civilization, and because only there was there a complete embrace of the Greek philosophical tradition. Elsewhere, in Islam, these disciplines were barred from the colleges (see Chap. 5), while in China likewise the study of science was a negligible part of the official teaching and examination system. For a description of the subjects covered by the Chinese encyclopedias designed to prepare for the examinations, see Etienne Balazs,

Development of Medicine as a Profession,[49] which starts with the Greeks, wholly omits any discussion of Arabic medicine, when, again, all European medical schools after the twelfth and thirteenth centuries were deeply indebted to the Greek and Arabic medical tradition, especially Avicenna's medical encyclopedia, the *Canon,* among many other works.[50] Many other omissions of this sort could be cited, but the point is that modern science is a product of multiple intercivilizational contributions and encounters, and only by considering that larger picture can we appreciate the magnitude of the achievements and the unique contributions that any society or civilization made to ecumenical science. It will not suffice to say that only Europe succeeded or that Arabic-Islamic civilization (or the Chinese) had no intention of contributing to modern universal science.[51]

Role-sets, institutions, and science

From a sociological point of view, it is evident that in order to go forward with our inquiry we need an analysis of the cultural and institutional contexts within which scientific activity was carried on in medieval Islam. The historians of science have rightly been totally immersed in studying the Arabic manuscripts and deciphering the inner meaning that such documents have in the narrow context of scientific problem solving. This is as it should be. Yet it is a sociological truism that in all societies men and women live and work within the confines of socially constructed roles and institutions. This is the perspective that we saw so carefully articulated by Robert Merton in the last chapter. It is the view which seeks to account for the consequences of the fact that people are always located in multiple roles and statuses with the result that attitudes, interests, and capacities, as well as limitations of one situation, often extend to another. This means that all social organizations and institutions are interdependent and that, therefore, "separate institutional spheres are only partially autonomous, not completely so."[52] Furthermore, "it is only after a typically prolonged development that social institu-

---

Chinese Civilization and Bureaucracy (New Haven, Conn.: Yale University Press, 1964), pp. 146ff.

[49] Vern Bullough, *The Development of Medicine as a Profession* (New York: Hafner, 1966).

[50] See Nancy Siraisi, *Avicenna in Renaissance Italy: The Canon and Medical Teaching in Italian Universities after 1500* (Princeton, N.J.: Princeton University Press, 1987).

[51] That Arabic science had a different orientation to knowledge in general than the "scientific" is the theme of S. H. Nasr, *Science and Civilization in Islam* (New York: New American Library, 1968). He claims that the "gnostic" and mystical tradition represents the true path to knowledge, especially in Islamic civilization, and that, therefore, the West is a deviation from that unified vision of knowledge.

[52] R. K. Merton, *Science, Technology, and Society in Seventeenth-Century England* (New York: Harper and Row, 1970 [1938]), p. x.

tions, including the institutions of science, acquire a significant degree of autonomy."[53] The connections between the political, religious, and other spheres of life are always present.

Thus, we remember that the great fourteenth-century Arabic astronomer, Ibn al-Shatir, was also the official timekeeper (the muwaqqit) in the Umayyad mosque in Damascus, and accordingly his role-set consisted of overlapping roles, that of a strictly religious official and that of a scientist. Similarly, Averroes was a great natural philosopher as well as *qadi,* that is, a judge specializing in religious law appointed first in Seville and then later in Cordoba. Accordingly, the "naturalization" of science in the present context refers to the domesticating of the foreign or ancient sciences, thereby incorporating them into an indigenous cultural and philosophical system, not to the process of institutionalizing them such that they carry their own specific gravity of autonomy and legitimacy, independent of the moral and religious scruples of the surrounding culture. For this is what did not happen in Islamic civilization. The real battle so far as modern science is concerned, then, is that whereby the sciences achieve their own autonomy, after having first been "Islamicized." So just as Merton was interested in the flow of influence that was deeply colored by religious conceptions arising from the dissenting churches in England which gave a new spur to scientific inquiry in the seventeenth century, it is pertinent to ask how religious and social involvements may have colored perceptions of scientific problems and set limits on possible solutions to them in medieval Islam.

Given the overlapping and often conflicting sources of motivation centered in different institutional spheres, it is important to ask, paraphrasing Merton, whether the shift in intellectual focus, especially in science, was a deliberate outcome of policy decisions and to what degree it was "the unanticipated consequence of value commitments among scientists" who were in various positions of power and influence.[54] Similarly, if one assumes that the motivations of social action are culturally conditioned, one could ask, "How does a cultural emphasis upon social utility as a prime, let alone an exclusive, criterion for scientific work variously affect the rate and direction of advance in science?"[55] In a general sense, how do the underlying cultural values of a society and civilization encourage or retard scientific inquiry? Thus both A. I. Sabra and George Saliba refer to a "naturalization" or "Islamization" process[56] whereby the ancient or Greek sciences became fully assimilated to

[53] Ibid.
[54] Ibid., p. ixf.
[55] Ibid.
[56] Sabra, "The Appropriation and Subsequent Naturalization of Greek Science"; Saliba, "The Development of Astronomy in Medieval Islamic Society."

the requirements of Islamic culture, including its religion. This was a system – the Islamic philosophical tradition – "in which truth was supposed to be [found] within a system that is consistent, harmonious, and well articulated, with religion having an essential position in that system."[57]

The task, in short, is to study the paths of mutual, but also conflicting influence that cultural institutions exert on scientific interests and activities. While Merton's project, located in seventeenth-century England, was "an empirical examination of the genesis and development of some of the cultural values which underlie the large-scale pursuit of science,"[58] we have reason to investigate the cultural values that may have stood in the way of continuing the "large-scale pursuit of science" in late medieval Islamic society. At the same time we must draw a distinction between the scientific revolution of the sixteenth and seventeenth centuries and the scientific movement – Merton's center of focus – that swept the Western world during the seventeenth century and later.

I must stress the fact that, from a sociological point of view, institutions are ideas, that is, social institutions are ideas that have been given paradigmatic expression so they are ready and available to Everyman in a particular society and civilization. Such ideas have been translated into sets of roles and role expectations, so that they have now become the legitimating and reigning directives for social action.[59] On the one hand, this takes the form of normativizing a set of values (or value patterns) in the moral and ethical spheres of social action, and on the other, a second level, it entails the institutionalization of these values by making them legal prescriptions and imperatives. Once values have been instituted as legal codes, they begin to have a life of their own.[60]

---

[57] Saliba, "The Development of Astronomy in Medieval Islamic Society," p. 222.

[58] Merton, *Science, Technology, and Society in Seventeenth-Century England*, p. xxxi.

[59] The connection between institutions and ideas is suggested by Ernst Gellner in "Concepts and Society," in *Rationality*, ed. Bryan R. Wilson (New York: Harper, 1970), pp. 18–49 at p. 18f. The technical sociological literature on this subject is not altogether satisfactory, but S. N. Eisenstadt's article is useful: "Social Institutions," *International Encyclopedia of the Social Sciences* 14: 409–21. It is among medieval historians that one gets the best sense of the connections between ideas, social structure, and institutions. The medieval historian Hastings Rashdall caught the essence of what is suggested when he wrote "Ideas pass into great historic forces by embodying themselves in institutions. The power of embodying its ideals in institutions was the peculiar genius of the medieval mind." *The Universities of Europe in the Middle Ages*, new ed. (New York: Oxford University Press, 1936), ed. F. M. Powicke and A. B. Emden, 1: 3. Mary Douglas, *How Institutions Think* (Syracuse, N.Y.: Syracuse University Press, 1986), goes in quite another direction.

[60] It seems to me that the history and sociology of law remains the best place to start working out these levels of institutionalization. One of the services of H. L. A. Hart's classic, *The Concept of Law* (New York: Oxford University Press, 1961), is to show that there are both general social norms, even rules, which are clearly not legal (though socially enforced) and

In the case of science, however, we must keep in mind that the modern scientific worldview is a unique metaphysical structure. This means that the modern scientific worldview rests on certain assumptions about the regularity and lawfulness of the natural world and the presumption that man is capable of grasping this underlying structure. In addition to subscribing to the notion of laws of nature, modern science is a metaphysical system which asserts that man, unaided by spiritual agencies or divine guidance, is single-handedly capable of understanding and grasping the laws that govern man and the universe. The evolution of this worldview has long been in process, and because we in the West simply take it for granted, we have not given the various stages of its evolution proper attention, above all in the context of the comparative sociology of science.

Accordingly, it is imperative that we view the problem that modern science arose in the West and not elsewhere as a set of intellectual struggles over these very issues. Above all they are intellectual struggles in the domain of moral decision. As the history of Western culture reminds us, people like Galileo had to join battle with established church authorities in order to warrant the claims they made for their scientific knowledge as well as their human capacity to achieve it. The rise of modern science was not just the triumph of technical reasoning but an intellectual struggle over the constitution of the legitimating directive structures of the West. Science conceived as an institutional structure is a new embodiment of roles and role-sets rooted in a peculiar intellectual ethos, as well as a legal context. Intellectually, modern science represents a new canon of proof and evidence, and institutionally it represents a new configuration of role structures.

From a sociological point of view, the decisive breakthrough to modern science should be sought in the social (and moral) sphere at the intersection of ideas and social roles. In order to decipher the fate of Arabic-Islamic science, we must attend to precisely these dimensions of the philosophical, religious, and legal reconstructions of social roles in the medieval period. A very good point for attacking this problem can be found in A. I. Sabra's recent thesis which suggests that it was precisely the naturalization of the foreign sciences – that is, the Greek or natural sciences – in Islam that led (paradoxically) to their decline. This is not, it should be observed, the thesis

---

also those rules and norms which are law. The distinction between the two is by no means obvious. Similarly, Paul Bohannan, "Law and Legal Institutions," *International Encyclopedia of the Social Sciences* 9 (1968): 73–8, goes to some length to argue that Malinowski's definition of custom, i.e., "a body of binding obligations regarded as right by one party and acknowledged as the duty by the other" (p. 75), is in effect a definition of law, not customs, since it represents a set of customs that have been "reinstitutionalized as law" and given that extra measure of force which only law can give.

that the pursuit of the natural sciences in fact became institutionalized in the sense I indicated above – that is, made autonomous – although Sabra does reject the view that scientists – "physician, astrologer, engineer, *muwaqqit* or *faradi*" – were marginal in Islamic society.[61] But before I consider that question, I will briefly sketch the cultural context – that is, the religious and legal background – within which medieval Arabic-Islamic science took place.

Social roles and cultural elites

In order to understand the evolution and social construction of the role of the scientist[62] we need to begin with an outline of the dominant status groups and intellectual roles in medieval Islam.[63] Among the intellectually salient groups in medieval Islam the *fuqaha* (the jurisconsults), the *mutakallimun* (the dialectical theologians), and the philosophers were the most significant.

The easiest to describe of these are the philosophers. These were individuals who fully embraced the substance and method of Greek philosophy and the rational or foreign sciences. While there were occasional currents of Neoplatonism, especially in al-Farabi and Ibn Sina, by and large the philosophers were most deeply impressed by and committed to the view of Aristotle. The most notable philosophers of Arabic-Islamic civilization were al-Kindi (d. ca. 870), al-Farabi (d. 950), al-Razi (d. ca. 925), Ibn Sina (d. 1037), al-Baghdadi (d. 1152), al-Biruni (d. 1048), and Ibn Rushd (Averroes; d. 1198). Although each of these men made outstanding contributions to philosophy, most of them earned their living as physicians or in other capacities as court-appointed officials, in law, for example, as in the case of Averroes. In the formative period of early Islam, the philosophers were tantamount to free thinkers in that they developed theories of knowledge by directly building on Greek philosophy. While they suggested that philosophical knowl-

---

61 Sabra, "The Appropriation and Subsequent Naturalization of Greek Science," pp. 229–36.

62 The classic treatment of this problem is Joseph Ben-David, *The Scientist's Role in Society* (Englewood Cliffs, N.J.: Prentice-Hall, 1971). See Chap. 1 above for the limitations of Ben-David's approach to our problem.

63 For a more detailed and inflected account of the many status groupings and overlapping levels of leadership in Islamic society, see Roy Mottahedeh, *Loyalty and Leadership in an Early Islamic Society* (Princeton, N.J.: Princeton University Press, 1980). Thomas Glick, *Islamic and Christian Spain in the Early Middle Ages* (Princeton, N.J.: Princeton University Press, 1979), chap. 4 (part 1) also contains a useful interpretation of Islamic stratification in the light of more Western conceptions of classes. For our purposes and the history of science, however, the groups I have identified are more useful.

edge was the highest and most noble, they also suggested that revealed religion was little more than superstition.[64] By the twelfth and thirteenth centuries, however, with the full development of Islamic dogmatics and the naturalization of Greek philosophy, philosophers such as Averroes were far more respectful of their religious tradition. They even criticized other philosophers for advancing arguments that might lead ordinary believers and even the undialectical religious scholars (the mutakallimun) astray. This is seen in the criticism of al-Ghazali by Averroes[65] as well as Ibn Taymiyya's (d. 1328) attack on philosophy.[66]

Medieval Islam produced many great philosophers whose intellectual enterprise was distinguished by its embrace of the problems and method of Greek philosophy. Quite in contrast to the philosophers, however, were the mutakallimun, the dialectical theologians, who used what methods of rational argument they could from the Greeks to enunciate and defend the First Principles of Islam. Laudable as this endeavor would seem to Christian theologians, from an Islamic point of view it was a highly dubious enterprise. This was so because of the unique status in Islam of the Quran and the teachings of the Prophet (the hadith), which were the alpha and omega of Islamic faith. The Quran and the hadith constituted the sacred law (the *Shari'a*) that established once and for all the patterns of conduct and the proper management of human affairs for all Muslims. This being the case, the specialists in fiqh (law) known as the fuqaha (jurisconsults) were the intellectually dominant actors in medieval Islam, above all, in the twelfth and thirteenth centuries. The dialectical theologians (the mutakallimun) from the outset – including Ash'ari, the founder of the orthodox school of sunni kalam – had conceded that their intention was only to defend the faith by establishing the philosophical foundations or the principles of religion, but otherwise all questions of moral, legal, and ethical conduct – the heart of Islamic belief – were left to the authority of the legists.[67] At best, the

---

[64] Shlomo Pines, "Philosophy," in *The Cambridge History of Islam* (New York: Cambridge University Press, 1970), 2: 780–823; Peters, *Aristotle and the Arabs*; R. Walzer, *Greek into Arabic*; Muhsin Mahdi, "Islamic Theology and Philosophy," *Encyclopedia Britannica* 9: 1012–25; and M. Fakhry, *A History of Islamic Philosophy*, 2d ed. (New York: Columbia University Press, 1983).

[65] Averroes, *Tahafut al-Tahafut*, trans. Simon Van den Bergh (London: Luzac, 1954); and Averroes, *On the Harmony of Religion and Philosophy*, reprint, ed. and trans. George Hourani (London: Luzac, 1976).

[66] See Ignaz Goldziher, "The Attitude of Orthodox Islam Toward the Ancient Sciences," in *Studies in Islam*, ed. Merlin Swartz (New York: Oxford University Press, 1981), pp. 185–215; Fazlur Rahman, *Prophecy in Islam* (Chicago: University of Chicago Press, 1979); and Pines, "Philosophy."

[67] See Rahman, *Prophecy in Islam*, pp. 123–4; Ash'ari, "The Elucidation of Islam's Foundation," in *The Islamic World*, ed. William H. McNeill and Marilyn R. Waldman (Chicago:

mutakallimun played a secondary role, and even that was frequently challenged by "the partisans of tradition" (*ahl al-hadith* or *ahl al-Sunna*), as in the thirteenth-century condemnation of theology by Ibn Qadama.[68] According to the latter, "no one is ever seen who has studied speculative theology, but there is a corrupt quality to his mind."[69] For such a person wallowing in theological speculation, the punishment cannot be too severe. Consequently Ibn Qadama cites a tradition from al-Shafi'i, asserting that, "My judgment with respect to the partisans of speculative theology is that they be smitten with fresh leafless palm branches, that they be paraded among the communities and tribes, and that it be proclaimed, 'this is the punishment of him who has deserted the Book and the Sunna, and taken up speculative theology.'"[70] From such a point of view, law, not theology, was the queen of the sciences, and only corruption could come from theological speculation. Furthermore, whenever challenges to the prevailing drift of Islamic thought arose, it was from the party of tradition (ahl al-Sunna), the religiously minded 'ulama', that the challenge arose.[71]

In general, the structure of thought and sentiment in medieval Islam was such that the pursuit of the rational or ancient sciences was widely considered to be a tainted enterprise. This has been shown most systematically in the work of Ignaz Goldziher.[72] The two most important charges brought forth by the orthodox were that the study of philosophy, logic, and the ancient sciences made one disrespectful of religious law and, insofar as such study did not pertain to strictly religious matters, it was "useless" and hence ungodly.[73] These sciences were also described as "repudiated sciences," as well as "wisdom mixed with unbelief."[74] This distrust of systematizing thought even extended to grammarians on some occasions, while the study of logic itself was often referred to as "forbidden."[75] Given this widespread hostility toward the rational sciences, Goldziher noted that it is "easily

University of Chicago Press, 1983), pp. 152–66; as well as Richard J. McCarthy, ed. and trans., *The Theology of Ash'ari* (Beirut: Imprimerie Catholique, 1953).

[68] On kalam, see Rahman, *Islam*, 2d ed. (Chicago: University of Chicago Press, 1979), chap. 5; Pines, "Philosophy"; and Pines "Introduction" to *The Guide of the Perplexed* (Chicago: University of Chicago Press, 1962); W. M. Watt, *The Formative Period of Islamic Thought* (Edinburgh: Edinburgh University Press, 1973); McCarthy, *The Theology of Ash'ari;* and L. Gardet and M. M. Anawati, *Introduction à la Théologie Musulmane*, 2d ed. (Paris: J. Vrin, 1970). On Ibn Qadama, see George Makdisi, ed. and trans., *Ibn Qadama's Censure of Speculative Theology* (London: E. W. J. Memorial Series, Luzac, 1962).

[69] Makdisi, *Ibn Qadama's Censure of Speculative Theology*, p. 12.

[70] Ibid.

[71] Rahman, *Islam*, pp. 131, 14; and Hodgson, *The Venture of Islam,* 1: 228f.

[72] Goldziher, "The Attitude of Orthodox Islam."

[73] Ibid., pp. 186–7.

[74] Ibid.

[75] Ibid.

understandable why people who wanted to protect their reputations concealed their philosophical studies and pursued them under the guise of some discipline that had better standing."[76] Likewise, a man who had books dealing with the foreign sciences in his home risked acquiring the reputation of being an impious person.

If we speak of the prevailing attitudes toward inquiry and learning in the various intellectual spheres during the medieval period, it must be said that those who pursued the religious sciences had the upper hand and periodically denounced those who pursued the foreign and ancient sciences. The intellectually distinct activities of the philosopher as well as those of the physician, astrologer, astronomer, and mathematician were all identifiable, but they were greeted with varying degrees of acceptance by the religious scholars. If we were to construct a status hierarchy, it would start with the legists, the fuqaha, at the top, below them the mutakallimun, and below them, the *faylasufs*, the philosophers cum natural scientists. Of course, at any point in time an outstanding intellect and philosopher might be highly regarded by the populace at large because of his notoriety, especially with regard to medical practice, but this could not be taken as an endorsement of his philosophical views. Still, the high reputation of physicians (whose training rested above all on the Greek classics and Galen) made them models of virtue, and they almost always served in high positions of political power and in local administrative offices.[77] The considerable freedom and resources that certain outstanding philosophers and mathematicians had to pursue their studies, however, was always contingent upon the official protection of local rulers. As Willy Hartner pointed out in the case of more than a dozen notable figures (such as al-Biruni, Ibn Sina, Abul Wafa, Ibn Yunus, and Ibn al-Haytham), royal patronage was a major element in their careers. Once that direct ring of protection and approval was withdrawn, scholars immersed in philosophy and the foreign sciences could easily fall into disrepute, as happened to Averroes, despite the fact that he had served as the chief qadi in Cordoba for years.[78] The possibility of opposition from the local 'ulama' was thus a constant threat and stemmed from the fact that the religious scholars – who orchestrated their own political constituency – and the local populace held views sharply different on a variety of theological, metaphysical, and scientific issues from the philosophers. These extended

[76] Ibid., p. 190.
[77] S. D. Goitein, *A Mediterranean Society*, 2 vols. (Berkeley and Los Angeles: University of California Press, 1967–71), 2: 240ff.
[78] Willy Hartner, "Quand et comment s'est arrêté l'essor de la culture scientifique dans l'Islam?" pp. 332–33. On Averroes, see A. Z. Iskandar and R. Arnaldez, "Ibn Rushd," *DSB* 12: 1–9; and Dominique Urvoy, *Ibn Rushd (Averroes)* (London: Routledge, 1991).

from contrasting notions about the creation of the world to whether natural necessity implied strict causality, whether man had free will, and whether moral principles could be reached strictly through the use of reason.[79] To the degree that the fuqaha and mutakallimun denied natural causality (by accepting the doctrine of Islamic occasionalism),[80] they denied that man had the power of reason to reach ethical as well as general truths unaided by revelation. In that regard they held views opposed to what we may call the metaphysics of modern science, or those assumptions without which no man is a scientist, to use Thomas Kuhn's phrase. It is fair to say, therefore, that during the early years of Islam there was a fundamental split between the philosophers and the religious scholars, and that in the long run, the question of the success of the scientific enterprise depended upon how these intellectual schisms were institutionalized. It depended upon whether the pursuit of the sciences was accorded a legitimate role in the structure of Islamic society, or whether this pursuit must always be a subsidiary inquiry carried on covertly. The hostility of the religious scholars did not extend equally to every branch of the ancient sciences, however, especially in regard to arithmetic, geometry, and astronomy. This was so because those sciences came to be religiously useful. That is, arithmetic proved to be indispensable in the division of inheritances, and the specialist who dealt with this, the *faradi*, was both a legal expert and a person trained in arithmetic. Likewise, it was essential for all Muslim believers to know the times of prayer as well as the direction to Mecca (the qibla). The establishment of the times of prayer and the direction of the qibla could only be exactly determined through the use of mathematics, geometry (later, trigonometry), and astronomy. Accordingly, these sciences eventually came to be highly developed by religiously committed individuals. As in the case of the muwaqqit, the timekeeper in the local mosque, these highly exact and scientific inquiries became incorporated into official religious roles. It is easy to imagine, therefore, that the pursuit of a religiously based science might transform itself into the pursuit of science for its own sake, just as in Europe in the twelfth and thirteenth centuries, the pursuit of theology as the queen of the sciences became transformed into the study of philosophy, logic, and science for its

---

[79] A good introduction to these philosophical issues can be found in Oliver Leaman, *An Introduction to Islamic Philosophy* (Cambridge: Cambridge University Press, 1985), as well as Pines, "Philosophy," Maimonides, *The Guide of the Perplexed,* and Ash'ari, "The Elucidation of Islam's Foundation."

[80] See Majid Fakhry, *Islamic Occasionalism and Its Critique by Averroes and Acquinas* (London: Allen and Unwin, 1958); and Barry S. Kogan, *Averroes and the Metaphysics of Causation* (Albany, N.Y.: SUNY Press, 1985).

own sake.[81] In short, the question of the pursuit of science and its long-range development depended on constructing a legitimate role for the scientist in the context of the prevailing religious views and within the legal framework allowed by Islamic law. In order to explore those questions we should examine another dimension of the social structure of Islam, that of education and learning.

Institutions of higher learning and research

Recent studies of the institutions of learning and instruction in medieval Islam, especially those of George Makdisi, have given us a greatly enlarged appreciation of their richness, diversity, and importance during an age that is too often said to be dark.[82] Makdisi also argues that both the scholastic method of disputation and the titles of academic positions in the European West were borrowings from Muslim educational practice and the Arabic language.[83] The fact is that the cultural elite of Arabic-Islamic civilization made an extraordinary commitment to all the forms of learning. This is manifested in the development and pursuit of all the institutions and techniques essential to the life of the mind, to the art and techniques of book writing and production, and to the gathering and transmission of knowledge among contemporaries and between generations.

On the most elementary level this commitment is seen in the cultivation of the arts of manuscript and book production. Here the first task was the production of suitable writing materials and, above all, the mass production of paper. Up until the middle of the tenth century, papyrus was the main source of writing material. Papermaking was first learned by the Arabs from

---

[81] See M. D. Chenu, *Nature, Man, and Society in the Twelfth Century,* selected, ed., and trans. Jerome Taylor and Lester K. Little (Chicago: University of Chicago Press, 1968); and William and Martha Kneale, *The Development of Logic* (New York: Oxford University Press, 1962); and David Knowles, *The Evolution of Medieval Thought* (New York: Vintage, 1962). More on this in Chap. 3.

[82] The most thorough study of the educational institutions of Islam are to be found in the following writings by George Makdisi, "Muslim Institutions of Learning in Eleventh-Century Baghdad," *Bulletin of the School of Oriental Studies* 24: 1–56 (1961); "Madrasah and University in the Middle Ages," *Studia Islamica* 32: 255–64; "On the Origin and Development of the College in Islam and the West," in *Islam and the Medieval West,* ed. Khalil I. Seeman (Albany, N.Y.: SUNY Press, 1980), pp. 26–49; and *The Rise of Colleges: Institutions of Learning in Islam and the West* (Edinburgh: Edinburgh University Press, 1981). Other useful introductions include Fazlur Rahman, *Islam,* pp. 181–92; and Johannes Pedersen and G. Makdisi, "Madrasa," in *Encyclopedia of Islam,* 2d ed. (Leiden: E. J. Brill, 1985), 5: 1123–34 (hereafter cited as *EI²*).

[83] *The Rise of Colleges,* chap. 4.

the Chinese as early as the eighth century in Samarqand. By the middle of that century there was a state-owned paper mill in operation in Baghdad, and by the middle of the tenth century the use of paper was so widespread that the manufacture and use of papyrus for writing materials had died out.[84]

The first paper manufacturing in Europe appears to have been done around 1150 by the Arabs in Spain. It seems that the Arabs developed and used paper for scholarly purposes several hundred years before the Europeans. In devising new methods of mass paper production some would say that the Arabs "accomplished a feat of crucial significance not only for the history of the Islamic book but also to the whole world of books."[85] Nevertheless, this art too was soon taken over by the West and by the thirteenth century there are reports of "paper from the Franks" being used in Egypt.[86]

Prior to the invention of the printing press the publication of an original work was generally a labor-intensive activity that consumed considerable amounts of time. This was so because of the Arab inclination to distrust the merely written word and to prefer oral testimonies that included the sources of the information, that is, the previous transmitters.[87] In part this stemmed from the religious context in which the traditions of the Prophet (hadiths) were carefully recorded and passed on through a chain of oral transmitters whose names and identities had to be recorded in order to authenticate the hadith. Since an act of forgery was always possible in the copying of a text – not to mention errors of copying – an oral tradition in which each transmitter of the text testified to its authenticity was an essential part of the authentication of any written source. Even today many Muslims claim that the Quran is a perfect, unchanged, and uncorrupted text because Muslims from the beginning committed every single word to memory, and thus no forgery of this living text was possible because many Muslims had memorized it. No account was thus taken of the possibility of faulty memories. What is more, a standard of moral and religious uprightness was often applied to the transmitter, and only a person considered to be morally worthy by the community could be relied upon. This same pattern of reliance on especially circumspect persons developed into the institution of the professional witness

---

[84] Johannes Pedersen, *The Arabic Book* (Princeton, N.J.: Princeton University Press, 1984; first pub. 1946), p. 62. The invention of papermaking among the Chinese is richly detailed in *SCC* 5/1 by Tsien Tsuen-hsuin (New York: Cambridge University Press, 1985). Although he purports to describe its transmissions to the West, his account omits the history of papermaking among the Arabs, leaving us in doubt about the actual path of transmission.

[85] Ibid., p. 59.

[86] Ibid., p. 64.

[87] This is the source of the concept of *ijaza*, the permission or authorization to transmit written texts. See George Vadja, "Idjaza," *EI²* (1971) 3: 1021.

(*amin*) who assists the judge, an institution that continues down to the present.[88]

Likewise, when an original work was presented to the learned world, this was accomplished orally as the author himself dictated the work word by word to students. After carefully checking the transcribed work, the author added his signature of authenticity along with his permission, his *ijaza*, "making the work lawful" and allowing the copyist to then transmit the work to still other auditors.[89] Of course, at some point the authentication of transmitters necessarily ceased, and scholars often went to great lengths to search out and personally copy versions of scholarly works they believed were genuine. Similarly, owners of books and manuscripts customarily inscribed their names as part of the tradition of authentic transmission.

It is therefore all the more remarkable that in the tenth and eleventh centuries there were hundreds of libraries scattered throughout the Middle East, usually attached to mosques or *madrasas* (colleges), in which thousands of hand-copied manuscripts were housed. For example, the palace library of the Fatimids in Cairo in the tenth century contained forty rooms, each full of books on different subjects. Among these were eighteen thousand volumes on the natural sciences, then referred to as the foreign or ancient sciences.[90] As we shall see in Chapter 8, this commitment to the creation of public libraries was far more extensive than anything found in China at the same time.

A still more impressive library was that of Shiraz in the tenth century, which was said to comprise 360 rooms surrounded by lakes and gardens. Here the books were housed in separate vaulted rooms with cabinets specially built for the books.[91] Aside from these private libraries, there were many public libraries attached to mosques and colleges. Although these schools taught only the religious sciences and excluded the natural sciences (except for arithmetic), they represent models of colleges at which provision was made for the teaching and housing of students who came from distant places.

In the city of Marv in eastern Persia, the historian Yaqut reported the existence of 10 wealthy libraries in the thirteenth century; while another report mentions 30 madrasas in Baghdad at about the same time, and each of these with its own library. In Damascus in 1500 there are reports of 150

[88] For an example in contemporary Morocco, see Lawrence Rosen, "Equity and Discretion in a Modern Islamic Legal System," *Law and Society Review* 15 (1980–1): 215–45.

[89] Pedersen, *The Arabic Book*, p. 31 and chap. 3 passim; and J. Pedersen and G. Makdisi, "Madrasa" in *EI²* (1985) 5: 1125; and see Makdisi, *The Rise of Colleges*, pp. 140ff.

[90] Pedersen, *The Arabic Book*, p. 116.

[91] Ibid., p. 123f.

madrasas and hence equivalent numbers of libraries.[92] Another madrasa in Egypt in the thirteenth century, founded by al-Qadi al-Fadil, is said to have received 100,000 volumes from its founder.[93] And "when the great college (madrasa) known as al-Madrasa al-Mustansiriya was founded in 1234, some of the books from the caliph's library were transferred to it – some 80,000 volumes."[94] By these standards Europe was extremely impoverished. For example, the Sorbonne Library of the University of Paris in the fourteenth century had a mere 2,000 manuscripts, while the Vatican Library in the fifteenth century had a paltry 2,257.[95] Even allowing for Arab exaggeration and the possibility that some European manuscripts might contain multiple works (as did the Arabic), the library resources of the Middle East were vastly superior to those of Europe.

From an early date the various Muslim sects recognized the value of libraries as means for spreading knowledge of Islam, both orthodox and sectarian. As a result, they created public libraries for this purpose.[96] Indeed, it sometimes occurred that those who were known to collect books and to have large libraries were suspected of being unorthodox, and their libraries were confiscated and destroyed. Yet mixed motives might enter the picture so that the confiscated library might serve to add riches to an existing library, as apparently happened with al-Kindi's library before its eventual return to its rightful owner.[97]

In short, libraries with very extensive holdings were scattered all over the Middle East during the height of Arabic-Islamic civilization from the ninth until the thirteenth century. While some of these were private collections, many others, above all those attached to mosques and mosque schools (*masjids*), were open to the public. Moreover, most of these had some provision for a staff of librarians and administrators.

While many lesser institutions of learning evolved in medieval Islam (including the mosque schools and the elementary schools), the dominating institution of higher learning was the madrasa, the prototype of the college (not the university) that developed later in the West. The madrasas began to

[92] Ibid., p. 128.
[93] Ibid., p. 119.
[94] Ibid., p. 115.
[95] John F. D'Amico, "Manuscripts," in *The Cambridge History of Renaissance Philosophy*, ed. Charles Schmitt and Quentin Skinner (New York: Cambridge University Press, 1988), pp. 11–24 at pp. 15ff; D'Amico follows K. Christ, *The Handbook of Medieval Library History*, trans. T. M. Otto (London: Methuen, 1984).
[96] Ruth S. Mackensen, "Moslem Libraries and Sectarian Propaganda," *American Journal of Semitic Languages and Literature* 51 (1935): 83–125.
[97] Ibid., p. 84f.

flourish in the eleventh century and, as the premier educational institutions of Islam, came to dominate a significant segment of intellectual life. Two aspects of the organization of the madrasas are of particular importance. The first is that the madrasas were established as charitable trusts: they were religious endowments that legally had to follow the wishes, religious or otherwise, of their founders. However, the law of trusts (the law of *waqf*) specifically forbade appropriating property and funds through the institution of waqf for purposes other than those sanctified by Islam.[98] This stipulation is, as we shall see, a major legal impediment to the unfolding of intellectual and organizational evolution in the Islamic world. Nevertheless, the founder of such a religious trust could retain proprietary rights in perpetuity for himself or his relatives and therefore could appoint himself (or his heirs) as administrator or professor of law in perpetuity. Secondly, the madrasas were schools of law (fiqh), and as such centered all instruction around the religious or Islamic sciences, to the exclusion of philosophy, the natural sciences, and theology as well.

In theory the curriculum of the madrasa focused on Quranic studies, hadith (the traditions of the Prophet), the principles of religion, and the principles and methodology of law.[99] The last of these included disputed questions in law and the principles of argument and disputation, as well as the practical point of view of the school of law to which the madrasa or the professor belonged. The deliberate exclusion of philosophy and the ancient sciences obviously stemmed from the suspicion with which such subjects were regarded by the religious scholars. Nevertheless, books on these subjects were often copied and made available in the libraries associated with the schools and mosques, and those law professors who became well versed in the foreign sciences gave private instruction (at home) on these subjects.

When in the eyes of the professor students had mastered the subjects taught in the madrasa – perhaps more accurately, mastered the manuscripts that were read, copied, and memorized – they were given an ijaza, an authorization to teach these matters to others. One could say that a license to teach these subjects was granted by the master-jurisconsult to the student. And here the stress is on the personal authorization that was involved:

When the master-jurisconsult, the *mudarris*, granted the license to teach law and issue legal opinions, he acted in his capacity as the legitimate and competent authority in the field of law. When he granted the license to the candidate he did so in his

---

[98] Makdisi, *The Rise of Colleges*, pp. 35ff; and Henry Catton, "The Law of Waqf," in *Law in the Near and Middle East*, ed. M. Khadduri and H. Liebesny (Washington, D.C.: The Middle East Institute, 1955), pp. 203–22.
[99] Makdisi, *The Rise of Colleges*, p. 80.

own name, acting as an individual, not as part of a group of master-jurisconsults acting as a faculty; for there was no faculty.[100]

It should be stressed that this sort of education was highly personalistic; the authorization or licensing was done by each professor, not by a group or corporate body, much less by a disinterested or impersonal certifying body. Likewise, neither the state, the sultan, nor the caliph had any influence over the recognition of educational competence.

Throughout its history down to modern times, the ijaza remained a personal act of authorization, from the authorizing 'alim [scholar] to the newly authorized one. The sovereign power had no part in the process; neither the caliph nor sultan nor amir nor wazir, nor qadi, nor anyone else could grant such a license. There being no church in Islam, no ecclesiastic hierarchy, no university, that is, to say, no guild of masters, no one but the individual master-jurisconsult granted the license . . . Islamic education, like Islamic law, is basically individualistic, personalist.[101]

To highlight the comparative aspects of this situation, it may be pointed out that educational certification in the Islamic world was the polar opposite of that in China: in the former case it was always (and only) the scholar himself who certified a student's competence, whereas in China it was always (and only) the state, never the corporate scholarly group, which conferred certification of educational competence.[102]

In short, education in the Islamic world, whether in the domain of the Islamic (religious) sciences or in the domain of philosophy and the foreign sciences, was a process of gathering permissions to teach (ijazas), and this was done, in the case of law, by attending the lectures of professors at one or more madrasas and, in the natural sciences, by apprenticing oneself to many different scholars in many different cities and accumulating ijazas. Even those who completed their religious and legal studies in a single madrasa received multiple ijazas from individual professors, not a degree from a specific college or university. Thus education in medieval Islam was centered around masters who tutored students according to their own wisdom, and whether or not the student completed his education by gathering ijazas at the college, or traveled around collecting them from private scholars, he collected personalized permissions from individual masters, not a certificate of mastery in a particular subject matter. It has been suggested, however, that in the case of law, the ijaza did imply an authorization to teach law as a subject matter and to issue legal opinions not just a sequence of books.[103]

---

[100] Ibid., p. 271.
[101] Ibid.
[102] See Chap. 7 for further discussion of traditional Chinese educational practice.
[103] According to Makdisi, *The Rise of Colleges*, p. 270, the license to teach law and the license

In the domain of the natural sciences, since all instruction occurred outside the colleges, specializing in a particular science clearly required that one travel around a good deal, from city to city, in search of scholars versed in the ancient sciences, in order to become a master of the current state of knowledge. Such a system clearly created institutional barriers to specialized scientific training and research.[104] This was less true in the case of medicine where self-instruction as well as tutoring by family members was a possibility.[105]

No doubt this personalized system of education had benefits for students who could freely choose their instructors, excepting the necessary problems of travel. While bright students could doubtless discern who the best and most knowledgeable professors were and apprentice themselves to these men, it was equally possible that students could avoid the best and even attack them through their own writings, as Ibn Salah did, when, after failing to learn logic under Kamal al-Din b. Yunus (fl. thirteenth century), he actually issued a *fatwa* (a legal opinion) condemning the study of philosophy and logic.[106] The lack of outside supervision, especially in the case of medicine, could lead to untoward consequences, above all the widespread prevalence of charlatanism and quackery.

Insofar as the development of science and scientific thought is concerned, such a system provided neither group support for philosophers cum scientists who held dissident views vis-a-vis the religious and political authorities, nor any mechanisms whereby received wisdom (as understood by the best and most competent experts, or as attested to by experiment) could be separated from the false and disproven. Likewise, the prohibitions against bringing it into the colleges perpetuated the personalized master-student pattern and prevented the efficient cumulation of knowledge by bringing scholars versed in the ancient sciences together in one place. While the medieval period is perhaps too early to expect the appearance of such vital

to issue opinions were combined into a single license: the *"al-ijaza li 't-tadris wa 'l-ifta'."* And see p. 343 n240 for Makdisi's correction of his earlier view.

104 In the late twelfth and thirteenth centuries, three madrasas designated for medical study appear to have been founded in Damascus. Makdisi, nevertheless, views this as an exception to the rule regarding the exclusion of the ancient sciences from the madrasas; *The Rise of Colleges*, p. 313 n38. One may wonder whether these madrasas were not actually set up for legists who also had been trained in medicine (self-taught or otherwise), and their religious/legal training may have provided the basis for their appointment. More on this in Chap. 5.

105 Gary Leiser, "Medical Education in Islamic Lands from the Seventh to the Fourteenth Century," *The Journal of the History of Medicine and Allied Sciences* 38 (1983): 48–75; Dols, Introduction to *Medieval Islamic Medicine*, pp. 3–73; and Goitein, *A Mediterranean Society*, 2: 240ff.

106 Goldziher, "The Attitude of Orthodox Islam," pp. 204–5.

elements of the scientific enterprise as the scientific journal and scientific guilds and societies, it may be suggested that the extremely personalistic structure of Islamic law and society contained built-in impediments to this evolutionary direction. Furthermore, movable-type printing had been invented in China in the eleventh century, and though the techniques of printing came to Islam earlier than they did to the West, nothing came of it.[107]

The personalistic emphasis is seen especially in law, where the individual believer may request alternative rulings on legal questions from several legal specialists, for "all legal opinions share this same quality of being 'right' for being the result of religiously executed effort," that is, the intellectual struggle (*ijtihad*) of the jurisconsult. Conversely, the "freedom of the mufti [legal consultant] in arriving at his personal opinion is matched by the freedom of the *mustafti* [the layman seeking an opinion] in following the opinion of his choice; for he may solicit as many opinions as he wishes, and may follow whichever he chooses."[108] This is obviously not the same situation as a contemporary Westerner asking for a legal opinion from a lawyer, since the *mufti* serves in the function of the adjudicating court, not as the surrogate petitioner as in the case of the lawyer.[109]

The inhibitive effect of these particularistic orientations on the development of modern science and institutions has been noticed by other students of the history of Arabic science. On the one hand, the extremely personalistic nature of human relations generally in the Middle East during the medieval period may be seen in the dominating influence of the *extended kin group*, whose influence on the development of the social structure of Islamic medicine has recently been stressed. Speaking of the traditional institution of the extended family as the essential social unit of the Islamic world, Lawrence Conrad notes that

it includes not only close relations, but also distant kin and even associates with no real blood ties to the group. But whether the justifying blood relationship is real or not, the bonds uniting the group are of extraordinary strength, and are constantly being renewed and reasserted through a continuous stream of favors and concessions granted and demanded as a matter of right and obligation by members of the

---

107 In a chapter on block printing in Egypt, T. F. Carter displays fragments of block-printed Arabic manuscripts which apparently were printed in the mid-fourteenth century. See *The Invention of Printing in China and Its Spread Westward*, chap. 18.
108 Makdisi, *The Rise of Colleges*, p. 277. See also F. Ziadeh, *Lawyers: The Rule of Law and Liberalism in Egypt* (Stanford, Calif.: The Hoover Institution, 1968), p. 9.
109 In fact, the situation seems to be no different when dealing with an actual qadi. Ziadeh, commenting on legal practice during the Ottomans, reports that if an individual were displeased with a ruling by a qadi and could find four muftis from each of the four legal schools who agreed (in opposition to the qadi's ruling), then the qadi would reconsider his opinion. Ziadeh, *Lawyers*, p. 9.

group. . . . Other groupings are recognized as valid and even important, but their claims on the individual are deemed secondary to that of the family, which regards other groupings as outsiders, and at all times seeks to ensure that such "foreign" groups do not encroach on its prerogatives and advantages, offend its dignity, or detract from its prestige.

Moreover, Conrad continues:

This pattern of an extended group of kin exchanging favors and demands and supporting each other in their relations with the "outside world," so to speak, repeated itself at all social, economic, and political levels. . . . On the most general level, the overwhelming dominance of the extended family was a major factor behind the fact that in medieval Islam there arose no corporate municipal institutions. That is, there were no guilds of physicians, surgeons, or druggists in medieval Islam, although these professions were highly developed. To the extent that medical practitioners *appear* to have been organized in some coherent way, this is often because of the tendency toward the gradual approximation of the limits of the family group to those of the profession, or some part thereof. But such groups had no legal or corporate identity, and no organized structure of leadership.[110]

Thus, the particularism of the extended kin group appears to have had a pervasive influence that worked against the development of scholarly guilds or any other autonomous groups – professional, legal, or corporate – which could sustain scientific inquiry and protect it from outside attack. In fact, the traditional, pre-Islamic extended family pattern was reinforced by Islamic law in that it did not recognize corporate entities. Islamic law does not recognize corporate personalities, which is why cities and universities and other legally autonomous entities did not evolve there.[111] Moreover, as I will have occasion to stress later, it was precisely the corporate (legally autonomous) nature of universities that gave them their dynamic thrust in the West and sharply set them apart from the madrasas of the Middle East.[112]

From another point of view, it may be said that the prevailing particularis-

---

[110] Lawrence Conrad, "The Social Structure of Medicine in Medieval Islam," *The Society for the Social History of Medicine Bulletin* 27 (1985): 11ff.

[111] On the corporate dimension in education, see Makdisi, *The Rise of Colleges*, pp. 237ff; as well as Joseph Schacht, *Introduction to Islamic Law* (Oxford: Oxford University Press, 1960), pp. 155ff; and Schacht, "Islamic Religious Law," in *The Legacy of Islam*, 2d ed. (New York: Oxford University Press, 1974), pp. 392–403. On cities and towns, see S. M. Stern, "The Constitution of the Islamic City," pp. 25–50 in A. H. Hourani and S. M. Stern, eds., *The Islamic City* (Philadelphia: University of Pennsylvania Press, 1970). Max Weber recognized this difference in *The City*, ed. and trans. Don Martindale and Gertrud Neuwirth (New York: Free Press, 1958). However, his articulation of the fundamental differences in urban structures between Islam and the West was often ambiguous. See H. J. Berman, *Law and Revolution: The Formation of the Western Legal Tradition* (Cambridge, Mass.: Harvard University Press, 1983), pp. 392–403, and my discussion below in Chap. 4.

[112] Makdisi, *The Rise of Colleges*, pp. 235–37, 289–90.

tic ethos of human relations, which was reinforced by an equally particularistic legal system, worked against the evolution of the scientific norm of
universalism: according to this norm, "truth-claims, whatever their source,
are to be subjected to preestablished impersonal criteria."[113] The idea of
objectivity, in other words, is bound up with the idea of impersonal criteria,
and these criteria are predetermined and universally applied by seekers after
the truth. However, the educational system of Islamic culture, whether
focused on law and the religious sciences or the foreign sciences, was highly
particularistic and personalized in its transmission, with all certification of
competence based on the authority of single individuals, not a faculty or
collegial judgment. The whole system of ijazas was modeled after the collection and transmission of hadiths – the religious sayings of the Prophet. The
need to certify each of these sayings (which were later assembled into large
written collections) produced the system of ijazas, which originally meant
the listening to, or audition, of the hadith from an authoritative source.[114]

In large part the problem of autonomy and universalism concerns the legal
concept of *jurisdiction*.[115] The point is that legal jurisdictions in the Islamic
world were never established because all Muslims are equally members of
the *umma*, the religious community, and it is impermissible to separate
Muslims into legally distinct groups. As a result, the domain over which a
mufti or a state official, the qadi, had jurisdiction – that is, legally authoritative voice – was never established. Furthermore, political rulers felt equally
capable of deciding for themselves what the law in their regimes was. In legal
terms, the "Shariah [religious law] had never developed the necessary procedures or writs that would bring the prince or executive power to account for
actions committed outside the law. Throughout Islamic history the judiciary,
composed of the 'ulama', exercised its functions as a result of delegation by
executive power and therefore was dependent on it."[116]

Moreover, as Harold Berman has shown, the separation of the ecclesiastical jurisdiction from the secular jurisdiction of emperors and kings was a
necessary precondition for the development of the modern European con-

---

[113] Robert K. Merton, "The Normative Structure of Science," chap. 13 in *The Sociology of Science: Theoretical and Empirical Investigations* (Chicago: University of Chicago Press, 1973), at p. 270. Further discussion of the sociological significance of the dichotomy be-tween universalism and particularism can be found in the many writings of Talcott Parsons, e.g., *Toward a General Theory of Action* (New York: Harper, 1951), pp. 81–2; and with very different stresses in Benjamin Nelson, *On the Roads to Modernity*, ed. Toby E. Huff (Totowa, N.J.: Rowman and Littlefield, 1981), pp. 184ff, 192, and passim.

[114] Makdisi, *The Rise of Colleges*, pp. 140–2.

[115] See Emile Tyan, "Judicial Organization," in *Law in the Middle East*, ed. Majid Khadduri and Herbert Liebesny (Washington, D.C.: The Middle East Institute, 1955), pp. 236–78.

[116] See Ziadeh, *Lawyers*, p. 149.

ception of corporate autonomy and, quite possibly, for the development of science as an autonomous enterprise as well.[117] Because Islamic law did not recognize corporate groups and entities, it prevented the evolution of autonomous status groups – professional as well as residential groups – within which legal privileges and universalistic norms, for example, the norms of science, objectivity, and impersonal justice, could be established and applied irrespective of personal, religious, or political considerations. Some students of the Islamic legal profession argue that even lawyers (qadis) were not an autonomous group until the introduction of the Western civil codes in the nineteenth and twentieth centuries.[118] From a sociological point of view, the decisive developmental breakthrough occurs when social groups, guilds, professional associations, and communities solidify and establish their own generalized rules of conduct that are recognized by the external authorities.

Following the criteria of Gabriel Baer, who denies the existence of guilds in the Middle East, it has been argued recently, nevertheless, that the Islamic schools of law (*madhhabs*) were guilds. According to Baer, "It would seem to us that one may be justified in speaking of the existence of guilds if all the people occupied in a branch of the urban economy within a definite area constitute a unit which fulfills at one and the same time various purposes such as economically restrictive practices, fiscal, administrative or social functions."[119] While Makdisi artfully shows that to some degree these conditions applied to the law schools centered in the madrasas from the tenth century onward, it remains the case that the madrasas were founded by single individuals (as required by the law of waqf), that the law professors had no special legal privileges vis-à-vis Islamic law (one mufti's interpretation was as good as another's), and even the ijazas (which they personally authorized) remained private authorizations that any learned scholar (*faqih*) could give whether or not he was a law professor or otherwise affiliated with a madrasa or masjid. What was unique about the European guilds, including cities and universities, was the fact that they were able to make their own laws – that is, internally valid rules and ordinances – which the church and political authorities recognized as valid. The phrase "city air makes one free" is the classic illustration of this level of social autonomy. Clearly this is not something that occurred in Islam.[120]

Still another aspect of the particularism of Islamic legal education was the

---

[117] Berman, *Law and Revolution*, especially pp. 151ff.
[118] Ziadeh, *Lawyers*, p. 149; and N. J. Coulson, *Conflicts and Tensions in Islamic Jurisprudence* (Chicago: University of Chicago Press, 1969), p. 66.
[119] G. Baer, "Guilds in Middle Eastern History," in *Studies in the Economic History of the Middle East*, ed. M. A. Cook (London: Oxford University Press, 1970), p. 12.
[120] I will have much more to say about this in Chap. 4.

fact that the madrasas as institutions of learning were exclusive, since they only admitted students from one legal tradition and excluded the rest.[121] What is important about this is that there was no thrust toward universalization, that is, toward developing a universal science of law, one which strove to synthesize the particularities and contradictions of law. Instead, the different legal traditions – the Hanbali, Maliki, Hanafi, and Shafi'i – maintained their separate identities and principles and thereby aborted the development of universal legal principles, as occurred in Western law.[122]

Finally, another aspect of the underlying ethos of medieval Islamic intellectual life is the sharp division between the knowers and the novices – the uninitiated – which has been termed "an impressive history of secrecy among philosophers."[123] George Hourani noted that this heritage from the Greeks flowed from Plato, through Galen, al-Farabi, Ibn Sina, al-Ghazali, Ibn Tufayl, to Averroes and, most famously, Maimonides. While their reasons for concealment vary, these philosophers all shared the sentiment that ordinary citizens (the masses) are not capable of grasping the higher truths of philosophy or, in the case of al-Ghazali, Tufayl, and Averroes, the "inner meaning" of the scriptures. In some cases it was simply asserted that if a person were "a believer he will know that to discuss those [philosophical] questions openly is forbidden by the Holy Law."[124]

The literary techniques for concealment and disclosure were well known and skillfully used by Averroes, both in his critique of al-Ghazali, that is, the *Tahafut al-Tahafut*, and in "The Decisive Treatise" (or *On the Harmony of Religion and Philosophy*). Such techniques included

> alluding to certain doctrines only symbolically; scattering or suppressing the premises of an argument; dealing with subjects outside their proper context; speaking enigmatically to call attention to significant points; transposing words and letters; deliberately using equivocal terms; introducing contradictory premises by which to divert the reader; employing extreme brevity to state the truth; refraining from drawing obvious conclusions; i.e., silence; and attributing one's own views to prestigious forebears.[125]

Such techniques, needless to say, run against the grain of the scientific ethos, which seeks brevity and clarity of expression, as well as the norms of univer-

---

[121] Makdisi, "The Guild of Law in Medieval Legal History," *Zeitschrift für Geschichte der Arabisch-Islamischen Wissenschaften* 1 (1984): 233–52 at p. 242.

[122] Berman, *Law and Revolution*, 140 and passim. More on this in Chaps. 4 and 6.

[123] Hourani, *Averroes: On the Harmony of Religion and Philosophy*, p. 106, n142.

[124] Averroes, *Tahafut al-Tahafut*, p. 430, as cited by Kogan, *Averroes*, p. 22.

[125] Kogan, *Averroes*, p. 21. For further illustration of the techniques, see Maimonides, *The Guide of the Perplexed;* and Leo Strauss, *Persecution and the Art of Writing*, reprint (Westport, Conn.: Greenwood Press, 1973).

salism and communalism.[126] This Averroist view, according to which one must use one form of expression for the masses and another for the cognoscenti, was later condemned by the Christian church as the doctrine of two truths. The ultimate demise of the view that ordinary believers had to be protected from the untoward consequences of deep knowledge occurred with the Protestant Reformation and the idea of the "priesthood of all believers" with its attendant notion of "the inner light," found also in Catholic medieval thought, especially in the context of discussions of conscience.[127] In addition, the advent of the printing press should be seen as an innovation that destroyed this elitist ethos. Nevertheless, in Islamic lands the printing press was banned for many centuries after its appearance in the West.[128]

In short, there were identifiable institutional patterns and cultural forces at work in medieval Islam that prevented the development of legally autonomous spheres of discourse and participation. These included the dominance of the traditional extended family, an ethos of secrecy in intellectual affairs, resistance to the formulation of generalized universal norms, as well as strongly particularistic legal norms. On one level it may be said that a breakthrough to associational autonomy and universalistic standards of thought and action would have required a revolution in Islamic law, something that actually happened only in the mid-nineteenth-century reforms of the Ottomans (the *tanzimat* reforms resulting in the *majalla*) and the "mixed courts" (of Egypt). Later in the twentieth century there was a wholesale adoption of Western legal codes, especially the civil code, in order to meet the legal needs of modernizing statehood.[129] On the other hand, the breakthrough to modern science required a breaking out of traditionalist patterns of human relations that were doubtless common in other parts of the world, to a greater or lesser degree.

[126] See Merton, "The Normative Structure."
[127] On these topics see Nelson, "Self-Images and Systems of Spiritual Direction," chap. 3 in *On the Roads to Modernity*, among others, and Chap. 3 below.
[128] Carter, *The Invention of Printing in China*; Pedersen, *The Arabic Book*, chap. 7; Hodgson, *The Venture of Islam*, 3: 123; and Stanford Shaw, *History of the Ottoman Empire and Modern Turkey* (New York: Cambridge University Press, 1976), 1: 235–8. It is interesting to note that a captured (and converted) Hungarian Calvinist (or Unitarian) was responsible for introducing the printing press to the Muslim community in Turkey early in the eighteenth century; ibid., p. 236.
[129] See the essays in Majid Khaddduri and Herbert Liebesny, eds., *Law in the Middle East*, especially S. S. Onar, "The Majalla," pp. 292–308, and H. Liebesny, "The Development of Western Judicial Privileges," pp. 309–33; as well as J. N. D. Anderson, *Law Reform in the Muslim World* (London: Athlone Press, 1976), and Majid Khadduri, *The Islamic Conception of Justice* (Baltimore, Md.: Johns Hopkins University Press, 1984). I have discussed some of these problems in "On Weber, Law, and Universalism: Some Preliminary Considerations," *Comparative Civilizations Review* no. 21 (1989): 47–79.

Institution building and the marginality problem

If we now place these considerations in the context of the question regarding the putative marginality of the natural sciences in medieval Islam, it appears that we are faced with a half-empty/half-full judgment. The formal exclusion of the teaching of philosophy, medicine, higher mathematics, optics, chemistry (alchemy), and astronomy from the madrasas suggests that the natural sciences were institutionally marginal in medieval Islamic life. In other words, scientific inquiry was generally tolerated, even sometimes encouraged by rulers for various short-lived periods of time, but in no case was it officially institutionalized and sanctioned by the intellectual elite of Islam:

The prevalence of a general picture of mild opposition or lack of encouragement is clearly reflected in the Islamic institutions of science and learning . . . the madrasa, the Islamic school of higher education, excluded systematic instruction in the secular sciences from its curriculum, and although exceptions to this general rule are found, these exceptions were short-lived and small in number. Thus, the observatory, the one [institution] among them which most closely related with the non-religious sciences, experienced the greatest difficulty in becoming an integral part of the Islamic civilization. Again the madrasa excluded systematic instruction in the secular sciences from its curriculum, and although exceptions to this general rule are found, these exceptions were short-lived and small in number.[130]

On the other hand, those such as A. I. Sabra, who reject the marginality thesis, base their judgments on the fact that even many law professors also privately taught philosophy and elements of the natural sciences, especially medicine; that scientific manuscripts were copied and were available in the libraries attached to the madrasas and other mosque schools; and that even logic was treated by many religious scholars as a necessary instrument, a balance for weighing arguments in all forms of intellectual discourse.[131] And finally, it is also noted that the development and use of higher mathematics (including algebra, geometry, and trigonometry) was most successfully carried forward in astronomy by religiously committed individuals, some of whom were timekeepers (muwaqqits) in the local mosques. And, as I noted at the outset of this inquiry, it was just such men who developed the mathematical planetary models that were at the heart of the Copernican revolution. In short, from this point of view, large segments of the rational sciences were *naturalized* (Sabra's term) by the Arabs by the twelfth century and as

---

[130] Sayili, *The Observatory in Islam* (Ankara: Turkish Historical Society Publication no. 38, 1960), p. 9.

[131] Sabra, "The Appropriation and Subsequent Naturalization," pp. 229–36. The case of medicine, which needs a somewhat different characterization, will be discussed in Chap. 5 in the section entitled "Islamic Protoscientific Institutions."

this occurred the Arabs achieved very significant heights of scientific creativity. Indeed, we have seen that the Arab astronomers criticized and perfected Ptolemaic planetary theory with the result that they produced greatly improved planetary models, models mathematically equivalent to those of the Copernican system – lacking of course the heliocentric orientation and the three simultaneous circular motions of the earth.

Even if we accept the view that by the twelfth and thirteenth centuries the foreign sciences had been thoroughly assimilated or naturalized within Islam, we must acknowledge that the natural sciences had not gained any institutional autonomy, a prerequisite for the transition to modern science. Moreover, Professor Sabra's naturalization (or Islamization) thesis clearly entails the logical consequence of the decline and fall of Arabic science, a deliberate reigning in of theoretical inquiry. This naturalization process has been cast into a three-stage model:

In the first stage we witness the acquisition of ancient, particularly Greek, science and philosophy through the effort of translation from Greek and Syriac into Arabic. . . . Greek science entered the world of Islam, not as an invading force setting off from a powerful stronghold in Alexandria, Antioch or Harran, but rather as an invited guest. The individuals who brought him in kept their reserve and aloofness with regard to the important matter of religion.[132]

However, during the second phase, this reserve and aloofness gave way to heightened curiosity and intellectual experimentation:

The guest quickly proved to hold an attraction for his hosts far beyond the promise of his practical abilities. His powers of persuasion can be seen in the unexpected but almost immediate and almost unreserved adoption of Hellenism by Muslim members of the household, like al-Kindî. But the real measure of his spectacular success is shown in the emergence, during the second phase, of a large number of powerful Muslim thinkers whose allegiance to a comprehensive Hellenistic view of the world of matter and thought and values can be described only as a thoroughgoing commitment. Those were the Fârâbîs, the Avicennas, the Ibn al-Haythams, the Bîrûnîs, and the Averroeses. I describe them as Muslims because they thought of themselves as such, and because they were attentive to problems generated by the collision between their religious beliefs and Hellenistic doctrines.[133]

In the third stage, we find the assimilation of philosophical inquiry within the bounds of religious prescription: the practice of *falsafa*,

the type of thought and discourse found in the writings of philosophers like Fârâbî and Avicenna, began to be practised in the context of *kalâm*; and in which the philosopher-physician (represented by Râzî) was replaced by the jurist-physician

---

[132] Sabra, "The Appropriation and Subsequent Naturalization," p. 236.
[133] Ibid.

(represented by Ibn al-Nafīs), the mathematician (*taʿlīmī*) by the *faraḍī*, and the astronomer-astrologer by the *muwaqqit*.[134]

In this last phase:

The carriers of scientific and medical knowledge and techniques now largely consisted of men who were not only Muslims by birth and faith, but who were imbued with Muslim learning and tradition, and whose conceptual framework had been produced in the process of forging a consciously Muslim outlook. No longer was the scientific scholar committed to the presuppositions of the earlier philosophers. Sometimes a scholar of this later breed distinguished himself equally in the religious and the rational sciences – such as Kamāl al-Dīn ibn Yūnus of Mawsil, and sometimes he held an office in a religious institution (like Ibn al-Shatir). In many cases he was an expert on *fiqh*, or grammar, or Qurʾānic science, or all of these. In almost every case he had undergone a thorough Muslim education.[135]

Given this course of assimilation and intellectual coloring, "the question . . . is not whether scientific education and practise mingled with traditional learning ·in this third stage; but how the process of combination developed and with what consequences for the character and progress of scientific thought."[136]

Here then the focus shifts to what might be called the spirit or ethos of intellectual and philosophic life after the naturalization had run its course. Sabra outlines a general observation that appears to have characterized the attitude of a great many prominent Muslim intellectuals from al-Ghazali to Ibn Khaldun. This is the strongly held view that the epitome of knowledge was that which brought one closer to his creator:

For the religiously committed Ghazâlî this means, not only that religious knowledge is higher in rank and more worthy of pursuit than all other forms of knowledge, but also that all other forms of knowledge must be subordinated to it. No occupation, no pursuit, however virtuous in itself, should be allowed to divert man from his ultimate goal. Thus, among the non-revealed forms of knowledge: medicine is necessary only for the preservation of health; arithmetic for the conduct of daily affairs and for the execution of wills and the division of legacies in accordance with the revealed law; astronomy, a science praiseworthy in itself but blameworthy in some of its implications, is useful in performing an operation legitimatized by the Holy Quran, namely the calculation of celestial movements; and logic is just a tool for weighing arguments in religious as well as non-religious branches of inquiry. . . . [T]here is only one principle that should be consulted whenever one has to decide whether or not a certain branch of learning is worthy of pursuit: it is the all-important consideration that "this world is the sowing ground for the next."[137]

134 Ibid.
135 Ibid., p. 237.
136 Ibid.
137 Ibid., p. 239.

The ultimate aphorism – and condemnation of idle curiosity – then is "May God protect us from useless knowledge." Thus within the Muslim world of the late Middle Ages, the utility and usefulness of knowledge is narrowly construed to mean knowledge useful in a strictly religious context. This religious utilitarianism may also be termed instrumentalism: "The final result of all this is an instrumentalist and religiously oriented view of all secular and permitted knowledge. This is the view that accompanied the limited admission of logic and mathematics and medicine into the *madrasa* and the conditional admission of the astronomer into the mosque."[138] What is more, Professor Sabra claims, what we have here "is not a general utilitarian interpretation of science, but a special view which confines scientific research to very narrow, and essentially unprogressive areas."[139]

From this point of view, the naturalizing and Islamicizing of the ancient sciences had the consequence of setting limits on intellectual innovation, in effect halting the free range of the imagination insofar as it impinged on the theoretical – and perhaps we should say the metaphysical – limits and assumptions of Islamic thought.

Summary

There were both internal and external impediments to the breakthrough to modern science in Arabic-Islamic civilization. If we confine our remarks mainly to astronomy, we see that the Arab astronomers from the eleventh to the fourteenth century established a broad-based research tradition aimed at reforming the Ptolemaic (geocentric) planetary model. These astronomers – in both Eastern and Western Islam – wanted a theoretical planetary model that conformed to what really is. In the thirteenth and fourteenth centuries the combined efforts of the Marâgha School of astronomers, capped by the work of Ibn al-Shatir, finally arrived at a planetary model mathematically equivalent to the Copernican model of a century and a half later. But having arrived there, Ibn al-Shatir and his successors failed to make the leap to the heliocentric view – the leap that distinguished the Copernican achievement – and thereby failed to achieve the philosophical and metaphysical transformation that we call the scientific revolution of the sixteenth and seventeenth centuries. It is clear that a significant number of astronomers, astrologers, and muwaqqits continued to work on this science after Ibn al-Shatir. David King, for example, lists seventy-five Mamluk astronomers between 1250 and

---

[138] Ibid., p. 240.
[139] Ibid., p. 241.

1517, fifty-seven of whom lived after al-Shatir.[140] Despite the fact that Shatir developed new planetary tables (*zij* tables) using his planetary constructions, "his models are not known to have had any influence in Islamic astronomy after his time."[141]

This suggests several inferences. First, consistent with Sabra's thesis of the naturalization of the ancient sciences in Islam by the High Middle Ages, there was no innovative spirit in astronomy thereafter that dared to break out of the received (religious) conception of the world's natural order. It might also be suggested that along with the established instrumentalist worldview there was an absence of the rationalist view of man and nature, most thoroughly exemplified in Plato's *Timaeus,* which played such an important role in the philosophical thought of the European Middle Ages.[142] Instead, the view that stressed the need to confine intellectual inquiry to those spheres that coincided with and aided the religious regulation of life carried with it the important theological view often referred to as Islamic *occasionalism,* a view which denied that the natural order was a rational order governed solely by laws of nature. The orthodox Ash'arite position was rather that the world was a continuous flux of moments, recreated each instant, but with a habitual pattern of continuity, knowledge of which was implanted in the believer's mind by God.[143] For anyone to declare otherwise would be foolhardy at best and life-endangering at worst. From the point of view of the dedicated Muslim believer who also functioned as an astronomer-muwaqqit, jurist-philosopher, or mathematician-faradi, there was no impetus to disrupt the received metaphysical conception of the natural order. If we accept Sabra's thesis – and the Arabic scientific writings that have so far been analyzed from the centuries after the fourteenth supports it – the free thinkers in the mold of Farabi, Ibn Sina, Razi, and Biruni were gone.

Second, if we consider the external influences on scientific and intellectual life, we see additional impediments. The first of these is the extremely personalistic nature of social relations in the medieval Middle Eastern world. On the one hand this is seen in the dominance of the extended kin family, which worked against the formation of guilds and associations of disin-

---

140 David King, "The Astronomy of the Mamluks," *Isis* 74, no. 274 (1983): 531–55; and for a useful discussion of the social, political, and intellectual life during the Mamluk rule of Syria in the thirteenth and fifteenth centuries, see N. A. Ziadeh, *Urban Life in Syria under the Early Mamluks,* reprint (Westport, Conn.: Greenwood Press, 1970).
141 King, "Astronomy of the Mamluks," p. 538.
142 See Chap. 3 for an elaboration of these themes.
143 See Kogan, *Averroes,* pp. 44, 135ff and passim; M. Fakhry, *Islamic Occasionalism and Its Critique;* and Harry Wolfson, *The Philosophy of Kalam* (Cambridge, Mass.: Harvard University Press, 1976).

terested, nonkin professionals. Likewise, the model of intellectual training and knowledge transmission was that of a personal authorization represented by the ijaza. What authenticated an intellectual tradition was not the wide acceptance of a body of knowledge by a group of experts, or an examination, but a personal authorization to teach a particular document or book. This is not to say that such a system of teaching and knowledge transmission could not produce highly skilled scholars – it clearly could. But it could not control the infinite proliferation of alternative interpretations that might be totally lacking in empirical grounding or conceptual sophistication. In addition, and most dangerously, it could not control or neutralize legally authentic opinions designed to attack progressive or innovative scientific theories, because such legal opinions were issued freely by legists and carried the weight of invincible religious authority. In such a system, each person's interpretation was as good as everyone else's, though the opinions of legists and the religiously learned carried the most weight, and there was no mechanism for selecting out the best and rejecting the worst, the implausible. In law this resulted in the formation of four mutually exclusive schools of law and a legal system in which one could request as many legal opinions as one desired on a question of law or morals and chose to follow the one most personally congenial. In science, an achievement as great as that of Ibn al-Shatir simply fell on deaf ears because it was not part of an ongoing educational system. Moreover, the original achievement of the Marâgha school appears to have been the product of just such a personalistic, extended kin grouping. Thus, al-Tusi, who organized the scholars of the Marâgha observatory, was very lucky to have a gifted student in the person of Qutb al-Din al-Shirazi. The transmission of the astronomical achievements of the Marâgha group to Ibn al-Shatir in Damascus seems somewhat surprising and anomalous. It may have had something to do, however, with the fact that al-'Urdi, another of the Marâgha astronomers and the son of al-'Urdi the builder and instrument maker engaged to build the Marâgha observatory, was also a Damascene.[144] But al-Shatir as a muwaqqit in the Damascus mosque appears not to have had any students to carry on his work, since, again, astronomy was not a subject taught in the madrasas, and hence the personal contact was lacking. Thus, Ibn al-Shatir's work seems not to have had any influence on later Arab astronomers.

These elements of particularism and personalism operated on the interpersonal level, but I should also stress that there were real legal impediments to the formation of social structures open to and supportive of science. This stemmed from a fundamental impediment in the nature of Islamic law, and

[144] Sayili, *The Observatory in Islam*, pp. 194, 220.

this defect, as we shall see, centered in the absence of a legal theory of corporations. Consequently it blocked all the avenues that could have opened up new forms of social action, of political consent, representation, and adjudication, as well as freedom of association and inquiry.

The foregoing description of the nature of medieval Arabic science and its cultural context can be taken as a starting point for a more systematic comparison of science and its institutionalization in the European West. For though the European West was definitely less rich and intellectually sophisticated prior to the twelfth and thirteenth centuries, thereafter it underwent a revolutionary transformation, thanks in part to the transmission of a wealth of Greek and Arabic scientific knowledge, which prepared it well for the revolution of modern science. The question, therefore, that cannot be avoided is, was the West really socially, legally, culturally, and institutionally different from the patterns we have seen in Arabic-Islamic civilization? In seeking an answer to that question, I will focus on the legal and forensic structures of the West, above all, on the manner in which they shaped social institutions and on the way they shaped Western man's image of his cognitive and intellectual powers.

3

Reason and rationality in Islam
and the West

In a broad sense one may say that the sources of reason and rationality in any civilization are to be found in its religion, philosophy, and law. These spheres of discourse and inquiry, before the emergence of autonomous science, interact to produce various amalgams of rational discourse based on the idioms, metaphors, and vocabulary of their domains. In some civilizations such as classical Greece, philosophy was undoubtedly the queen of intellectual life. This has prompted many observers to note that wherever Greek thought prevailed it shaped images of man and his capacities in the most rational of directions, and this influence has been felt even down to the present day.[1]

The strictly religious sources of rationality, however, as Max Weber so acutely saw,[2] are scarcely to be overlooked. For once images of the proper aims of the religious life have been conjured up, they establish, to use Clifford Geertz's formulation, "powerful, pervasive, and long-lasting moods and motivations in men by formulating conceptions of a general order of existence and clothing these conceptions with such an aura of factuality that the moods and motivations seem uniquely realistic."[3] In the Western world these religious images created an unprecedented faith in reason and the rational ordering of the natural world. This rationalist metaphysic has continued to undergird the scientific worldview ever since the Greeks.

[1] This is the thesis of A. C. Crombie's new work; see "Designed in the Mind: Western Visions of Science, Nature, and Humankind," *History of Science* 26 (1988): 1–12. The great impact that Greek philosophy had on early Christianity is richly revealed in Edwin Hatch's classic, *The Influence of Greek Ideas on Christianity*, reprint (Gloucester, Mass.: Peter Smith, 1970).

[2] Weber, "Religious Rejections of the World and Their Direction," in *Essays from Max Weber*, ed. by Hans Gerth and C. W. Mills (New York: Oxford University Press, 1948), pp. 323–59.

[3] Clifford Geertz, "Religion as a Cultural System," in *The Interpretation of Cultures* (New York: Basic Books, 1973), p. 90.

In addition to the strictly religious order of things, one needs to consider the legal conceptions that in many ways have become the operative mechanisms whereby the more narrowly religious moods and motivations have become ensconced in an institutional order. For it would be unduly restrictive to overlook the independent influence that legal canons and methods of procedures, indeed, the legal mind, have had on the construction of authoritative modes of reason and rationality in the daily practice of dispute resolution.

Historians of science, on the other hand, have sought more narrowly to find the sources of scientific rationality in the arts and crafts, that is, in the prevailing technology of artisanry. Even Max Weber took this path at various points in his thinking. Weber insisted that it was from the renaissance arts that "the method of experiment" arose.[4] Likewise, in *The Religion of China*, he wrote that "the 'experimenting' great art of the Renaissance was the child of a unique blend of two elements: the empirical skill of occidental artists based on craftsmanship, and their historically and socially determined rationalist ambitions. They sought eternal significance for their art and social prestige for themselves by raising art to the level of 'science.'"[5] Similarly, Edgar Zilsel,[6] Joseph Needham, and many others down to the present have attributed the rise of various forms of experimentalism to the crafts and even the arts. But their remarks were by and large made in ignorance of the history of experimental practice in Arabic science. Joseph Needham, while better informed, lays heavy stress on the influence of the "higher artisanate"[7] because of his early Marxist leanings, leanings shared by Zilsel. All such discussions overlook the clear statements of the logic of experimental science set out by the Arabs in at least three different sciences long before the Renaissance. Furthermore, discussions of these techniques were passed to the West in the twelfth and thirteenth centuries.[8] In short, the influence of art-and-craft-based experimentalism on the development of scientific thought and practice must take second place to articulated statements of experimentalism provided by natural philosophers and scientists themselves.

[4] *The Protestant Ethic and the Spirit of Capitalism* (New York: Scribners, 1958), p. 13.
[5] *The Religion of China* (New York: Free Press, 1951), p. 151.
[6] Edgar Zilsel, "The Sociological Roots of Science," *American Journal of Sociology* 47 (1942): 544–62; and "The Origin of William Gilbert's Scientific Method," *Journal of the History of Ideas* 2 (1941): 1–32.
[7] Needham, *SCC* 3: 154ff, 159, 160, 166, as well as *Clerks and Craftsmen in China and the West* (Cambridge: Cambridge University Press, 1970), and *GT*.
[8] See for example, A. C. Crombie, "The Significance of Medieval Discussions of Scientific Method for the Scientific Revolution," in *Critical Problems in the History of Science*, ed. Marshall Clagett (Madison, Wis.: University of Wisconsin Press, 1959), pp. 79–102; and "Avicenna's Influence on the Medieval Scientific Tradition," in *Avicenna*, ed. G. Wickens (London: Luzac, 1952), pp. 84–107.

For example, A. I. Sabra has traced the concept of *experiment* in Arabic science back to the eleventh-century optical work of Ibn al-Haytham (d. ca. 1040). Although al-Haytham's use of the term experiment and its cognates was something of a novelty, Sabra observes that al-Haytham's medieval Latin translator "did not hesitate to render *i'tabara* by *experimentare* (or *experiri*) *'itibar* by *experimentum* (or *experimentatio*), and *mu'tabir* by *experimentatar*."[9]

Accordingly, our tack will be one that casts a broader net to capture the sources of reason and rationality in the wellsprings of religion and sacred law. Given the unique centrality of sacred law (the Shari'a) in Arabic-Islamic civilization, we might suppose that Islamic law is an extreme illustration of the general impact of law on social and institutional life. It would be a mistake, however, to suppose that law in the West was any less implicated in the domains of social, economic, political, and intellectual life. To put it differently, every civilization constructs its own metaphysical geometry with its attendant sociological consequences. The outcomes of legal development have indeed been different in the two civilizations, but rather than supposing that the effects of law are generically different, we may suppose that law in both civilizations played a central role in the evolution of social and cultural life and that the law of the West, once it had undergone the legal revolution of the twelfth and thirteenth centuries, was a far more fertile soil than could be found elsewhere – including the civilizations of Islam and China – for the growth of free institutions and autonomous spheres of scientific discourse. Our first order of business, therefore, is to compare the alternative anthropologies of man embedded in the two civilizational centers and to unfold their differing conceptions of reason and rationality. Later we will also explore the role of law in China, and in all three cases we will be concerned to elucidate the mutually reinforcing impacts that these alternative conceptions of reason, rationality, and law had on each other, as well as their effects on the general social order.

The Islamic legal background

Islamic law, as we have seen, is a preeminently sacred law that coalesced around the Quran and the collected sayings of the Prophet Muhammad as compiled by later followers. As a sacred decree, Islamic law was presumed to

---

[9] A. I. Sabra, "The Astronomical Origins of Ibn al-Haytham's Concept of Experiment," *Actes du XIIe Congrès International d'Histoire des Sciences*, Tome IIIa (1971), pp. 133–6 at p. 133; and *The Optics of Ibn al-Haytham: Books I–III on Direct Vision* (London: The Warburg Institute, 1989) 2: 10–19. For further discussion of the medieval sources of experimentation, see below Chap. 6, section entitled "Internal Factors."

be fully complete, perfect, and unchanging. While first-rate legal minds saw that this could not be literally true, they nevertheless proceeded on this assumption and set about developing a legal canon and a set of intellectual devices that could be employed to extend the Shari'a to those cases for which exact parallels were missing or in which circumstances were unlike anything actually discussed in the Quran or the *Sunna* (the traditions of the Prophet).

There is a deeply felt tension and paradox here. On the one hand, the Shari'a, the sacred law of Islam, is the command of God, a fait accompli, and the task of the jurist and believer is to understand this command. Law or jurisprudence (fiqh) is the science of understanding, and the roots of law (the *usul al-fiqh*) comprise the material to be understood and the intellectual means by which this is to be achieved.[10] On the other hand, as N. J. Coulson concluded, "the Qur'ân and the *sunna* taken together in no sense constitute a comprehensive code of law. The legal material they contain is a collection of piecemeal rulings on particular issues scattered over a wide variety of different topics; far from representing a substantial corpus juris, it hardly comprises the bare skeleton of a legal system."[11] That being the case, Islamic jurists were perpetually faced with the task of deriving legal rulings from a sacred writ for novel situations, when no such anomalies were supposed to exist.

According to classical Islamic legal theory, the roots or sources of law are four: the Quran, the traditions of the Prophet (the Sunna), analogical reasoning (*qiyas*), and the consensus (*ijma'*) of the scholarly community. The theory asserts that God's law (God's command) is completely encompassed within the Quran and the oral tradition of the Prophet (written down in various hadiths). Although God would never allow his community of believers to wander from the straight path, and, therefore, within the Shari'a there must be an appropriate tradition for guidance on all occasions, an intellectual struggle (ijtihad) is necessary to understand the whole Shari'a and to apply it to all the complex situations encountered by Muslims.

In the course of the development of this legal theory, the acceptable modes of reasoning were increasingly narrowed to strict analogy (qiyas), that is, finding the similarity, and shunning the idea of personal opinion (*ra'y*) and especially personal discretion (*istihsan*).[12] The figure who did the most to lay these foundations and to systematize Islamic legal thought was al-Shafi'i (d.

[10] N. J. Coulson, *A History of Islamic Law* (Edinburgh: Edinburgh University Press, 1964), p. 75ff.; and Fazlur Rahman, *Islam*, chap. 4.

[11] Coulson, *Conflicts and Tensions in Islamic Jurisprudence* (Chicago: University of Chicago Press, 1969), p. 4.

[12] See Schacht, *Introduction to Islamic Law* (Oxford: Oxford University Press, 1964), p. 37.

820).[13] Shafi'i redefined the permissible modes of reasoning in legal thought, reducing intellectual struggle (ijtihad) to qiyas, that is, reasoning by analogy. Shafi'i wrote:

Analogy is of two kinds: the first, if the case in question is similar to the original meaning [of the precedent], no disagreement in this kind [is permitted]. The second, if the case in question is similar to several precedents, analogy must be applied to the precedent nearest in resemblance and most appropriate. But those who apply analogy are likely to disagree [in their answers].[14]

In principle, the operation of analogical reasoning seeks a similarity between some established aphorism, injunction, or narrative in the Quran and a new situation.[15] Shafi'i was aware that there are many different forms of analogy and spoke of the Arabic tongue as "the tool by means of which [one] applies analogy."[16] He also spoke of the strongest form of analogy as something like a deduction "from an order or prohibition by God or the Apostle involving a small quantity, which makes equally strong or stronger an order or prohibition involving a great quantity, owing to the [compelling] reason in the greater quantity. Similarly the commendation of a small act of *piety* implies the presumably *stronger* commendation of a greater act of piety."[17]

Thus guidance in these cases is based upon some indication of a similarity that allows one to draw a rightly guided conclusion. Shafi'i made it clear, furthermore, that "on matters touching the [life of a] Muslim there is either a binding decision or an indication as to the right answer. If there is a decision, it should be followed; if there is no indication as to the right answer, it should be sought by *ijtihad*, and *ijtihad* is *qiyas* (analogy)."[18] The capstone of this thinking led to a complete subordination of reason to the divine law:

On points of which there exists an explicit decision of God or a *sunna* of the Prophet, or a consensus of the Muslims, no disagreement is allowed; on the other points scholars must exert their own judgment in search of an indication in one of these three sources. . . . If a problem is capable of two solutions, either opinion may be held as a result of systematic reasoning; but this occurs only rarely.[19]

In short, the rationalizing impulses in early Islamic legal thought, which sought to shape law into a systematic and coherent body of knowledge and

---

[13] See Schacht, *Origins of Muhammadan Jurisprudence* (Oxford: Oxford University Press, 1950), pp. 269–82; 283–8; 315ff.

[14] Majid Khadduri, ed. and trans., *Islamic Jurisprudence: Al-Shâfi'i's Risâla* (Baltimore, Md.: Johns Hopkins University Press, 1961), p. 209.

[15] Cf. Bernard, "Qiyas," *EI²* 5: 238–42.

[16] Al-Shafi'i, *Risâla*, p. 307.

[17] Ibid., p. 308.

[18] Ibid., p. 288, pa. 493.

[19] Shafi'i as cited in Schacht, *Origins of Muhammadan Jurisprudence*, p. 97.

thus to prevent believers from straying from the straight path, achieved the results of eliminating the role of reason as a source of law. The actual reasoning of Muslims in particular cases could take many different forms, "but whatever form it took, juristic speculation in classical times was not regarded as an independent process which created a field of man-made law alongside the divine ordinances."[20] Shafi'i took the position that the Muslim community as a whole had preserved all the divinely inspired traditions of the Prophet, none having been lost. It was further stipulated that "the consensus of the community could not contradict the *sunna* of the Prophet . . ." and, therefore, "no room for discretionary exercise of personal opinion" was left. "Human reasoning had to be restricted to making correct inferences and drawing systematic conclusions from the traditions."[21]

By this means the gates of intellectual struggle (ijtihad) were closed,[22] and no new legal principles could be added. This did not mean that qadis and jurists would stop issuing legal opinions (fatwas) to resolve questions in individual cases, but it did mean that no new legal principles and conceptions could be added to the canon of Islamic law – for these were given once and for all by God in the Quran and the Sunna vouchsafed by the scholarly consensus. This outcome explains why so many elements of a complete legal code are missing in Islamic law; it explains why legal personalities and institutions such as corporations do not exist, why the idea of personal liability and the concept of negligence are unknown to Islamic law,[23] why the rules of evidence are hardly developed at all,[24] and why Islamic penal law, as well as Islamic codes of public administration, are completely inadequate for a modern state.[25] In later times a conception of public interest did

[20] Coulson, *Conflicts and Tensions*, p. 19.
[21] Schacht, *Introduction to Islamic Law*, p. 47f.
[22] Ibid., pp. 69ff.
[23] Schacht, *Introduction to Islamic Law*, p. 182.
[24] Regarding rules of evidence, see Coulson, *Conflicts and Tensions*, pp. 61–6; as well as Schacht, *Introduction to Islamic Law*, pp. 151, 192ff; and *Islamic Criminal Law and Procedure*, by M. Lippman, S. McConville, and M. Yerushalmi (New York: Praeger, 1988), pp. 59–77.
[25] On penal law, see Schacht, *Introduction to Islamic Law*, pp. 175ff; as well as Lippman et al., *Islamic Criminal Law and Procedure*, and the essays in M. Cherif Bassiouni, ed., *The Islamic Criminal Justice System* (New York: Oceana, 1982). Also see Majid Khadduri, *The Islamic Conception of Justice* (Baltimore, Md.: Johns Hopkins University Press, 1984); on law and public administration see M. Khadduri, ed., *Major Middle Eastern Problems in International Law* (Washington, D.C.: American Enterprise Institute for Public Policy, 1972); *War and Peace in the Law of Islam* (Baltimore, Md.: Johns Hopkins University Press, 1955); and *The Islamic Law of Nations: Shaybani's Siyar*, trans. Majid Khadduri (Baltimore, Md.: Johns Hopkins University Press, 1966); N. J. Coulson, "The State and the Individual in Islamic Law," *International and Comparative Law Quarterly* 6 (1957), pp. 49–60; and Martin Shapiro, "Islam and Appeal," *California Law Review* 68 (1980): 350–81. For the

develop but in the limited sense of "government in accordance with the revealed law" (*siyasa shar'riyya*),[26] which served mainly to grant discretionary powers to the temporal ruler while placing no legal constraints on him, since the idea of discretionary powers transcends all such limitations.[27]

From a certain point of view, these early developments in Islamic law hint at the idea of precedent and of following a path marked out by juridical consensus, but one must not read Western conceptions into Islamic reality. While Shafi'i sanctified scholarly consensus, in both theory and practice this idea lacked explicitness. Since there were no centralized courts, and since judges were not even an autonomous profession until the late nineteenth century,[28] there was no institutional mechanism by which a history of precedents and judicial rulings could be put into place. Of course there are collections of sayings attributed to the Prophet (for example, those of Bukhari), but these lacked universal assent as well as systematic organization as legal principles. These manuals "were organized around such headings as faith, purification, prayer, alms, fasting, pilgrimage, commerce, inheritance, wills, vows and oaths, crimes, murders, judicial procedure, war, hunting, and wine."[29] As such they may be considered the first step in the development of a legal system, but without the further elaboration and systematization required of a legal system, they constitute a congeries of religious, ritual, liturgical, moral, and customary, as well as legal, precepts. In fact, both of these ideas, precedent and binding man-made judicial rulings, are un-Islamic, since anything strictly legal would have to be found in the Quran or

---

problems Islamic law confronted in the nineteenth century, when it encountered Western legal conceptions, and the adjustments it had to make, see the essays in Khadduri and Liebesny, eds., *Law in the Middle East* (Washington, D.C.: The Middle East Institute, 1955), especially those by Liebesny (on restoring Western legal privileges), Onan (on the Ottoman legal reforms of the nineteenth century resulting in the *majalla*), and Tyan (on judicial organization and the *mazalim* courts, and the absence of the notion of jurisdiction in Islamic law).

From the perspective of the development of universal legal standards within the Western legal tradition and their borrowing by several countries of the Middle East, see T. E. Huff, "On Weber, Law, and Universalism," *Comparative Civilizations Review* no. 21 (1989), pp. 47–79.

[26] Coulson, *A History of Islamic Law*, pp. 129ff.

[27] Ibid., pp. 132ff. There is another concept of public interest (*maslaha*); see Khadduri, "The *Maslaha* (Public Interest) and *'Illa* (Cause) in Islamic Law," *New York University Journal of International Law and Politics* 12 (1979): 213–17. But it remains a philosophical ideal that has yet to be shown to be a legal principle in Islamic law (apparently because the term *maslaha* does not appear in the Quran); see M. Khadduri, *The Islamic Conception of Justice* (1984), pp. 137f.

[28] See F. Ziadeh, *Lawyers: The Rule of Law and Liberalism in Egypt* (Stanford, Calif.: The Hoover Institution, 1968); and Coulson, *Conflicts and Tensions*, pp. 68f.

[29] Lapidus, *A History of Islamic Societies* (New York: Cambridge University Press, 1988), p. 102.

the actual sayings of the Prophet. In other words, the record was complete at the time of the death of the Prophet, and therefore it makes little sense to speak of precedent outside the Quran and the Sunna. It was not permissible to imagine that a judicial ruling, per se, established a precedent: it was simply an application of the existing law. To codify actual judicial rulings, fatwas, into a codebook of legal canon, would be to usurp the rightful place of the Quran and the Sunna.

In addition, the four major schools of law – the Shafiʻi, Hanafi, Hanbali, and the Maliki (to which other dissenting schools were later added) – remained separate domains, laws unto themselves. They were not taught as a single entity in the madrasas[30] but as exclusive systems that applied individually to a situation because the litigant belonged to that legal school, or because one or more of the parties to the legal dispute belonged to another school. Thus Herbert Liebesny refers to the introduction of precedent into Islamic law as a result of the European incursions into India and the Middle East in the seventeenth and later centuries.[31] The very different nature of legal evolution and the uses of analogical reasoning in Western law are to be seen in the standard works on this subject.[32] It is clear from Edward Levy's and Melvin Eisenberg's accounts that legal change is brought about by finding legal rules and then finding higher principles, which frequently expand legal concepts and categories and then subsume the older rules. Nothing like that occurred in Islamic law. Islamic practice in this or that town, city, or village might understandably diverge from the ideal, but its primary thrust did not give birth to a major systematization of law equivalent to the canon law, nor did it give birth to the revolution in law witnessed in the West – as we will see in Chapter 4. Islamic law's rejection of the cumulation of precedents is actually closer to the theory and ideology of contemporary Continental civil law,[33] but it should be remembered that classical Islamic law

---

[30] Makdisi, *The Rise of Colleges*, p. 304 and passim.

[31] Herbert Liebesny, "English Common Law and the Islamic Law in the Middle East and South Asia: Religious Influences and Secularization," *Cleveland State Law Review* 34 (1985/6): 19–33.

[32] Edward H. Levy, *An Introduction to Legal Reasoning* (Chicago: University of Chicago Press, 1949); and Melvin A. Eisenberg, *The Nature of Common Law* (Cambridge, Mass.: Harvard University Press, 1988); as well as Ruggero J. Aldisert, *Logic for Lawyers* (New York: Clark Boardman, 1989). The first articulation of a legal theory of precedent based on previous judicial decisions in common law – that is, actual case law – goes back to Sir Henry Brackton, the great thirteenth-century English jurist.

[33] John Henry Merryman, *The Civil Law Tradition*, 2d ed. (Stanford, Calif.: Stanford University Press, 1985); and Mary Glendon, W. M. Gordon, and Christopher Osake, eds., *Comparative Legal Traditions* (St. Paul, Minn.: West Publishing, 1985); as well as Arthur von Mehren and James Gordley, eds., *The Civil Law System*, 2d ed. (Boston: Little Brown, 1977).

precluded substantive and procedural change by means of statute or parliamentary enactment. Indeed, it is because of the presumed perfection and unchangeability of the Shari'a that modernizing Islamic countries have had to radically restrict the application of the Shari'a to the family and inheritance while adopting Western civil codes (sometimes with modifications) for application in all other domains.[34]

Reason, man, and nature in Europe

From the point of view of the centuries-long gestation of the central cultural forms of Islamic civilization, Europe in the eleventh century without that tradition was as fresh, young, and naive in comparison with Arabic-Islamic civilization as the United States was in comparison to Europe in 1776. For while it had a thousand-year-old religious tradition, it had lost much of Rome's heritage, especially the Roman legal tradition, as well as the major portion of Greece's heritage, and it had failed to establish major intellectual traditions outside the church. It is not surprising, therefore, that when the European translators, such as Adelard of Bath (fl. 1116–42), Gerard of Cremona (ca. 1114–87), and Michael Scot (1217–35), among others, began to encounter the rich intellectual heritage of the Middle East (largely in Spain), they quickly became enthusiasts of and promoters of the wisdom of their "Arab masters."[35]

It has long been understood by medievalists that the recovery of the Roman legal heritage, along with the transmission of the long-lost Greek tradition, through the new contacts with Arabic-Islamic culture in the twelfth century produced a renaissance in Europe. This new burst of energy and creativity affected virtually every sphere of intellectual activity: it was evident in law, philosophy, theology, and scientific inquiry. It was marked equally by the founding of colleges and universities as well as new cities and towns.[36] Indeed, there was a new sense of well-being brought on by a

---

[34] See J. N. Anderson, *Law Reform in the Muslim World* (London: Athlone Press, 1976); the essays in D. Khadduri and Liebesny, eds., *Law in the Middle East;* as well as M. Khadduri, *The Islamic Conception of Justice.* Also Huff, "On Weber, Law, and Universalism," *Comparative Civilizations Review* no. 21 (1989): 47–79.

[35] For an account of this transmission process see Haskins, *The Renaissance of the Twelfth Century* (New York: Meridian, 1957 [1927]), chap. 9, as well as the more recent account by David C. Lindberg, "The Transmission of Greek and Arabic Learning to the West," in *Science in the Middle Ages,* ed. David C. Lindberg (Chicago: University of Chicago Press, 1978), pp. 52–90.

[36] Haskins, *The Renaissance;* Rashdall, *The Universities of Europe* (Oxford: Oxford University Press, 1964); Alexander Murray, *Reason and Society in the Middle Ages* (Oxford: At the Clarendon Press, 1978); M.-D. Chenu, *Nature, Man, and Society in the Twelfth Century* (Chicago: University of Chicago Press, 1968); Harold Berman, *Law and Revolution: The*

heightened economic development that served to support this new soaring of reason and the imagination. One may say that it was this new spirit of human agency and empowerment that produced one of the sharpest contrasts between the thinkers of Europe and those of the Middle East during this period. Furthermore, it was among the Christian religious elite that this attitude was most pronounced. It was seen equally in the work of the canonists (the students and architects of ecclesiastical law), the Romanists (the students of the revived Roman law), and the theologians cum philosophers, such as Peter Abelard, William of Conches, Thierry of Chartres, and many others. It manifested itself in a jubilant new spirit of reason that saw the marks of rationality and planned order in every domain. It so pervaded all aspects of intellectual life that it was as if the thinkers of the age wore reason-colored glasses.

The most significant source of this rationalist impulse was the *Timaeus* of Plato, the one work of Plato and the Greeks that had survived the lapse of learning after the decline of the Roman Empire. Translated into Latin toward the end of the third century A.D. by Chalcidus, the *Timaeus* was first embraced by Augustine and thereafter taken up with enthusiasm by the *moderni* of the twelfth and thirteenth centuries.[37] Although the Arabs knew of the *Timaeus*, it did not attain the popularity nor receive the enthusiastic embrace the Christians of the West gave it.[38] What most impressed the European thinkers of the early modern period about the *Timaeus* was the image of nature as an orderly, integrated whole. The natural world was portrayed as a rational order of causes and effects, while man, as part of the rational order of things, was elevated by virtue of his reason. The historic passage that has been the source of citation, commentary, and inspiration is the following:

Formation of the Western Legal Tradition (Cambridge, Mass.: Harvard University Press, 1983), among others.

[37] Chenu, *Nature, Man, and Society*, pp. 60ff; and see Stiefel, *The Intellectual Revolution in Twelfth-Century Europe* (New York: St. Martins, 1985), p. 22, for the usage of the term *moderni*; and Edward Grant, "Science and the Medieval University," in *Rebirth, Reform, Resilience: Universities in Transition, 1300–1700*, ed. James M. Kittelson and Pamela J. Transue (Columbus, Ohio: Ohio State University Press, 1984), pp. 68–102, at p. 85.

[38] See Richard Walzer, *Greek into Arabic* (Columbia, S.C.: University of South Carolina Press, 1962); and F. E. Peters, *Aristotle and the Arabs* (New York: New York University Press, 1968); as well as Shlomo Pines, "Philosophy," in *The Cambridge History of Islam* 2: 780–823. Although paraphrases of it appear in the writings of some Arab philosophers, so fragmentary are references to the *Timaeus* among Arab philosophers that one has the impression that it was never translated into Arabic. This is the impression left both by Pines, "Introduction," to *The Guide of the Perplexed*, and Peters in *Aristotle and the Arabs*, as well as *Allah's Commonwealth* (New York: Simon and Schuster, 1973).

Now everything that becomes or is created must of necessity be created by some cause, for without a cause nothing can be created. . . . Was the heaven then or the world . . . always in existence and without beginning? or created, and had it a beginning? Created, I reply. . . . But the father and maker of all this universe is past finding out; and even if we found him, to tell of him to all men would be impossible. And there is still the question to be asked about him: Which of the patterns had the artificer in view when he made the world – the pattern of the unchangeable, or that which is created?[39]

It is difficult to overestimate the impact of the elements of Platonic thinking on the Christian medievals of this period prior to the arrival of the "new" Aristotle. For this Platonism infected every area of inquiry, including the study of nature as well as of Scripture.[40]

A not inconsiderable achievement of this period in the West, when it discovered nature, was the separation and elaboration of the spheres of nature and supernature, that is, the separation of the miraculous from the forces of nature. William of Conches (d. 1154) was perhaps the preeminent architect of this protoscientific philosophy of nature. In his commentary on the *Timaeus* he asserts:

Having shown that nothing exists without a cause, Plato now narrows the discussion to the derivation of effect from efficient cause. It must be recognized that every work is the work of the *Creator* or of *Nature,* or the work of a human artisan imitating nature. The work of the Creator is the first creation without pre-existing material, for example, the creation of the elements or spirits, or it is the things we see happen contrary to the accustomed course of nature, as the virgin birth, and the like. The work of nature is to bring forth like things from like through seeds or offshoots, for nature is an energy inherent in things and making like from like.[41]

This image of nature entailed both the notion of orderliness (or hierarchy) and lawfulness. The medievals had discovered the world as a universe, a cosmos of connected and interlocking parts. Thus, Honorius of Autun speaks of the supreme artisan who "made the universe like a great zither upon which he placed strings to yield a variety of sounds," who divided the world into two kinds of complementary parts, the spiritual and the material:

for he divided his work in two – into two parts antithetical to each other. Spirit and matter, antithetical in nature yet consonant in existence, resemble a choir of men and boys blending their bass and treble voices. . . . Material things similarly imitate the distinction of choral parts, divided as things are into genera, species, individuals,

---

[39] Plato, *The Timaeus* (Jowett trans.), as cited in Chenu, *Nature, Man, and Society,* p. 57 n15.
[40] For its influence on poetry, see W. Weatherby, *Platonism and Poetry in the Twelfth Century* (Princeton, N.J.: Princeton University Press, 1972).
[41] As cited in Chenu, *Nature, Man, and Society,* p. 41.

forms and numbers; all of these blend harmoniously as they observe with due measure the law implanted within them and so, as it were, emit their proper sound.[42]

Repeatedly one finds a stress on the idea of a unified cosmos, ordered and regulated and in which the laws of nature and the forces of nature are presumed to operate autonomously. Another expression of this view is seen in the writings of Hugh of St. Victor (d. 1141), speaking of the idea of the world as an ordered unity (*universitas*): "The ordered disposition of things from top to bottom in the network of this universe . . . is so arranged that, among all the things that exist, nothing is unconnected or separated by nature, or external."[43] Likewise, Abbot Thierry of Chartres (d. ca. 1156) asserts that "the world would seem to have causes for its existence, and also to have come into existence in a predictable sequence of time. This existence and this order can be shown to be rational." Peter Abelard (d. 1142), in like manner, sought to explicate the separation between the autonomous forces of nature and those of the divine and preferred naturalistic explanations whenever they could be worked out:

Perhaps someone will ask, too, by what power of nature this came to be. First, I reply that when we require and assign the power of nature or natural causes to certain effects of things, we by no means do so in a manner resembling God's first operation in constituting the world, when only the will of God had the force of nature in creating things. . . . We go on to examine the power of nature . . . so that the constitution or development of everything that originates without miracles can be adequately accounted for.[44]

Furthermore, this image of the world as an ordered place gave rise to the idea of the world as a machine, as witnessed in the writings of Hugh of St. Victor: "As there are two works, the work of creation and the work of restoration, so there are two worlds, visible and invisible. The visible world is this machine, this universe, that we see with our bodily eyes."[45] This idea of a universal machine is likewise the title of one of Robert Grosseteste's scientific treatises and is an image witnessed repeatedly in the writings of the age.[46]

In short, the Platonisms of the twelfth century came to form a model of

---

[42] As cited in ibid., p. 8.
[43] As cited in ibid., p. 7 n10.
[44] Ibid., p. 17 n34.
[45] Ibid., p. 7 n10.
[46] Benjamin Nelson stressed the importance of this notion of a world machine in the medieval period in several of his papers. See *On the Roads to Modernity* (Totowa, N.J.: Rowman and Littlefield, 1981), especially pp. 190 and 197 n6. Also see Lynn White, Jr., *Machina Ex Deo* (Cambridge, Mass.: MIT Press, 1968); and White, *Medieval Technology and Social Change* (Oxford: Oxford University Press, 1962), pp. 105 and 174 n5.

inquiry such that everything – natural and supernatural – was examined for the purpose of finding the causes and giving the reasons thereof.[47] It was fully recognized that this way of proceeding was difficult and likely to provoke opposition on the part of strict religious enthusiasts. Nevertheless, Adelard of Bath, adopting this new naturalistic philosophy of nature, set out an agenda, which was to be the domain of the philosopher:

For the functioning and the interconnection between all the senses are manifest in all living things . . . but which forces come into play in what connections with which method or mode, none except the mind of a philosopher can make clear. For the effects of these interactions are most subtly connected with their causes, and the relationship between the causes themselves is also so very intricate that knowledge of these matters is often concealed from philosophers by nature herself.[48]

William of Conches put it even more boldly when he affirmed that "it is not the task of the Bible to teach us the nature of things; this belongs to philosophy."[49] How strongly parallel this seems to Galileo's statement in the seventeenth century that "the intention of the Holy Ghost is to teach us how one goes to heaven, not how heaven goes."[50]

But these *moderni* did not stop here: they went on to examine and even criticize the Bible and to suggest that if passages of the Bible contradict reason and the natural order, it should not be taken literally. Thierry of Chartres was perhaps not willing to go so far; yet, writing a commentary on the book of Genesis he says, "This is an exegetical study of the first portion of Genesis from the point of view of an investigator of natural processes

---

[47] The view that the Western world had to wait for the sixteenth and seventeenth centuries for these ideas, especially the idea of laws of nature, is now outdated in the light of more recent scholarship. Cf. Zilsel, "The Genesis of the Concept of Physical Law," *The Philosophical Review* 51 (1942): 245–79, especially pp. 255–8; and "The Sociological Roots of Science," *American Journal of Sociology* 47 (1942): 544–562; with the account in Stiefel, *The Intellectual Revolution in Twelfth-Century Europe;* Chenu, *Nature, Man, and Society,* chaps. 1 and 2; and even Dijksterhuis, *The Mechanization of the World Picture* (London: Oxford University Press, 1961), pp. 119–25. Zilsel omits any discussion of the Platonists of Chartres as well as such important figures as Abelard, Adelard of Bath, and Alain of Lille. For a more recent account of the development of the idea of laws of nature in Greek thought, see Helmut Koester, "Nomos and Physeôs: The Concept of Natural Law in Greek Thought," in *Religions in Antiquity: Essays in Memory of E. R. Goodenough,* ed. Jacob Neusner (Leiden: E. J. Brill, 1968), pp. 521–41.

[48] As cited in Stiefel, "The Heresy of Science: A Twelfth-Century Conceptual Revolution," *Isis* 68 (1977): 355.

[49] Chenu, *Nature, Man, and Society,* p. 12.

[50] Galileo, "Letter to the Grand Duchess Christina," in M. Finocchiaro, *The Galileo Affair: A Documentary History* (Berkeley and Los Angeles: University of California Press, 1989), p. 96. Galileo was seeking authorization for his inquiries apparently by quoting Cardinal Baronius, an associate of Cardinal Bellarmine.

(*secundum physicam*) . . . and of the literal meaning of the text."[51] But
William of Conches takes it further by asserting the priority of physical
reasoning: "And the divine page says, 'He divided the waters which were
under the firmament from the waters which were above the firmament.'
Since such a statement as this is contrary to reason let us show how it cannot
be thus."[52] It would appear that these medievals were pioneering that approach to the Scripture and the historical record which places reason and the
logical tools of inference above the literal word. This is, in other words, the
beginning of higher criticism in biblical studies.[53]

Just as the universe itself was conceived to be a unified whole, so too man
was presumed to be part of this rational whole. As such he was thought to be
endowed with reason and thereby enabled to read and decipher the patterns
of the universe, that is, to read "the book of nature."[54] One finds this
rationalist philosophy (or anthropology) of man articulated by many writers
including Adelard of Bath:

Although man is not armed by nature nor is [he] naturally swiftest in flight, yet he
has that which is better by far and worth more – that is, reason. For by possession of
this function he exceeds the beasts to such a degree that he subdues them. . . . You
see, therefore, how much the gift of reason surpasses mere physical equipment.[55]

Thus it is that we find a variety of preeminently modern ideas – about the
constitution of nature, about the role of philosophy as opposed to religion
and the Scriptures, and about the rationality of man – appearing in the
renaissance of the twelfth century. All of this was to undergo strong modification with the arrival of the new Aristotle and other translations from
Greek and Arabic, yet Platonic rationalism did lay a foundation regarding
the physical (and metaphysical) nature of the universe as a rational and
coherent whole. Most important of all, it established a firm belief in the
rational capacity of man to understand and explain nature and, equally
important, to interpret and explain the Scriptures. Accordingly, Christian

---

[51] Cited in Stiefel, "Science, Reason, and Faith in the Twelfth Century: The Cosmologists'
Attack on Tradition," *Journal of European Studies* 6 (1976): 7.

[52] Ibid.

[53] See Beryl Smalley, *Study of the Bible in the Middle Ages* (Oxford: Oxford University Press,
1952). More recent examples would be Sir Edwyn Hoskins and Noel Davey, *The Riddle of
the New Testament* (London: Faber and Faber, 1958). Richard Popkin, "Bible Criticism and
Social Science," in *Methodological and Historical Essays in the Natural and Social Sciences*
(Boston Studies in the Philosophy of Science, vol. 14), ed. R. S. Cohen and Marx Wartofsky
(Dordrecht: Reidel, 1974), pp. 339–60, would seem to place these beginnings much later.

[54] On the significance of the idea of the "book of nature," see B. Nelson, "Certitude, and the
Books of Scripture, Nature, and Conscience," in *On the Roads to Modernity*, chap. 9.

[55] As cited in Tina Stiefel, "Science, Reason, and Faith in the Twelfth Century: The Cosmologists' Attack on Tradition," p. 3.

philosophy and theology in the twelfth and thirteenth centuries unequivo-
cally declared man to be the possessor of reason, and this capacity enabled
him to decipher the most mysterious puzzles of God's creation. It also
enabled man to decipher the mysteries of the divine word itself – unaided by
revelation and without the need for prevarication.

This outlook stands very much in contrast to Islamic thinking of the time,
whether one speaks of the philosophers or the theologians. For among the
latter, it is clear that the Ash'arite view of man and nature, based on Islamic
atomism (known as occasionalism),[56] was very much opposed to the well-
ordered, even mechanistic and physically determined conception of the natural order that evolved in the writings of the Christian theologians of the
twelfth and thirteenth centuries. While al-Farabi and Ibn Sina had been
deeply influenced by Plato, they did not find the inspiration that the Latins
found in the *Timaeus*. To be sure, these Arab philosophers did develop
Platonist philosophical views that were offensive to the religious elite of
Islam, but they did not elaborate the rationalistic or mechanistic worldview
that the European Platonists of the twelfth century built on Plato's edifice.
Even more in contrast, the dialectical theologians of Islam, the mutakallimun, could not embrace the naturalistic image of nature composed of
causal forces of nature and, above all, would not tolerate the idea that events
described in the Quran could be explained by naturalistic accounts, as Thierry and William of Conches had attempted in the case of the Christian
Scriptures in twelfth-century Europe. Even in the twentieth century, H. A.
R. Gibb relates the case of an Egyptian sheikh, Muhammad Abu Zaid, who
in 1930

published an edition of the Quran with annotations, criticizing the old commentaries
and interpreting supernatural references in simple naturalistic ways. Although the
purpose of the work was to encourage the younger generation to study the Quran,
the book was confiscated by the police, and an injunction was secured to prevent the
writer from preaching or holding religious meetings.[57]

These same sentiments were at work in the Islamic condemnation of Salman
Rushdie's novel, *The Satanic Verses*, in 1989. One might say that the tradition of biblical criticism (sometimes called higher criticism) is a uniquely
Western religious tradition that originated in the twelfth and thirteenth centuries and was to undergo vigorous development during the Enlightenment.[58] Conversely, such a tradition never developed in Islamic thought and

[56] M. Fakhry, *Islamic Occasionalism and Its Critique by Averroës and Aquinas* (London: Allen and Unwin, 1958).
[57] H. A. R. Gibb, *Modern Trends in Islam* (Chicago: University of Chicago Press, 1947), p. 54.
[58] See note 53 above.

its absence is a contemporary source of objections to fictionalizing the people and events of the history of Islam.[59]

Without developing the details further, it may be suggested that the twelfth-century renaissance in Europe laid the foundations for a scientific research program (subsequently transformed and embellished). It was developed by the religious elite who dominated intellectual thought and it contained, according to Tina Stiefel, the following assumptions:

1. That a rational and objective investigation of nature in order to understand its operation is possible and desirable.
2. That such an investigation might make use of techniques of mathematics and deductive reasoning.
3. That it should use empirical methodology – i.e., evidence based on sense-data, where possible.
4. That the seeker for knowledge of nature's operations (a "scientist") should proceed methodically and with circumspection.
5. That the scientist should eschew all voices of authority, tradition and popular opinion in questions of how nature functions, except to the extent that the information is rationally verifiable.
6. That a scientist must practise systematic doubt and sometimes endure a state of prolonged uncertainty in his disciplined search for an understanding of natural phenomena.[60]

Accordingly, one may argue, as Tina Stiefel does, that this intellectual departure constituted a scientific revolution in that it clearly laid out a research program within a metaphysical context that assumed that the events of nature are rationally explicable by men using the tools of logic and the God-given agency of human reason. This research plan, furthermore, accepted the postulate of the causal ordering of nature so clearly stated in the *Timaeus* and other sources. If it had a place for miracles, that place was delimited within increasingly narrowed boundaries.

Given the naturalist, rationalist, and even mechanistic elements in this new philosophy of nature, it is not surprising that the whole movement, once it had been fortified by the new Aristotle, provoked a severe reaction on the part of the more traditionalistic officials of the Christian church in the thirteenth century. This took the form of the famous condemnation of 1277 (by the Bishop of Paris, Stephen Tempier) of a wide variety of philosophical assumptions. Yet, on balance, the condemnation did little to curb academic

---

[59] See Daniel Pipes, *The Rushdie Affair: The Novel, the Ayatollah, and the West* (New York: Birch Lane Press, 1990); and Lisa Appignanesi and Sara Maitland, eds., *The Rushdie File* (Syracuse, N.Y.: Syracuse University Press, 1990), as well as the many other discussions of the controversy spawned by Rushdie's fiction.

[60] Stiefel, *The Intellectual Revolution*, p. 3.

freedom, to prevent the teaching of Aristotle, or to inhibit scientific think-ing. On the contrary, the teaching of the natural books of Aristotle – those dealing with physical nature, with plants, animals, and meteorology – was well established in the major universities of Europe and continued into the seventeenth century.[61]

Moreover, it should not be overlooked that the effort to make theology itself a science was a fundamental movement during this period of time[62] and that it inexorably pushed intellectual life in the direction of systematization and self-conscious methodological reflection. The same may be said of the study of law (as we will see in Chapter 4). Once this step had been taken and an equally strong metaphysical commitment had been made to the idea that man had reason – an idea embedded in philosophical as well as strictly theological thought – the movement toward a science of nature, freed from religious constraints, was virtually inevitable, whatever setbacks might be encountered on the way. But there is still another source in the West of the metaphysical belief that man has reason and rationality: the idea of con-science.

Reason and conscience

While classical thought in Islamic law sought to tightly circumscribe and eliminate human reason as a source of law, European and Western law gener-ally developed along a path that took the opposite direction. As we have seen, European medievals in philosophy, theology, and natural science were wholly committed to the idea that the universe is a rationally ordered cosmos and that man, as a co-creator of this order, was endowed with reason. This metaphysical commitment was in the first instance an inheri-tance from the Greeks, especially through the vehicle of the *Timaeus*. There was also a more strictly religious and Christian channel for this idea, how-ever, the notion of *synderesis* or conscience, an idea that can be traced back to the Bible, especially the New Testament.[63]

---

[61] Among others see Edward Grant, "Science and the Medieval University," pp. 68–102; and "The Condemnation of 1277, God's Absolute Power, and Physical Thought in the Late Middle Ages," *Viator* 10 (1979): 211–44; as well as James A. Weisheipl, "The Curriculum of the Faculty of Arts at Oxford in the Early Fourteenth Century," *Medieval Studies* 26 (1964): 143–85. More on this in Chap. 5.

[62] See Chenu, *Nature, Man, and Society*.

[63] The literature of this subject is very extensive. A very selective bibliographic overview would include the following: Eric D'Arcy, *Conscience and Its Right to Freedom* (New York: Sheed and Ward, 1961); Michael Baylor, *Action and Person: Conscience in Late Scholasticism and the Young Luther* (Leiden: E. J. Brill, 1977); K. E. Kirk, *Conscience and Its Problems: An Introduction to Casuistry* (London: Longmans, Green, 1927); John McNeill, *A History of*

One finds in the writings of St. Paul the word *conscience* associated with the idea of "an interior witness and judge of one's past actions and motives, which can be a source of comfort or remorse."[64] Christian commentators, moreover, have been keen to point out that conscience, whatever its precise status as a faculty, a habitual faculty, or a cognitive agency, is a faculty both for prescribing the right course of action as well as for judging and censoring actions taken in the past. Thus conscience is not just the presence of a feeling of remorse, that is, pangs of conscience, in the familiar psychological (and Freudian) idiom of today, but a far more complex agency of the soul that is capable of discernment. According to Paul Tillich:

In principle Christianity has always maintained the unconditional moral responsibility of the individual person in the Pauline doctrine of conscience. Aquinas states that he must disobey the command of a superior to whom he has made a vow of obedience, if the superior asks something against his conscience. And Luther's famous words before the emperor in Worms, insisting that it is not right to do something against the conscience . . . are based on the traditional Christian doctrine of conscience.[65]

This contrast is further accentuated in those writings where Paul distinguished between the law as the written word of God, that is, the Torah, versus the law of God written in men's hearts. Paul writes, "When Gentiles who have no law obey instinctively the Law's requirements, they are a law to themselves, even though they have no law; they exhibit the effect of the Law written on their hearts, their conscience bears them witness, as their moral convictions accuse or it may be defend them."[66] It might also be noted that the term conscience (synderesis) in Paul's thinking did not arise from the Old Testament and the Jewish tradition, from which the term appears to be absent, but from popular and Hellenic thought.[67]

The church fathers, for example, St. Ambrose (d. 397), Basil (d. 379), and Origen (d. ca. 253/4), all issued significant commentaries on this cognitive cum spiritual faculty. But it was the famous "Gloss of St. Jerome"[68] that both broadened and deepened the concept, making it a persistent source of

the *Cure of Souls* (New York: Harper, 1964); M.-D. Chenu, *L'éveil de la conscience dans la civilisation médiévale* (Paris: J. Vrin, 1969); and D. E. Luscombe, "Natural Morality and Natural Law," in *Cambridge History of Later Medieval Philosophy* (New York: Cambridge University Press, 1982), pp. 705–19. A useful overview of the issues is in Benjamin Nelson, "Casuistry," *Encyclopedia Britannica* 5 (1968): 51–2.

64 D'Arcy, *Conscience and Its Right*, p. 8.
65 Tillich, "The Transmoral Conscience," reprinted in *The Protestant Era* (Chicago: Phoenix Books, 1957), p. 139.
66 *Romans* 2:14–15 (Moffat trans.) as cited in Tillich, *The Protestant Era*, p. 139.
67 C. A. Pierce, *Conscience in the New Testament* (London: SCM Press, 1955), pp. 52–4.
68 See Baylor, *Action and Person*, pp. 24ff; and D'Arcy, *Conscience and Its Right*, pp. 15–19.

philosophical and theological speculation throughout the medieval period. This state of affairs was wrought by Jerome's Latin rendering of the Greek word *synteresis* as conscience (*conscientia*).[69] This was a significant innovation, "for the Latin notion was very much broader and more indefinite than its Greek equivalent, if indeed, they are equivalents. . . . More broadly, *conscientia* in Latin meant consciousness of knowledge in general, especially in the sense of experience or perception, but also in the sense of knowing facts, or that a state of affairs prevails."[70]

It was of course the moral dimension of conscience that became the central core of the medieval Christian doctrine of conscience. Prior to this the Stoics had elaborated on the idea of conscience, "and for them it carried the connotation of man's awareness of the natural moral law and his awareness of the correspondence or lack of correspondence of his own actions to the law."[71] Thus Seneca "regarded the conscience as a divine guardian within men."[72] It was St. Jerome's suggestion, however, following Platonic interpreters of the dream of Ezeckiel, that synderesis was a fourth and additional component of the conscience, and it was this idea that set the philosophers and theologians to work trying to reconcile all the philosophical possibilities raised by this quartet of elements. According to St. Jerome:

There are Reason, Spirit, and Desire; to these correspond respectively the man, the lion, and the ox. Now above these three are the eagle, so in the soul, they say, above the other three elements and beyond them is a fourth which the Greeks call synderesis. This is that spark of the conscience which was not quenched even in the heart of Cain when he was driven from paradise.[73]

For the next two centuries the philosophical and conceptual problems raised by Jerome's gloss were a source of critical discussion; they were finally resolved in the thirteenth century by the work of Thomas Aquinas (1225–74).

Without going into the intricacies of this debate any further,[74] it may be said that the Christian medievals ascribed to man a conscience that implied the existence of an inner cognitive agency which allowed the individual to arrive at moral and ethical truths and to judge moral states of affairs. They attributed to man a rational faculty that was especially able to wrestle with

---

[69] Baylor, *Action and Person*, p. 24. There are many spellings of this word, including "synderesis" and "syneideresis."

[70] Ibid., p. 24.

[71] Ibid., p. 25.

[72] Ibid.

[73] As cited in D'Arcy, *Conscience and Its Right*, pp. 16f, 26.

[74] Today in theological circles, *conscience* is defined as "the proximate rule of right reason in the moral sphere"; see Benjamin Nelson, *On the Roads to Modernity*, p. 72, and his "Casuistry."

moral and ethical dilemmas. We can conclude that in the European West the
reigning philosophical and theological images of man accentuated the rationality of man. Both the clerics who drew on Greek philosophical thought
and those who drew on the more scriptural and theological sources ascribed
to man complex capacities of reason and rationality. This rational capacity
extended to the understanding of nature as well as the grasping of ethico-
religious truths. Precisely because man had a conscience, an unquenchable
moral-rational agency, he could arrive at moral truths unaided by revelation.
But even if he fell into error he was compelled to follow his conscience.
Having a good conscience became for many Christians the mark of having
executed one's Christian duties with intelligence and in good faith. In 1215
the forum of conscience became universally available and all Christians were
enjoined to go to confession, precisely because they were in possession of a
conscience and would want to rid themselves of moral error.[75] However
restrictive one might construe this effort to institutionalize the regulation of
inner moral affairs, an irrevocable metaphysical image of man had been
created: man had reason and conscience and there was no escaping it. Such
an agency, once officially recognized, increasingly became a standard against
which moral and legal precepts had to be weighed. Individuals in all sorts of
stations and stages of grace would have to exercise it. The Lutheran Reformation, from this point of view, was a revolution that fully unleashed the
conscience, making it the supreme arbiter of even scriptural truth. In effect,
religious orthodoxy lost the battle, and reason and conscience had to be free
to do their work.[76]

By way of contrast, we have seen that the Islamic theologians, as well as
the legists, constructed a different anthropology of man. Their view stressed
the inherent limits of man and his intellect. The world is too complex for
such a mortal being to fully understand and, therefore, the uses of reason
have to be carefully circumscribed. In the writings of the chief architect of
the main Sunni school of law, al-Shafi'i, the various modes of reason (ra'y),
personal discretion (istihsan), and personal preference were assimilated to
analogy (qiyas) or simply abandoned, and the legists were left with only a
narrowly conceived reasoning by analogy. After al-Shafi'i, it was unthinkable that anyone could come up with new principles of law, principles that
subsume the application of particular rules under higher or broader con-

---

[75] See Nelson, *On the Roads to Modernity,* pp. 45, 224. The older sources still include Henry
Lea's classic, *A History of Auricular Confession and Indulgences,* 3 vols., reprint (New York:
Greenwood Press, 1968).
[76] For Luther's contribution, see Baylor, *Action and Person.* But also see Rosenstock-Huessy,
Out of Revolution (New York: Argon Books, 1969), chap. 7, for a legal view of the Lutheran
revolution.

cepts, whether they pertain to evidence and procedure or substantive law. God's command had been given once and for all: it was a perfect, complete, uncorrupted work of God, and man's task was to understand it, not to add principles or doctrines. Furthermore, the consensus (ijmaʿ) of the scholars had worked out all the foundational problems and there was nothing left for future generations of scholars to uncover; they might, perhaps, deepen their understanding of the sacred law but they could not add to it. The Greek and Christian idea of conscience (synderesis) was unknown to the orthodox Islamic legists as well as to philosophers.[77] Nor did they embrace the rational conception of man and nature to be found in the *Timaeus,* if we assume that it was available.[78] Within Islamic thought, to recognize theology (kalam) at all was a major battle, as Islamic legists were prone to condemn it out of hand. There could be no thought of kalam as the queen of the sciences, since it was a suspect enterprise.[79] Likewise, without a philosophic view of man as a rational being possessed of reason, who could arrive at ethical and moral truths unaided by revelation, there could be no thought of philosophy as the handmaiden of theology.[80]

During the formative period of Islam,[81] there were in fact Islamic rationalists, the *Muʿtazilites.*[82] These religious philosophers were prepared to grant man the full powers of reason and even to claim "parity for reason and revelation."[83] For them, reason was the attribute that made man "the creator of his acts" and the judge of what is good and what is bad.[84] This position attributed to man "an innate power to act, and an innate understanding of

[77] Cf. L. Gardet and M.-M. Anawati, *Introduction à la Théologie Musalmane,* 2d ed. (Paris: J. Vrin, 1970), p. 348.

[78] See note 38 above.

[79] F. Rahman, *Islam,* p. 123; and G. Makdisi, ed. and trans., *Ibn Qadama's Censure of Speculative Theology* (London: Luzac, 1962).

[80] Although there is no word in classical Arabic for "conscience," the Arabs did have many other terms to refer to various aspects of the intellect, reason, and discourse, many of which were indebted to the Greeks. ʿAql seems to be the Arabic translation of "nous," the "active intellect" of Aristotle. See Fazlur Rahman, *Prophecy in Islam,* reprint (Chicago: University of Chicago Press, 1979); and his article on "ʿAql" in *EI²* 1:341–2. Also see E. J. Rosenthal, *Knowledge Triumphant* (Leiden: E. J. Brill, 1970).

[81] See Watt, *The Formative Period of Islamic Thought* (Edinburgh: Edinburgh University Press, 1973).

[82] In addition to Watt, *The Formative Period,* chaps. 7 and 8, also see his *Islamic Philosophy and Theology: An Extended Survey* (Edinburgh: Edinburgh University Press, 1985); as well as R. Frank, "Some Fundamental Assumptions of the Basra School of Muʿtazila," *Studia Islamica,* 33 (1971): 5–18; George Hourani, *Islamic Rationalism: The Ethics of ʿAbd al-Jabbar* (Oxford: The Clarendon Press, 1971); and F. E. Peters, *Aristotle and the Arabs,* pp. 136–46.

[83] Rahman, *Islam,* p. 90.

[84] Gardet and Anawati, *Introduction à la Théologie Musalmane,* p. 347.

the fundamental criteria of good and evil."[85] At the same time this thinking led them to believe that "God *cannot* do the unreasonable and unjust."[86] But this rationalist thrust was defeated by the orthodox triumph that resulted from the teachings of al-Ash'ari (d. 935) in the tenth century. Ash'ari, like most other dialectical theologians, granted full authority to the Shari'a, so that "all the practical issues including law and ethics, that impinge on actual concrete life were given to the authority of the shari'a."[87] This, as Fazlur Rahman points out, led to a divorce between theology on the one hand and law and ethics on the other. In effect law and ethics were frozen so that ethical thought was no longer free to work out the practical solutions to the dilemmas of life – a living casuistry – and to create a more informed vision of ethical life. Al-Ash'ari made it clear that there was a sharp division between the domain of law and that of religious thought:

The [early generations] have discussed and disputed about such matters as arose at that time concerning Din [religion] from the side of the shari'a (i.e., law) . . . like the legal duties such as punishments and divorce which are too numerous to mention here . . .

Now these are legal matters which pertain to details of life which they (i.e., the early generations) brought under the shari'a which concerns itself with the detailed conduct of life . . . and hence can never be comprehended except on the authority that comes from the Prophets. But as for the matters that arise in the field of the principles for the determination of questions (of Faith) every intelligent Muslim must refer these to the general and agreed principles founded upon reason, sense-experience and immediate knowledge, etc. Thus the questions of details of the shari'a (i.e., law), which are based on traditional authority must be referred to the principles of the shari'a whose source is traditional authority, whereas questions arising out of reason and experience must be referred to their own bases, and authority and reason must never be mixed up.[88]

While Ash'ari makes it clear that the authority and the domain of tradition (as represented by the Shari'a) is never to be confused with or subordinated to reason, it should be noted that Ash'ari's conception of reason (perhaps better translated as intelligence or intellect) is not the same as the philosophers' active intellect, nor is it similar to the Western notion of inner light. As al-Farabi (d. 950) put it, "As for the intellect that the *Mutakallimun* are always talking about, when they say about a thing, 'This is necessitated, or denied, or accepted, or not accepted by the intellect,' they mean thereby something that is universally accepted by the first [reflections] of the opinion

[85] R. Frank, "Some Fundamental Assumptions," p. 7.
[86] Ibid., p. 102.
[87] Ibid., p. 123.
[88] Ash'ari, as cited in Rahman, *Islam*, p. 105.

of everybody. For they designate as intellect the first [reflections] of common opinion [professed] by everybody or by most people."[89] Although the theologians, he continues, "believe that the intellect that they talk about among themselves is the intellect that is mentioned by Aristotle. . . . If you examine the premises they make use of, you will find that all of them without exception derive from the first [reflections] of common opinion. Accordingly they point out one thing and make use of another."[90] In short, Ash'ari had completely rejected the rationalism of the Mu'tazilites.

But in developing his position, Ash'ari fashioned the formidable occasionalist and determinist view, according to which God holds the world together from moment to moment by willing it. Moreover, man's agency – both his reasoning from beginning to end and his actions – is determined by his *acquisition*[91] from God of this capacity to complete them. The doctrine of acquisition goes back to the Mu'tazila of the seventh and eighth centuries, and in Ash'ari's hands, it becomes a powerful bulwark establishing the agency and determinacy of God in all matters. According to this view, "The acts of men are created and . . . a single act comes from two agents, of whom one, God, creates it, while the other, man 'acquires' it (*iktasabu-hu*); and [according to this view] God is the agent of the acts of men in reality, and . . . men are the agents of them in reality."[92] This doctrine of acquisition became thereafter a major theological conception in kalam, and it is repeated with various nuances by all the major Ash'arites from the tenth century through to the fifteenth-century culmination of systematic Muslim theology.[93] In other words, all acts of man originate with God, but man somehow acquires the will and power to carry out his acts, which thereby remain the will of God.

The study of the evolution of this doctrine by such scholars as Montgomery Watt and the late Harry Wolfson, among others, shows that the mutakallimun were greatly perplexed by the doctrine's claim that both man and God were responsible for human acts. In the end, God's omnipotence is preserved while the question of human responsibility (as a legal matter) becomes blurred. Men have to be accountable for acts committed against God and others, but the line was drawn before a theory of negligence could

[89] As cited by Shlomo Pines in "Translator's Introduction" to Maimonides, *The Guide of the Perplexed* (Chicago: University of Chicago Press, 1963), p. lxxxiii.
[90] Ibid.
[91] For the doctrine of *acquisition*, see "Kasb," *EI*² 4:690–94. Watt, *The Formative Period*, pp. 192ff and passim; and Harry Wolfson, *The Philosophy of Kalam* (Cambridge, Mass.: Harvard University Press, 1976), pp. 663–719.
[92] Watt, *The Formative Period*, p. 192.
[93] See Wolfson, *The Philosophy of Kalam*, pp. 663–719; as well as Gardet and Anawati, *Introduction à la Théologie Musalmane*.

be fashioned, before a theory about responsibility for acts of omission that cause harm to others, as well as a systematic theory of criminal acts, could be fashioned.[94] Ash'ari's blunt position was that "in the case of everything concerning which God is described as having power to create it as an acquisition, God has the power to force men to it."[95] But if there is compulsion, then man loses his responsibility, and God's other attribute, that of justice, falls into question, both with regard to the fate of men in the courts of trial and earthly punishment as well as their fate in the court of divine retribution. For if man is not responsible for his acts, then it is unjust to punish him for them, and conversely, if some men experience suffering and privation, this is an injustice visited by God himself. The theodicy problem in effect receives no solution, and Islamic ethics and moral philosophy are unable to move forward toward a systematic theory of intention, such as Peter Abelard began to work out in the twelfth century in the West.[96]

The ultimate Islamic formulation of this problem, which completely undid a metaphysics of causality in human affairs, was set out by al-Ghazali. A rebuttal of it was attempted by Averroes, but to no avail. In Ghazali's words:

According to us the connection between what is usually believed to be a cause and what is believed to be an effect is not a necessary connection; each of the two things has its own individuality and is not the other, and neither the affirmation nor the negation, neither the existence nor the non-existence of the one is implied in the affirmation, negation, existence, and non-existence of the other – e.g., the satisfaction of thirst does not imply drinking, nor satiety eating, nor burning contact with fire, nor light sunrise, nor decapitation death, nor recovery the drinking of medicine, nor evacuation the taking of purgative, and so on for all the empirical connections existing in medicine, astronomy, the sciences, and the crafts. For the connection of these things is based on a prior power of God to create them in successive order, though not because this connection is necessary in itself and cannot be disjointed – on the contrary, it is in God's power to create satiety without eating, and death without decapitation, and to let life persist notwithstanding the decapitation, and so on with respect to all connections.[97]

The psychological effect of such a theological position can be seen in the description of the physician dispensing drugs and prescriptions, recorded in

---

[94] On the Islamic definition of crime and criminal acts, see Lippman, McConville, and Yerushalmi, *Islamic Criminal Law and Procedure*, pp. 45–56.

[95] As cited in Wolfson, *Philosophy of Kalam*, p. 552.

[96] See Nelson, *On the Roads to Modernity*, pp. 223ff, and passim.

[97] As cited in Averroes, *Tahafut al-Tahafut*, trans. Simon Van den Bergh (London: Luzac, 1954), p. 316. The firm embrace of this doctrine is still to be seen in such contemporary Islamic writers as Seyyed Hossein Nasr, e.g., in *Science and Civilization in Islam* (New York: New American Library, 1968), pp. 307ff, and in his other writings. Likewise, throughout Islamic countries today, including Indonesia, it is a matter of habit to qualify anticipations of one's future actions with the ritual prescription "Inshallah" ("God willing").

the Geniza records of Cairo of this period. Professor S. D. Goitein reports that "we are moved by the expressions of piety which are rarely absent from a prescription. It would be superscribed with the words: 'To be taken with God's blessing' or with the Muslim formulae, 'In the name of the Merciful, the compassionate,' and end almost invariably with remarks such as 'It will help, if God wills,' or 'Thanks are due to God alone.' "[98]

Theology after this period, according to Fazlur Rahman, "split into two distinct types, the dogmatic and formally rational theory of *kalam*, and the speculative theology of Sufism."[99] The dogmatic element of the former retained its occasionalist determinism so that man had no capacity for original thought or action. Kalam also wanted "to show that reason in fact yields no universal principles," and as al-Ghazali put it, "no obligations flow from reason but from the shari'a."[100] In the final analysis, according to Fazlur Rahman, theology "monopolized the whole field of metaphysics and would not allow pure *thought* any claim to investigate rationally the nature of the universe and the nature of man."[101]

Although al-Ghazali's far-reaching discussions of philosophy forced later theologians to become familiar with the functions and uses of philosophy and logic as a balance for weighing arguments,[102] it must also be said that Ghazali destroyed any idea that philosophy could be a demonstrative science. Likewise, "his occasionalist interpretation of the empirical premises of demonstrative science"[103] severely limited the value of natural sciences in the eyes of the orthodox. Still, philosophical speculation of sorts lived on in Islam, in Sufi mysticism and even in small schools of philosophers cum theologians. Fazlur Rahman suggests that it was in the work and spirit of Fakhr al-Din al-Razi (d. 1209) that the tradition of philosophical speculation was carried on after the twelfth century. He argues that there "was an astonishingly wide scope . . . offered for the exercise of speculative reason"[104] in al-Razi's work. Nevertheless, Razi and the writers who arose in the thirteenth and later centuries, while attempting to defend philosophy, "nevertheless accepted orthodox positions" on the most central issues of theology and dogma.[105] The tensions between philosophy and the party of

---

[98] Goitein, *A Mediterranean Society* 2:254.
[99] Rahman, *Islam*, p. 107.
[100] As cited in ibid., p. 106.
[101] Ibid., p. 107.
[102] Sabra, "The Appropriation and Subsequent Naturalization of Greek Science in Medieval Islam: A Preliminary Assessment," *History of Science* 25 (1987): 232.
[103] M. Mamura, "Ghazali's Attitude Toward the Secular Sciences," in *Essays on Islamic Philosophy and Science*, ed. G. F. Hourani (Albany, N.Y.: SUNY Press, 1975), p. 108.
[104] Rahman, *Islam*, p. 121.
[105] Ibid., p. 122.

tradition continued, often erupting into "fierce outbursts of the traditional-ist attacks on intellectualism," and culminated in an imposing anti-intellectualist attack by Ibn Taymiyya (1263–1328). His basic view, Rahman suggests, "consists of a restatement of the shari'a and vindication of religious values."[106] His view of kalam, F. E. Peters writes, "is even dimmer than al-Ghazali's."[107] When he set out to analyze the "Harmony Between the Truth of Tradition and Evidence of Reason," he "severely criticized both the phi-losophers and the theologians,"[108] thus perpetuating the fierce antagonism between philosophy and religion. One concrete result was the "extradition from the syllabus" of philosophy at the great Al-Azhar College in Cairo, from which it remained banned until the arrival of modernism in the late nineteenth century.[109]

Conclusion

I have proposed in this chapter that to understand the evolution and devel-opment of systems of scientific thought, we have to consider the broader metaphysical picture within which patterns of discourse are carried on. I have shown, moreover, that a major source of the images of man's rational capacities (as well as of nature itself) are to be found in the religious and legal doctrines of a civilization. Such conceptions profoundly shape man's self-image and either enhance or constrain his rational powers. In the case of Arabic-Islamic civilization, the architects of both law and theology tightly circumscribed the rational capacities of man. Both theology and law specifi-cally rejected the idea of a rational agency attributable to all men in favor of the view that man should follow the path of tradition, of traditional author-ity (*taqlid*), and not attempt to fathom the mysteries of external nature or sacred writ. Islamic theology and law promoted the view that both the wisdom of God and the consensus of the scholars was superior to human agency and declined to endorse the idea that human reason could be a source of law and ethics. This stance seems related to the fact that the logic of intention is not developed in Islamic law (though religious intentions are recognized in matters of ritual), and thus the shades and grades of legal liability that form the backbone of the law of negligence did not develop in Islamic law. The closest Islamic thought came to developing a rationalist

---

[106] Ibid., p. 111.
[107] Peters, *Aristotle and the Arabs* (New York: New York University Press, 1968), p. 201.
[108] Rahman, *Islam*, p. 123.
[109] Ibid. To have philosophy taught at all in an Islamic college was unusual in itself, and this action seems mainly to bring al-Azhar in conformity with most other colleges. Cf. Makdisi, *The Rise of Colleges*, pp. 77ff.

perspective is in the work of the Islamic rationalists, the Mu'tazilites. But human agency was denied the power to innovate in religion or ethical thought, with the resultant closing of the gates of ijtihad. The greatest philosophical thinkers in Arabic-Islamic civilization after al-Ghazali never failed to cast doubt on the powers of human reason and to disparage the virtues of demonstrative logic; they insisted instead on the priority of faith (fideism) or on the unsurpassed authority of tradition (the Shari'a and the Sunna). Reason for the orthodox was little more than common sense, and there was no acknowledgment of the idea that reason could reach new truths unaided by revelation. Innovation, in matters of religion, was equivalent to heresy.[110] Christianity and the West were not without fears about heretical innovation and the means to eradicate it, but insofar as this applied to statements about the natural world, a new, powerful, and increasingly autonomous zone of intellectual freedom had been carved out.

In contrast to their Muslim counterparts, the European medievals felt especially enabled to study and decipher nature, to systematize, to organize, and to rationally evaluate the merits of the religious and legal texts that lay before them as the renaissance of the twelfth century unfolded. In their view, reason was precisely the instrument that set men off from the lower animals and allowed rational inquiry in all domains. This view did not proclaim a holiday of unfettered free thinking in all the realms, but it did lay the groundwork for intellectual autonomy. Not least of all, it assimilated man's rational capacities to those of God by asserting that man's rational powers were a gift from God given for his glory. Furthermore, there were multiple sources that extolled the rational ordering of nature and the rationality of man. On the one hand, the religious scholars, the theologians, and canonists embraced the rationalist images of man and nature implied by Plato and the *Timaeus*. By adopting this metaphysics as their own, the European medievals of the twelfth and thirteenth centuries became the architects of a uniquely rationalist conception of man and his powers. What is more, the Christian clerics had the tradition – or, rather, recaptured and transformed the tradition – of the Bible according to which man has an unquenchable rational agency that enables him to make informed judgments on moral and ethical affairs. While the informed Christian might go astray – and confession before a superior would remedy that possibility – it was undeniably accepted that man was a rational creature who could and must exercise his rational faculties in the spheres of moral action.

In the hands of powerful thinkers such as Peter Abelard, the beginnings of

[110] B. Lewis, "Some Observations on the Significance of Heresy in the History of Islam," *Studia Islamica* 1 (1953): 43–63.

a whole new system of moral accounting – casuistry – was worked out. Within the context of this new system, and from the point of view of intention, it could even be argued within Christian dogmatics, as Abelard did, that lacking an intention to harm God, the Jews of the New Testament could not be charged with the crime of crucifying Christ.[111] In short, the European medievals had fashioned an image of man that was so imbued with reason and rationality that philosophical and theological speculation became breathtaking spheres of inquiry whose outcomes were far from predictable, or orthodox – to the consternation of all. Furthermore, this theological and philosophical speculation was taking place within the citadels of Western learning, that is, in the universities. Christian theology had indeed clothed man with a new set of moods and motivations, but it had also attributed to him a new set of rational capacities that knew no bounds.

We turn, therefore, to the ways in which these putative rational faculties were applied to the study of law, as well as nature, and to the ways in which these conceptions became institutionalized as legal concepts, powers, and agencies.

[111] See Nelson, *On the Roads to Modernity*, p. 223; as well as D. E. Luscombe, *Peter Abelard's Ethics* (London: Cambridge University Press, 1976).

4

<====================================>

The European legal revolution

Ever since the appearance of Charles Homer Haskins's classic study, *The Renaissance of the Twelfth Century*,[1] scholars have known that the twelfth and thirteenth centuries experienced an extraordinary efflorescence of creativity and new cultural forms.[2] Charles Haskins, Hastings Rashdall, F. W. Maitland, and other scholars were certainly aware of the revival of the study of law and its impact on the development of the university, and even the impact of legal studies on the church and canon law. But with the publication of Harold J. Berman's book, *Law and Revolution*,[3] we have been reminded as perhaps never before of the extraordinary revolutionary nature of the legal and institutional reforms that erupted and swept across Europe during this period. Professor Berman's fresh account, based on the harvest of legal scholarship since the 1930s, brings to light the centrality of the sweeping legal reforms, indeed, the revolutionary reconstruction, of all the realms and divisions of law – feudal, manorial, urban, commercial, and royal – and therewith the reconstitution of medieval European society. It is this great legal transformation, I shall argue, that laid the foundations for the rise and autonomous development of modern science.

At the center of this development one finds the legal and political principle of treating collective actors as a single entity – a *corporation*. Some social theorists have recognized that the existence of these "new corporate actors"[4]

[1] Charles Homer Haskins, *The Renaissance of the Twelfth Century* (New York: Meridian, 1957).
[2] This medieval portrait has now been updated by the essays in *Renaissance and Renewal in the Twelfth Century*, edited by Robert Benson and Giles Constable (Cambridge, Mass.: Harvard University Press, 1982).
[3] Harold J. Berman, *Law and Revolution: The Formation of the Western Legal Tradition* (Cambridge, Mass.: Harvard University Press, 1983).
[4] James Coleman, *Foundations of Social Theory* (Cambridge, Mass.: Harvard University Press,

changes the nature of social action, creating new social and economic dynamics that must be accounted for by revised social, economic, and political theories. The emergence of corporate actors was unquestionably revolutionary in that the legal theory which made them possible created a variety of new forms and powers of association that were in fact unique to the West, since they were wholly absent in Islamic as well as Chinese law. Furthermore, the legal theory of corporations brings in its train constitutional principles establishing such political ideas as constitutional government, consent in political decision making, the right of political and legal representation, the powers of adjudication and jurisdiction, and even the power of autonomous legislation. It seems to me that, aside from the scientific revolution itself, and perhaps even the Reformation, no other revolution has been as pregnant with new social and political implications as the legal revolution of the European Middle Ages. By laying the conceptual foundations for new institutional forms in legal thought, it prepared the way for the other two revolutions.

As I sketch some of these developments in the West it is useful to remember that the twelfth and thirteenth centuries were graced by the presence of great intellects in both Islam and Europe, so I am by no means suggesting that the great men of Arabic-Islamic civilization were gone. Insofar as natural science is concerned, it is probably true, as George Sarton concluded, that after the twelfth century, the number of significant scientific scholars in the Arabic world fell behind that of Europe.[5] It may also be true that scholarship in general declined absolutely in the Middle Ages after this time.[6] Nevertheless, for the twelfth and thirteenth centuries, one can find virtually an equal number of great scholars in both civilizations, as well as Chinese civilization.

Accordingly, I am suggesting that the great men of Arabic-Islamic civilization deliberately chose to follow another path, with all the consequences for mankind which that decision entailed. For example, Peter Abelard's life (1079–ca. 1144) parallels that of the great al-Ghazali (1058–1111), and John of Salisbury (1120–80) was a contemporary of Ibn Rushd (Averroes, 1126–98). While Avicenna died in 1037, the important legist, philosopher, and astronomer Nasir al-Din al-Tusi (d. 1274) was a contemporary of Thomas

---

1990), pp. 531f, and passim. Likewise, Peter Drucker has suggested that the idea of corporations was the most potent social innovation since the Middle Ages.

[5] George Sarton, *Introduction to the History of Science*, 3 vols. in 5 parts (Baltimore, Md.: Williams and Wilkins, 1927–48), 2/2. For an interesting attempt to quantify the results of Sarton's work, see Pitirim Sorokin and Robert Merton, "The Course of Arabian Intellectual Development, 700–1300 A.D.," *Isis* 22 (1935): 516–24.

[6] This is the impression given by Sorokin and Merton's study cited above.

Aquinas (1225–74). And the arch traditionalist Ibn Taymiyya (1273–1328), along with the Egyptian theologian al-Iji (d. 1355), was a contemporary of Marsilius of Padua (ca. 1280–ca. 1343) as well as William of Ockham (ca. 1285–ca. 1349). Clearly, powerful intellects were busy in both civilizations following their unique but culturally conditioned agendas. While Arabic-Islamic civilization was undoubtedly intellectually richer at the outset of the High Middle Ages, at the conclusion of them the West had effected a radical transformation that marked a decisive turning away from the political, legal, social, and institutional forms that prevailed in the Islamic Middle East.

The development of modern Western law

Although there were parallels in the organization of state and society between the civilization of Islam and that of the European West, there were also fundamental differences not only in law but also in custom and tradition. In the Islamic setting there was no question that the law of the land was the sacred law of Islam, the Shariʿa, and there were no competing sources of law. While the Roman law could not have been unknown to the Muslims and Turks who traded with the Italians and Byzantines, there is no evidence that Islamic legists desired or actually ever borrowed the legal principles, concepts, or practices of the Justinian Code. Unlike the European medievals, they never considered the possibility that Roman law as codified in the ancient law books could be a source of legal principles and ideas, since the Quran and the Sunna were the complete record of God's commands. While local custom in the Middle East was granted tacit recognition by the Sunna, the lawyers and legists (qadis and fuqaha) were clear that the operative law was that of the Shariʿa. But most important of all, the temporal ruler of Islam, whether caliph – successor to the Prophet – or emir was assumed to be the ruler of the whole Muslim community and the enforcer of the Islamic legal order. There was no conceptual distinction between the sacred and the secular, the spiritual and the temporal. The law of the realm consisted precisely of those commands the believer must follow if he was to pass the reckoning on the day of judgment.

In the civilization of the West, the situation was significantly different. The kings of Europe and their defenders by custom and legal precedent had often asserted their claim to be a source of law as well as guardians of the spiritual body of Christ, the Christian community of the faithful. Since the adoption of Christianity by the Romans, however, the church had its own hierarchy of officers and officials who claimed exclusive monopoly of the religious realm, and this extended in various ways into the temporal order. From biblical times the clash between religion and world, between the demands of Christ and the demands of Caesar,[7] had been acknowledged, debated, and repeatedly fought over. On the eve of the legal renaissance of the twelfth and thirteenth centuries, the great legal historian Ernst Kantorowicz pointed out, "the king was the fountain of justice; he was supposed to interpret the law in the case of obscurity; the courts were still the 'king's courts' and the king was still considered the judge ordinary of his realm whereas the judges, who derived their power from him, acted only as delegate judges."[8] From Roman law the legists were all too familiar with the stock phrase, "What has pleased the Prince, has the force of law,"[9] although Professor Berman suggests that the Justinian Code contained an equally pertinent, if not so well known phrase suggesting that rulers also ought to be obligated to follow the law: "It is a statement worth of the majesty of the reigning prince," the code reads, "for him to confess to be subject to the laws; for Our authority is dependent upon that of the law."[10]

The final drama of this period, which ultimately established the institutional fabric of modern society, was therefore the great clash between church and state in the investiture controversy (1050–1122), largely won by the papal revolution. Above all else this battle was an intellectual and legal contest that produced the first modern system of law assumed to be universal in scope. This was the canon law, whose first definitive statement was the *Decretum,* issued by the Italian monk, Gratian, in 1140. To produce this monument of legal scholarship and institutional architectonics, Gratian had to work through a massive collection of disparate documents whose legal import were both questionable and frequently contradictory.

The body of materials that lay at hand in Europe between the fifth and tenth centuries was derived from many different sources. Much of church law was derived from the Bible, both Old Testament and New. But since the adoption of Christianity by the Emperor Constantine in 313, Christianity had become wholly enmeshed in the Roman legal order, a practical and administrative order very different from the Semitic sources of Judaism and early Christianity, as well as Hellenism. While the Roman Empire had indeed collapsed in the fifth century, Professor Berman points out that "in the clan-dominated culture of Western Europe the Church was considered to be a bearer of Roman law, and the eighth-century 'code' of the Ripuarian

---

[7] "Render, therefore, unto Caesar the things which are Caesar's and unto God the things that are God's" (Matthew 22:21).

[8] "Kingship under the Impact of Scientific Jurisprudence," in *Twelfth-Century Europe and the Foundations of Modern Society,* ed. M. Clagett, Gaines Post, and R. Reynolds (Madison, Wis.: University of Wisconsin Press, 1966), p. 93.

[9] Ibid., p. 96.

[10] *Justinian Codex* 1.14.4, as cited in Berman, *Law and Revolution,* p. 585 n58.

Franks, the Lex Ribuaria, contained the provision: *Eccelesia vivit jure Romano* ('the Church lives by Roman Law'). This meant that to the extent each person carried the law of his clan with him, and was to be judged according to it wherever he went, the church was deemed to carry with it the Roman law."[11]

Church law itself, because it had a dual charge of administering both the temporal and the spiritual domains, contained regulations regarding church finances and property, rules for specifying ecclesiastical authority, regulations governing sacred and secular interactions, regulations for dealing with crime and punishment, as well as rules for marriage and family life.[12] In addition, it contained elements of German folk law as well as various rulings of early church councils and statements (decretals) by church fathers. Not surprisingly, there were also elements of church liturgy and theology interlaced within this whole body of material, which could hardly be called a system of law prior to the twelfth century. As Berman points out, "There were no professional judges or lawyers. There were no hierarchies of courts."

Also lacking was a perception of law as a distinct "body" of rules and concepts. There were no law schools. There were no great legal texts dealing with basic legal categories such as jurisdiction, procedure, crime, contract, property, and other subjects which eventually came to form structural elements in western legal systems. There were no developed theories of the sources of law, of the relation of divine and natural law to human law, or ecclesiastical law to secular law, of enacted law to customary law, or of the various kinds of secular law – feudal, royal, urban – to one another.[13]

Perhaps the greatest spur to the development of the systems of modern law was the discovery in Italy of a manuscript containing the Justinian civil law – the *corpus juris civilis* – toward the end of the eleventh century.[14] Although Roman civil law had long since ceased to be the law of everyday practice, the shear magnificence, the range of issues, and the integral construction of the code outshone anything available in Europe during this time. Consequently it quickly attracted a great deal of attention. By 1087 the great Romanist Irnerius was ensconced at the University of Bologna and was regularly commenting upon and teaching the Roman civil law. What is also highly significant is the fact that Bologna was largely a lay institution of higher learning (since it had been founded by students). It was a place

[11] Ibid., p. 200.
[12] Ibid.
[13] Ibid., p. 85.
[14] Haskins, "The Revival of Jurisprudence," in *The Renaissance of the Twelfth Century*, chap. 7. And Berman, *Law and Revolution*, p. 122.

dominated by legal studies carried on by laymen, not canonists,[15] for the canon law had to wait until 1140 for the appearance of Gratian's "Concordance of Discordant Canons" (*Concordia discordantium canorum*) to give it a self-definition as coherent as the Roman law. The birth of the new science of law involved three separate elements: a body of legal materials to work with; a new method of analysis; and a place in which to carry on these new legal studies, that is, the universities.[16] It is also useful to note here a major deviation from the path of intellectual, legal, and institutional development in the Islamic world: the commitment in Bologna at the outset to teach the secular law, that is, the Roman *corpus juris civilis,* was a major concession to the authority of the secular law. Only later did Bologna become a center recognized equally for the teaching of ecclesiastical (canon) law.[17]

In the Islamic world, by way of contrast, aside from being pious endowments (and lacking the legal attributes of a corporation), the Islamic colleges (madrasas) allowed only the teaching of Islamic law (the Shari'a) – not Roman, Greek, Judaic, or customary law – and only one school (madhhab) of Islamic law was permitted in a single madrasas.[18] Consequently, the study and teaching of the complete body of Islamic law (in the sense of the four schools and their divergent interpretations) was never brought together in one place where a reconciliation of discordant opinions (fatwas), legal principles, and practices could be fashioned into a uniform and universal system of law.

What took place in the eleventh, twelfth, and early thirteenth centuries in Western Europe was a radical transformation that created, among other things, the very concept of a legal system with its many levels of autonomy and jurisdiction and its cadres of legal experts. A profound change took place that dramatically altered "the very nature of law as a political institution and as an intellectual concept."[19] This change was so dramatic, so complete and far-ranging that it "can only be called a revolutionary development of legal institutions." It was, moreover, "not only an implementation of policies and theories of central elites, but also a response to social and economic changes 'on the ground.'"[20] That is, it was not only an intellectual

---

[15] Alfred Cobban, *The Medieval Universities: Their Organization and Development* (London: Methuen, 1975), chap. 2.

[16] Berman, *Law and Revolution,* p. 123.

[17] Berman suggests that Roman law as well as natural law and divine law were treated as sacred law; *Law and Revolution* p. 146. While I do not dispute this, it would not occur to Muslims to put such inherited, foreign, and man-made laws on the same footing as the revealed law of the Scriptures.

[18] George Makdisi, *The Rise of Colleges: Institutions of Learning in Islam and the West* (Edinburgh: Edinburgh University Press, 1981), pp. 10, 304 and passim.

[19] Berman, *Law and Revolution,* p. 86.

[20] Ibid., pp. 86, 87.

revolution, but a social, political, and economic revolution whereby new legal concepts, entities, procedures, powers, and agencies came into being and transformed social life.

The papal revolution

At the heart of this transformation is what has been called the papal revolution (ca. 1072–1122). This was the struggle by means of which the papal authority of the Christian church declared itself free from secular control, above all, free from interference in the appointment and governance of the clergy. Prior to this time, clergy at all levels, located in religious offices, in cloisters, abbeys, and monasteries scattered all over western Christendom, had often been chosen and appointed by secular and local officials.[21] With the papal revolution all these extrareligious influences were drastically curtailed. Put differently, the papal revolution withdrew the spiritual authority that emperors, kings, and princes had previously claimed.[22] Though this seems a rather minor adjustment, the fact is that the papal revolution, by working out a new legal system deeply indebted to Roman concepts (reworked in the light of church law and European customary law), placed restraints on the prerogatives of the secular authorities and, in the process, created the "first modern Western legal system."[23] What is more, this revolutionary adjustment "gave birth to the modern Western state – the first example of which, paradoxically, was the church itself."[24] This outcome is paradoxical because we are accustomed to thinking of the modern state as a secular entity. Nevertheless, Professor Berman argues that the church from this time forward exercised all the legal functions that we attribute to the modern state.

[21] Berman, *Law and Revolution*, p. 88.
[22] Several legal historians have put this change appositely: they suggest that the secular order became sacralized and thereby forced rulers to conform to natural law. Thomas Aquinas affirms that while the prince can make law, he is at the same time "bound to the *vis directiva*, the directive power of the natural law to which he should submit voluntarily." Hermann Kantorowicz, "Kingship under the Impact of Scientific Jurisprudence" in Clagett, Post, and Reynolds, eds., *Twelfth-Century Europe and the Foundations of Modern Society* (Madison, Wis.: University of Wisconsin Press, 1966), p. 97. Cf. Berman, *Law and Revolution*, pp. 114ff; and Brian Tierney, *Religion, Law, and the Growth of Constitutional Thought, 1150–1650* (New York: Cambridge University Press, 1982), chap. 3.
[23] Berman, *Law and Revolution*, chap. 5.
[24] Ibid., p. 113. Cf. Joseph Strayer, *On the Medieval Origins of the Modern State* (Princeton, N.J.: Princeton University Press, 1970); Tierney, *Growth of Constitutional Thought 1150–1650*, p. 22; and Garrett Mattingly: it is "intelligible . . . that the medieval church should foreshadow and as it were recapitulate in advance the development of the modern state," as cited in Tierney, *Growth of Constitutional Thought*, p. 12.

It claimed to be an independent, hierarchical, public authority. Its head, the pope, had the right to legislate, and in fact Pope Gregory's successors issued a steady stream of new laws. . . . The church also executed its laws through an administrative hierarchy, through which the pope rules as a modern sovereign ruler through his or her representatives. Further, the church interpreted its laws, and applied them, through a judicial hierarchy culminating in the papal curia in Rome.

Given all these functions exercised by the church, it can be said that it "exercised the legislative, administrative, and judicial powers of a modern state,"[25] including the levying of taxes in the form of tithes and other fees.

For our purposes the most significant result of this revolution that separated the religious and secular domains was the declaration of the church's legal autonomy, thereby creating the very idea of separate and autonomous legal jurisdictions. Put differently, when, through the battles of the investiture controversy, the church in the person of Pope Gregory VII (r. 1073–85) "declared the legal supremacy of the clergy, under the pope, over all secular authorities,"[26] it created a new and autonomous legal order. It asserted a right to jurisdiction, a right to hear all cases within its domain, a right to legislate new laws, and a commitment to conduct its affairs according to law. In effect, it took a giant step toward becoming the first *Rechtsstaat*, a state ruled by law.[27] While its zeal in the service of the faith was not always contained within the freely consensual framework of the principles and procedures of the canon law, it did leave a "legacy of governmental and legal institutions, both ecclesiastical and secular, for resolving . . . tensions and maintaining an equilibrium throughout the system."[28] It established the model by which secular states could organize their affairs, establish courts, elect officials, and enact their own laws, in order to govern their political, economic, and social domains. Indeed, the papacy's legal separation from the secular domain set the stage and encouraged the development of parallel secular legal structures. While papal authority had a tendency to expand its domain and to assert its dominion over large areas of civil and domestic affairs, for example, over marriage and the family, inheritance, divorce, and the like, authority for control of such matters rested more on customary practice and Roman law than on biblical prescription; hence, papal authority would eventually be forced to rescind them to secular authorities. As another church historian has put it, "In spite of the persistent tendency toward papal centralization, the whole Church, no less than the secular states,

---

[25] *Law and Revolution*, pp. 113–14.
[26] Ibid., p. 94.
[27] Ibid., pp. 215, 292–4.
[28] Ibid., p. 115.

remained in a sense a federation of semi-autonomous units, a union of innumerable greater or lesser corporate bodies."[29]

These developments in law and legal theory put European life on an entirely new footing. As an intellectual innovation, the canonists – above all, Gratian – had taken a very bold step in applying reason and logic to the great body of legal materials that they fashioned into a new legal system – the canon law.[30] Instead of insisting on the priority of or superior sacredness of any one part of the inherited legal traditions, Gratian (and others before him, such as Ivo and Irnerius) proceeded as if there were a natural harmony of law and legal reasoning in the world. Thus, Bishop Ivo of Chartres (in 1095) self-consciously sought to unite the rules of the church "into one body," thereby becoming "one of the first to set forth conflicting passages in the authorities and to suggest some standard by which they could be reconciled."[31] The task of the legist, they assumed, was to find the harmony of sources in precept and in principle that would unite and integrate the existing laws. It fell to Gratian to carry out this task and to succeed in a fashion that established a new standard that lasted for centuries. He collected and studied about thirty-eight hundred canonical texts from various periods of time and set about organizing them into new divisions and categories. In the second part of his work, he discussed specific legal issues in an effort to work out the general principles that should serve as foundations for a working legal system. As examples he analyzed thirty-six complex cases "by presenting patristic, conciliar, and papal authorities pro and con, reconciling the contradictions where possible or else leaving them unresolved, offering generalizations and sometimes harmonizing the generalizations."[32] Not only did the cases themselves present complex moral and legal issues, but Gratian and others found it necessary to work out a harmony between divine law, natural law, customary, national, and enacted law. In doing so, Gratian worked toward establishing a hierarchy of legal sources, in effect, a theory of legal sources:

He started by interposing the concept of natural law between the concepts of divine law and human law. Divine law is the will of God reflected in revelation, especially the revelations of Scripture. Natural law also reflected God's will. However, it is found in both divine revelation and in human reason and conscience. From this Gratian could conclude that "the law [*leges*] of princes [that is, of the secular authorities] ought not to prevail over natural law [*jus naturale*]." Likewise ecclesiastical

---

[29] Brian Tierney, *Conciliar Theory*, as cited in Berman, *Law and Revolution*, p. 215.
[30] Berman, *Law and Revolution*, chap. 5.
[31] Ibid., p. 144.
[32] Ibid.

"laws" may not contravene natural "law." "Ius," he wrote, "is the genus, *lex* a species of it."[33]

It is, perhaps, only in the context of the Islamic concepts of law and legal reasoning that we can appreciate the revolutionary quality of these innovations of Gratian and the canonists. For, as Professor Berman remarks, "the theory that custom must yield to natural law was one of the greatest achievements of the canonists."[34] This was so because it established a new standard by which not only customary but also ecclesiastical law could be judged for its justice and its fittingness. In the first case, "the theory of Gratian and his fellow canonists provided a basis for weeding out those customs that did not conform to reason and conscience."[35] To accomplish this end, the canonists worked out elaborate legal tests, many of which are still in use today, to determine the validity of a custom. These included "its duration, its universality, its uniformity of application, its reasonableness."[36] Such tests clearly pushed toward the establishment of the idea of the relativity of legal rules.

On the other hand, if such tests were to be applied to all laws, then the ecclesiastical laws might also be challenged by the test of natural law. Indeed, Gratian wrote that "enactments, whether ecclesiastical or secular, if they are proved to be contrary to natural law, must be totally excluded."[37] This was an intellectual revolution of surpassing achievement in three senses.

The breakthrough in inherited logics

First of all, the canonists had produced a new system of law, a system that harmonized numerous legal traditions and enunciated new principles for its foundation. Secondly, this process had created a new science, the science of law, a new model of intellectual achievement. In Professor Berman's view, this new science of law was a prototype of modern science in the generic sense.[38] As such, the new science of law may be seen as a protoscience of the modern type, a substantive discipline meeting certain methodological requirements. These requirements included the following elements: "(1) an integrated body of knowledge, (2) in which particular occurrences of phenomena are systematically explained, (3) in terms of general principles or truths ('laws'), (4) knowledge of which (that is, of both the phenomena and the general principles) has been obtained by a combination of observation,

---

[33] Ibid., p. 145.
[34] Ibid.
[35] Ibid.
[36] Ibid.
[37] Ibid., p. 147.
[38] Ibid., pp. 115–64.

hypothesis, verification, and to the greatest extent possible, experimentation."[39] In legal science, it may be said, "the phenomena studied were the decisions, rules, customs, statutes, and other legal data promulgated by church councils, popes, and bishops, as well as emperor," and so forth, and these items constitute the data to be explained, analyzed, categorized, systematized, and put into a logically coherent conceptual pattern. These findings have also to be tested and verified through further inquiry. Accordingly, Berman argues, the twelfth-century jurists were "the first scholars to see and develop not only empirical tests of the validity of general principles but also empirical uses for such principles."[40]

Thirdly, and most important in my view, as an act of intellect and imagination, the breakthrough to modern law established the principle of the authority and legitimacy of reason over discordant authorities. The act of innovation in its totality established the principle that men could discover new harmonies in the world order; they could find and enunciate new principles that would place sacred and scriptural sources on an entirely new footing. If one could exercise reason and conscience in the domain of the inherently sacred, then the metaphysical bonds which declared that there was only one interpretation of the domain of the sacred law was destroyed. Likewise, the exercise of reason and conscience denied the argument that man's intellectual resources were too meager to affect changes in his understanding of the order of man, society, and Scripture. If freedom of reason and conscience was now certified in the domain of law, it would be exceedingly difficult to constrain it in any other domain where scared claims to authority would be less well grounded. In Benjamin Nelson's words, this was a breakthrough in "the received logics of decision,"[41] one that opened up vast new intellectual possibilities – new realms of intellectual freedom. It was that sort of breakthrough in the following ways.

In the first instance, the new science was the product of a new method, the method of *dialectic*, which Peter Abelard and the Scholastics developed. Underlying this was a new mode of analysis and synthesis that was first applied to law and theology. This method "presupposes the absolute authority of certain books," which are taken to be fully complete, "but paradoxically it also presupposes that there may be both gaps and contradictions" in the text, the solution to which is attained in the "resolutio" of the dialectical reasoner.[42] This is the dialectical method which "seeks the reconciliation of opposites." In full form, the method entails

---

[39] Ibid., p. 152.
[40] Ibid.
[41] Nelson, *On the Roads to Modernity*, p. 72.
[42] Berman, *Law and Revolution*, p. 131.

a *questio* relating to contradictory passages in an authoritative text, followed by a *propositio* stating authorities and reasons in support of one position, followed by an *oppositio,* stating authorities and reasons for the contrary view, and ending with a *solutio* (or *conclusio*) in which it is shown either that the reasons given in the oppositio are not true or that the propositio must be qualified or abandoned in light of the oppositio.[43]

In developing dialectical logic in this way, the legists (canonists and Romanists) went beyond the inherited logical standards of both the Greeks and the Romans. On the one hand, the classical Greek forms of reasoning drew a sharp distinction between *apodictic* reasoning – proceeding from universally valid premise to a certified conclusion – and *dialectical* reasoning. In the latter no assurance of certainty was given; one had only probabilities, since one proceeded from a selection of cases and sought to "induce" the generalized premise that would cover the initiating examples. The medieval canonists, however, went beyond this, seemingly "turning Aristotle on his head."[44] Peter Abelard contributed significantly to this new mode of reasoning by giving examples of reasoning from species to genus. The covering principle, "the Maxim," he wrote, "contains and expresses the sense of all such consequences and demonstrates the mode of inference common to the outcome."[45]

On the other hand, the ancient Roman lawyers (in contrast to the canonists) were extremely conservative and their efforts were directed "not to theoretical synthesis, but to the consistent and orderly treatment of individual cases."[46] Professor John Dawson argues that

their whole impulse was toward economy, not only of language, but in ideas. Their assumptions were fixed, the main purposes of the social and political order were not to be called in question, the system of legal ideas was too well known to require much discussion. They were problem-solvers, working within this system and not called upon to solve the ultimate problems of mankind's needs and destiny. They worked case by case, with patience and acumen, with profound respect for inherited tradition.[47]

The Roman judges had insisted upon restricting the application of particular rules to particular contextual situations. Yet "the jurists of Bologna, contemporaries of Abelard, induced universal principles from the implications of particular instances."[48] This was just the opposite of the older Roman con-

---

[43] Ibid., p. 148.
[44] Ibid., p. 140.
[45] As cited in *Law and Revolution,* p. 140.
[46] John Dawson, *The Oracles of the Law* (Cambridge, Mass.: Harvard University Press, 1968), p. 114.
[47] Ibid., p. 115.
[48] Berman, *Law and Revolution,* p. 140.

cept of a rule as merely "a short account of matters."[49] In sum, the European jurists of the twelfth and thirteenth centuries launched a bold new program, reversing and overcoming the limits of previous models of logic and inference, and set about building a new legal system. With bold strokes they assumed "that the whole law, the entire jus, could be induced by synthesis from the common characteristics of specific types of cases."[50] They had irrevocably broken out of the limitations of method and logic as well as custom and tradition that would confine analysis to the local, the particular, and the ethnic, as well as the exclusively religious. They fashioned a new method of analysis and synthesis and this empowered them to create a new legal system that presumed to be universally applicable, since, above all else, it rested on the standard of reason and natural law. This did not mean that the Scriptures or the other components of the ecclesiastical law were to be abandoned; it only meant that they had to meet new tests of validity and legitimacy, tests spelled out in the very idea of natural law that was effected through reason.

Not only did canon law transcend the logical and methodological limits of the philosophical and legal traditions of Greece and Rome, it differed profoundly in both method and spirit, from Islamic law. In Islamic law the method of developing the legal system was based on the authority of the sacred sources. There was no question of reason or conformity to natural law, nor was there any thought of developing a universal system of law applicable to all peoples of non-Islamic persuasion. Islamic law was particularized for Muslims, for those who submit to the will of Allah. The view was that God's command, the Quran, has been given in completed form directly to man through the Prophet Muhammad. This was not all there was to the sacred law (the Shari'a), as Muslim jurists deferred to the idea that tradition, that is, the traditions of the Prophet (Sunna) was an equally valuable part of the Shari'a. These two parts of the law, these two roots of law (usul al-fiqh), had to be accepted as completely authentic and irrevocable, incapable of being overridden by future generations. While the Quran was assumed to be authenticated by its direct transmission through the Prophet Muhammad, the traditions were more troublesome. "The jurists therefore landed on the idea of certifying traditions (hadiths) only if one could trace the saying directly back to the Prophet or his companions through an unbroken chain of transmitters. This line of transmission was called the *isnad*."[51] To this end, many famous collections of these traditions have been assembled, with individual hadiths in the collections numbering as many as

---

[49] Ibid.
[50] Ibid., p. 140.
[51] Joseph Schacht, *Origins of Muhammadan Jurisprudence* (Oxford: Oxford University Press, 1950), pp. 36ff; and Fazlur Rahman, *Islam* (New York: Doubleday, 1968), pp. 56ff, 69ff.

six to eight thousand. While these are taken as authentic by Muslims and the various schools of law, Western scholars do not find them supported by historical evidence.[52] Indeed, so dubious does the legal scholar Joseph Schacht find these historical legends that, in referring to particular traditions whose origins are suspect, he uses such locutions as "put in the mouth of the Prophet," or "put into circulation a tradition to the effect . . ."[53] The point is not that the Muslims were deliberately duplicitous about these sources, though some may have been, but that tracing back such sayings to an original source two or three centuries earlier is a very dubious enterprise in a search for legal authenticity.

After imposing the requirement of the *isnad* for each tradition (hadith), the great ninth-century legist al-Shafiʻi (d. 820) set about systematically organizing the science of law, its sources, materials, and permissible modes of reasoning. He wanted to make sure that in the process of Islamicizing the whole body of practiced law in Middle Eastern society, the methods and materials used were authentically Islamic. In this, and especially in his systematizing of the legitimate modes of legal reasoning, his efforts add up to "a ruthless innovation"[54] that solidified Islamic law and its sources so that further innovation was ruled out. That result was achieved by adopting a very strong traditionalist bias such "that nothing can override the authority of a former tradition from the Prophet." He also "recognized in principle only strictly analogical and systematic reasoning . . . to the exclusion of arbitrary opinion and discretionary decisions."[55] The end result was that Shafiʻi "cut himself off from the natural and continuous development of doctrine in the Ancient schools."[56] Furthermore, the capstone of Shafiʻi's system was the doctrine of consensus (ijmaʻ), according to which God would never allow his community to fall into error and, therefore, once a scholarly consensus had been reached, it could never be overridden. In the end, the legal system stagnated, since the methodological procedures put into effect "could hardly be productive of progressive solutions" to legal questions.[57]

The early Muslim legists had also formulated various legal maxims such as "the child belongs to the marriage bed," "there is no divorce and no manumission under duress," or "profit follows responsibility,"[58] but these also (after the work of Shafiʻi) could only enter the canon of law (the usul al-fiqh)

---

[52] See Schacht, *Origins*, pp. 138–76, and Rahman, *Islam*, pp. 47ff and 69ff.
[53] *Introduction to Islamic Law*, pp. 38, 40, and passim.
[54] Ibid., p. 48.
[55] Ibid., p. 46.
[56] Ibid.
[57] Ibid., p. 47.
[58] Ibid., p. 39.

as elements of specific traditions (hadiths), along with their isnads, which comprise the Sunna. That is, these aphorisms could not be turned into authentic legal principles that controlled adjudication, because this would be to innovate, to add to the already completed work of God, and that was formally a heresy. Such aphoristic traditions, therefore, carried no specific weight, and no effort was made (such as Gratian and the canonists had exerted in Europe) to turn the maxims into logical and rational legal principles and concepts with their own rationale. The Islamic root of law designated by tradition means precisely that: a tradition of the Prophet, authenticated by the isnad (the names of its transmitters back to the originator), no matter how illogical or inconsistent it might be with respect to other practices, principles, or traditions. There was no imperative to achieve a hierarchy of explicit principles, nor any overt logical imperative to eliminate all traditions that implied contradictory outcomes in actual practice.

If we compare this outcome to the spirit that animated the lawyers and jurists who fashioned the papal revolution in Europe, we have to say that for the latter "there was a dynamic quality, a sense of progress in time, a belief in the reformation of the world. . . . Law in the West in the late eleventh and twelfth centuries, and thereafter, was conceived to be an organically developing system, an ongoing growing body of principles and procedures, constructed – like the cathedrals, over generations and centuries."[59] Whereas the Western legal systems had adopted reason and conscience as well as the idea of natural law as the ultimate standards for accepting or rejecting a specific legal practice or principle, Islamic law opted for tradition and the scholarly consensus. "In the end," F. E. Peters writes, in Islamic law

consensus turned out to be a far more potent force than al-Shafi'i could have foreseen. Consensus, with its attendant notion of infallibility . . . was not only the guarantor of the other three sources of law [Quran, hadith, and qiyas]; it served as the base and justification of important elements in Islam that were either unsanctioned by the Quran and the sunna, like the institution of the Caliphate, or were expressly opposed there, like the cult of saints. . . . Once the four schools assented to their natural claim to authority and, in effect, agreed to disagree on the rest of the details, no other basic approach to the law could be permitted except there be another consensus, a possibility that was ruled out. It was as later affirmed, "the closing of the gates of independent thought."[60]

But the European revolution in the overthrow of inherited logics of decision was accompanied by equally revolutionary new institutional arrangements. As Hastings Rashdall pointed out long ago, the European medievals

---

[59] Berman, *Law and Revolution*, pp. 118, 119.
[60] F. E. Peters, *Allah's Commonwealth* (New York: Simon and Schuster, 1973), pp. 244–5.

were masters at institutionalizing great ideas and ideals, and the new social order – above all, the universities – was a product of this institutionalizing impulse.[61] This institutional revolution was suggested above in the context of the breakthrough to the modern state. But this was only one aspect of the substantive legal revolution that centered in the idea of treating collectives as single entities, that is, corporations. Even more revolutionary was the adjunct idea that such entities had greater or lesser amounts of jurisdiction.

Corporations and jurisdiction

At the heart of this revolution was the legal idea of treating a collective body of people as a unit, a whole body or corporation. This was a product of what has been called the "communal movement" of the Latin Middle Ages.[62] This had the most far-reaching consequences for social, political, and legal organizations, the effects of which even today are far from settled.

The idea of a corporation centers on the principle that collective actors may be treated as a single person or agent.

Legally a corporation (*universitas*) was conceived of as a group that possessed a juridical personality distinct from that of its particular members. A debt owed by a corporation was not owed by the members as individuals; an expression of the will of a corporation did not require the assent of each separate member but only of a majority. A corporation did not have to die; it remained the same legal entity even though the persons of the members changed.[63]

On the one hand, these principles established the existence of fictive personalities that are treated as real entities in courts of law and in assemblies before kings and princes. Such a device allows the treatment of the actions of a plurality of individuals as a single outcome, a single will. Such a group of collective actors might be economic actors, for example, a guild or business enterprise; an educational institution, for example, a university; a religious order or chapter; or indeed a state. In each of these cases the collective actions of the group are granted a unitary legal status.

There was a sense among the medievals that groups of individuals had legitimate purposes which brought them together and that these interests established a right to be represented as a group in the life of the community.[64] There were many sources of the idea that collectivities ought to be

---

[61] Rashdall, *The Universities of Europe in the Middle Ages,* new ed. (New York: Oxford University Press, 1936), 1:3.

[62] Pierre Michaud-Quantin, *Universitas: Expressions du mouvement communautaire dans le moyen-âge Latin* (Paris: J. Vrin, 1970).

[63] Tierney, *Growth of Constitutional Thought, 1150–1650,* p. 19.

[64] Michaud-Quantin, *Universitas,* part 1, chap. 1; part 2, chap. 1; and passim.

treated as unities. In his review of all the sources and meanings of the term universitas, Michaud-Quantin found them among the many collective groupings of the church (for example, clerical associations, congregations, convents, and chapters); the communal aggregations represented by geographical and territorial subdivisions in and around cities (for example, urbs, municipalities, burgs, communes, and villages); as well as ethnic enclaves formed into communes and communities. So also, there are various forms of fraternities, confraternities, and charitable associations. Not of least importance were those groups denoted by the Latin words *societas* and *collegium*. In the end, it was but an accident of history that the Latin universitas (corporation or whole body) came to refer exclusively to the places of higher learning that retain the name universities.[65]

The medievals came together to form more or less permanent collectives for a great variety of purposes – religious, economic, communal, educational, professional – and the canon law recognized these collectives as legitimate legal entities with the rights of assembly, ownership, and representation (both internal and external).[66] All of these agglomerations of individuals laid claim to collective interests, and once recognized by law as entities, as whole bodies, their collective existences were transformed into legal personalities bundled with legal rights: to own property, to have representation in court, to sue and be sued, to make contracts, to be consulted when one's interests were affected by actions taken by others, especially kings and princes. It is here that one finds the application of the famous Roman maxim "What touches all should be considered and approved by all."[67] By the thirteenth century this idea had been elevated to a major principle of corporate and communal representation, a "principle of due process in the court . . . an integral part of the *rationale* of the representation of individuals and corporate rights, before the king and his court and council in assembly."[68]

The principle of treating collective actors as a single entity carried with it the principle of *election by consent*. If the corporation, the collective group of actors considered as a whole, is to be represented by a single voice, for example, in a court of law, then it must elect such a person. Furthermore,

---

[65] Rashdall, *Universities* 1: 4–5.
[66] Pierre Gillet, *La personnalité juridique en droit ecclésiastique* (1927), comes to the same conclusion: the term corporation was applied to a great variety of associations and these were conceived as entities "formed for preserving to each his justice." As cited in Berman, *Revolution*, p. 606 n40.
[67] Gaines Post, *Studies in Medieval Legal Thought: Public Law and the State, 1100–1322* (Princeton, N.J.: Princeton University Press, 1964), i.e., *Quod omnes tangit omnibus tractari et approbari debet*, chap. 3, and passim; and Berman, *Law and Revolution*, p. 608 n54.
[68] Ibid., p. 90.

the idea of acting with full powers of attorney (*plena potestas*) was also clearly articulated. During the High Middle Ages these agents had many titles such as proctor, syndic, actor, and even economus. As Gaines Post tells us, "Whatever his title, this representative of a corporation was chosen by the whole *universitas* or by the *maior et sanior pars* [the greater and sounder part] of at least two-thirds of the members in assembly."[69]

In this procedure we can see the beginnings of a formalized process of representative government. In its very nature, the legal idea of a corporation is a major institutional location of the principle of constitutional limitations and governance. As Post puts it, from this idea of legal representation there "developed in the 13th century the idea that English shires and towns could be represented in Parliament; and indeed modern representation of communities is based on this principle."[70] Likewise, Joseph Strayer stressed the point: "The idea of political representation is one of the great discoveries of medieval governments." While the Greeks and Romans took steps in that direction, they lacked fundamental legal and philosophical presuppositions. "In medieval Europe, on the other hand, representative assemblies appeared everywhere: in Italy, Spain, and southern France early in the thirteenth century; in England, northern France and Germany anywhere from fifty to one hundred years later."[71] From the idea of treating collectives of individuals as separate legal entities we get the first realization of group representation. Since this takes place within the corporate and legal framework of acknowledged rules, it signifies the beginnings of constitutional government.[72]

From a point of view within the comparative sociology of law, there is an even more revolutionary aspect of the legal status of corporation, the principle of *jurisdiction*, that is, the concept of legitimate domains of legal action.[73] This principle cuts many ways. On the one hand, corporate entities in the

---

[69] Post, *Studies*, p. 51f.

[70] Gaines Post, "Ancient Roman Idea of Laws," *Dictionary of the History of Ideas* (New York: Scribners, 1973) 2: 686.

[71] Joseph Strayer, *On the Medieval Origins*, pp. 64 and 65.

[72] Tierney, *Growth of Constitutional Thought*, pp. 19ff.

[73] As I shall argue in some detail below, there are good grounds for believing that the distinction in canon law between ownership and jurisdiction created a powerful opportunity for separating the public and private domains. For the theory sharply distinguishes between obligations owed to other persons versus obligations to the *corpus*, the whole body of the corporation. Thus the property of the corporation is not owned by the individual members but by the whole group; likewise in the case of debts. Correspondingly, there is a sharp distinction drawn in legal theory between ownership and jurisdiction: those who serve as agents of adjudication (exercising jurisdiction) are not synonymous with the owners of the corporate assets; and likewise the agent of ownership is acting as a fiduciary and is not necessarily the agent exercising legal powers of jurisdiction, that is, deciding matters of

twelfth and thirteenth centuries were permitted to enact their own ordinances and statutes; stated differently, they could be a source of new laws and regulations, the purpose of which was to regulate and control the corporate members. The most obvious example of this is the set of rules and regulations enacted by the medieval universities. At the University of Paris, for example, in the twelfth and thirteenth centuries there were rules and regulations enacted for the admittance as well as the expulsion of students. There were also rules that prescribed the conduct of faculty and established the courses of instruction and their sequence. Not least of all, there was the *licentia docendi,* a legally exclusive right to teach (a license) that was awarded only by the chief official of the university – not by the state and not by individual scholars. By 1215 at the latest, Gaines Post asserts, the scholars at the University of Paris were firmly established as a corporate body, "a *universitas magistrorum et scholarium*" (a corporation of masters and scholars) who "could enact statutes and enforce obedience to them."[74]

So it was with each collectivity in this period that achieved corporate status. Each enacted laws to govern its members and thereby whole new systems of law – for example, urban law, merchant law, royal law – developed that served to counterbalance jurisdictions and prevent the monopoly of power and authority over the whole realm. Thus guilds, associations of merchants, and various assortments of workers and tradesmen became lawmaking bodies. They enacted ordinances to regulate their membership, to fix prices, control trade, and standardize business transactions. The heads of the guilds in many cities also "became the magistrates of the communes."[75] In Milan consuls of merchants were authorized to establish courts and to adjudicate all merchant cases within their urban jurisdiction.[76]

law. Such a theory must have served to create a separation between the public and the private, that is, the family and the enterprise. More on this distinction below.

In terms of commercial relationships, the *societas* and the *companie* grew out of families and brotherhoods, where partners in several are directly responsible for the profits and losses of the enterprise. Yet in the *corporation* there is an abstract level of ownership that clearly distances the members of the corporation from direct ownership. These legal precepts provide a strong rationale for overcoming the bonds and limits of kinship relations, which as we saw in Chap. 2 remained so very strong in Arabic-Islamic culture and civilization and also in all forms of kin-based societies, such as traditional China. Accordingly, societies and civilizations lacking such a distinction (the Islamic Middle East as well as China and Southeast Asia) would not find it problematic (i.e., corrupt) to use public funds to reward family and friends. The theory of corporations, therefore, may be seen as one of the historic developments that provided a means for breaking out of the parochialism of kin, clan, and caste, something Weber saw repeatedly as a stumbling block to economic development.

[74] Post, *Studies,* p. 37.
[75] Berman, *Law and Revolution,* pp. 391 and 392.
[76] Ibid.

Likewise, European cities of this period were constituted and acted as modern states, "just as the church of that period was a modern state – in the sense that they had full legislative, executive, and judicial power and authority, including the power and authority to impose taxes, coin money, establish weights and measures, raise armies, conclude alliances, and make war."[77] In this manner the legal concept of the corporation created a whole new array of social actors and other realms of social existence. They were, to be sure, abstract realms; nevertheless, all such actors and entities had the right to be represented before official bodies, and this was acknowledged by kings and princes and eventually parliaments.

In this social movement leading to the recognition of collective actors or whole entities and legal personalities, we see a reconstruction of the institutional framework of European society and civilization. Once the church had declared itself legally autonomous from the secular order, the stage was set for the recognition of all the secular states – the national as well as city and communal states – as autonomous legal bodies, bound by their own laws. Moreover, the canonists and Romanists had in theory worked out the many complications implied by this new order of things. In the first place, the head of a corporate body had both the wisdom and the power (jurisdiction) to make the laws of the corporate body: he had jurisdiction within the domain of the corporate group and could serve as judge of all cases. This did not mean, however, that the head owned the corporation or its property, for clearly there was a distinction between ownership and jurisdiction. Among others, John of Paris (d. 1306) asserted, "To have proprietary right and ownership is not the same thing as having jurisdiction over it. . . . Princes have the power of judging even though they do not have ownership of the property in question."[78]

But having jurisdiction specified that the agent in question had the legitimacy as well as the power to make and carry out the ordinances. Furthermore, from the time of the investiture controversy onward, there was an implied hierarchy of jurisdictions both within the church and between the secular and the religious legal orders. Within the church, the hierarchy of authority flowed down from the pope, through the cardinals, to the archbishops, bishops, and so on, down to the local chapters and individuals within. In effect, these organizing principles served to establish boundaries of legitimacy within which all the major territorially bounded corporate actors could establish laws and ordinances, hold court, and issue rulings.

If, on the other hand, we reflect on the Arabic-Islamic situation, it is

---

[77] Ibid., p. 396.
[78] As cited in Tierney, *Growth of Constitutional Thought*, p. 32.

evident that no such legal or organizational revolution occurred until the nineteenth century when, in response to the European presence in the Middle East, new forms had to be created, and European civil law codes were either borrowed wholesale or used as a foundation for creating a new legal system that at least acknowledged Islamic principles.[79] But there was no separation of church and state, the sacred and the secular. According to medieval Islamic theory, the temporal ruler was the successor of the Prophet, and he was charged with carrying out the commands of God. It is true that various Muslim rulers, especially among the Abbasids, introduced adjunct administrative structures even though they lacked a foundation in religious law. The so-called courts of complaints, the *mazalim* courts, were extended to cover virtually anything that theoretically could come before a qadi, the legitimate religious judge appointed by the ruler. But there were "no texts, no written or customary rules . . . [to] define and limit the categories of litigation which . . . [came] under the jurisdiction of the mazalim."[80] As a result, the conflict between "ideal and reality," the religious ideal of Islam and the practical reality of ruling a society, put Islamic political life in the worst of all possible worlds. For lacking any ability, that is, legal legitimacy, to create and enact new laws and to effect a separation between church and state, Islamic rulers often took matters into their own hands, though not without some intellectual props. The courts of complaints were established by the temporal rulers who sought both to achieve "government in accordance with the precepts of divine law (*siyasa shar'iyya*)"[81] and to curb or correct any abuses of the objective rules of the Shari'a by the qadis. Hence the mazalim courts were also charged with administering all cases that resulted from "erroneous or faulty application of the objective rules of law, either in specific cases or in a general way."[82] From the eleventh century onward, N. J. Coulson notes, Islamic "writers on constitutional law . . . assert that while the Shari'a doctrine embodies the ideal order of things, the overriding duty of the ruler is to protect the public interest" and, therefore,

---

[79] For this development, see Khadduri and Liebesny, eds., *Law in the Middle East* (Washington, D.C.: The Middle East Institute, 1955), especially Herbert Liebesny, "The Development of Western Judicial Privileges," pp. 309–33; J. N. D. Anderson, *Law Reform in the Muslim World* (London: Athlone Press, 1976); and M. Khadduri, *The Islamic Conception of Justice* (Baltimore, Md.: Johns Hopkins University Press, 1984), pp. 205–16. On the development and use of precedent, see Liebesny, "English Common Law and Islamic Law in the Middle East and South Asia: Religious Influences and Secularization," *Cleveland State Law Review* 34 (1985–6): 19–33.

[80] Tyan, "Judicial Organization," in Liebesny and Khadduri, eds., *Law in the Middle East*, p. 263.

[81] N. J. Coulson, *Conflicts and Tensions in Islamic Jurisprudence* (Chicago: University of Chicago Press, 1969), p. 68.

[82] Tyan, "Judicial Organization," p. 263.

"in particular circumstances of time and place the public interest might necessitate deviations from the strict Shari'a doctrine."[83] In other words, a secular ruler must do what he will to preserve the public order and there are no particular bounds on his actions. The "political ruler is recognized as the fount of all judicial authority, with the power to set such bounds as he sees fit to the jurisdiction of his various tribunals, including the Shari'a courts."[84] Or, put negatively, since the ruler could set whatever bounds he chose, there was no concept of jurisdiction. For "in Islamic law there is no distinction between various degrees of jurisdiction."[85] Part of that absence stemmed from the fact that "the qadis had never been an independent judiciary body in the true sense of that word. Appointed by the political ruler and subject to dismissal by him, they exercised their judicial office as his delegates." When their "administration of justice proved defective" because of their idealistic conception of the Shari'a, "the ruler simply appointed other delegates."[86]

From the point of view of the history of Western law, including the contributions of Roman civil law and the canon law, the crucial ingredients missing here are the concept of jurisdiction and the idea of legitimate interest groups (independent of the family and kin group), that is, wholes or corporations, whose existence demands acknowledgment of their legal rights. No such acknowledgment was possible within the Islamic sacred law. This conception of a corporate legal personality – as well as legal rights apart from those of family members – did not emerge in Islamic law. Without the concept of legitimate group boundaries and the attendant concepts of jurisdiction (*imperio*) or sovereignty, there could be no basis for constructing politically autonomous groups for whom the Roman maxim "What touches all should be considered and approved by all" (*Quod omnes tangit, omnibus tractari et approbari debet*) could apply. As Joseph Schacht put it, "The whole concept of an institution is missing" in Islamic law.[87]

Furthermore, a significant portion of the Shari'a enjoins all believers to "encourage good and discourage evil." This duty was given officially to the market inspector (*muhtasib*). When this was done by the Abbasids, the official duties of the inspector entailed not just enforcing the rules of traffic, sanitation, and proper weights and measures, but also "summary punishments which came to include the flogging of the drunkard and the unchaste

[83] Coulson, *Conflicts and Tensions*, p. 68.
[84] Ibid., p. 69.
[85] Tyan, "Judicial Organization," p. 241.
[86] Coulson, *Conflicts and Tensions*, p. 66.
[87] Schacht, "Islamic Religious Law," in *The Legacy of Islam*, 2d ed. (New York: Oxford University Press, 1974), p. 398.

and even the amputation of the hands of thieves caught in the act."[88] All this discretionary power issues directly from the fact that anything which violates the spirit, not just the letter, of Islamic teachings is a crime, the punishment for which is discretionary, that is, not spelled out in the Quran.[89] Simply put, the spirit as well as the letter of Islamic and Western law are profoundly different. In Chapter 7 I will review traditional Chinese law from the same points of view.

Revolution and the parting of the ways

In this chapter I have suggested that legal systems constitute one of the most powerful and enduring elements of the social structure of societies and civilizations. By their very nature they create structures of action and agency through their enactments, codes, prescriptions, and procedures. In the twelfth and thirteenth centuries in Europe legal systems underwent radical reconstruction, and this created a new social order with new and expanded concepts of agency, reciprocity, responsibility, and representation. While it is of course true that law and legal codes are one thing and social action another, we have seen that the new legal ideas (created out of the fusion of Roman precept and the new legal reasoning) did play a large part in creating a new order of things, social, economic, and political. The broader effects of the legal revolution on urban law and economic action have been addressed by others, especially Professor Harold Berman in *Law and Revolution*, among others.[90] It is not my intention to rehearse all those changes here, but only to consider the ways in which the new legal conceptions shaped the directive structures of medieval society and provided a fertile ground for the development of modern science.

I have also stressed the need to examine the symbolic elements of society that, embedded in law, religion, and philosophy, create a new anthropology of man. By this I mean to suggest that different societies, but even more so, different civilizations composed of two or more societies sharing the same religion, law, and knowledge base, rest on alternative conceptions of who man is and what his capacities for reason and rationality are. In the present comparison of such images of man in Western European civilization and Arabic-Islamic civilization, we have seen that legal thought during the

---

[88] Schacht, *Introduction to Islamic Law*, p. 52.
[89] Among others, see Lippman et al., *Islamic Criminal Law and Procedure* (New York: Praeger, 1988), chaps. 3, 4, and 5.
[90] Also see Robert Lopez, *The Commercial Revolution of the Middle Ages, 950–1350* (New York: Cambridge University Press, 1977).

twelfth and thirteenth centuries had a profound effect on the conceptions of the powers of reason and rationality which were thought to be indwelling in man. Mankind, in the European view, is possessed of reason and may use that reason to weigh the validity of custom, tradition, and religious authorities, and even Scripture itself. Furthermore, man is possessed of conscience, that unquenchable agency which allows men to discern right and wrong in moral affairs. So vigilant and acute is this agency that even if a superior, a prince, should enjoin a person to do something against conscience, the civilian actor should resist.

In contrast, Islamic legal and religious thought, which is far more integrated and unitary than the religious and legal spheres in the West, insists above all that the powers of man's reasoning are too limited and too uncertain to be a guide in moral, religious, and legal affairs. The commands of God were given by him in a written form, the Quran, and the believer may use all his powers of linguistic and grammatical analysis, and even analogical reasoning, to understand the Scriptures and the traditions of the Prophet. But the believer is in no way empowered, much less authorized, to add anything to (or subtract anything from) this body of knowledge, for the consensus of the scholars has already perfected our knowledge of fiqh and the Shari'a. It is not up to mankind to innovate (which in religious affairs is equivalent to heresy)[91] where understanding fails. At that point the believer should turn to taqlid, acceptance of the authority of tradition. Most important of all, God would not allow his people to roam, to wander unguided, so there is always guidance for the believer. Conscience, as the medieval Europeans understood it, is unknown to Islamic law, ethics, and religion, and classical Arabic lacked such a term. At best one may acquire insight or understanding of the mysteries of God's revelation, but that acquisition[92] is the way God's mysterious and unknowable and all-powerful will is activated in the world, that is, by allowing the believer to think he has achieved great insight, when in reality it is God's active presence in the world that allows us to discern anything at all.

The dichotomy of spirit between the two anthropologies of man are reflected in the writings of the two contemporaries, Peter Abelard (d. ca. 1144) and al-Ghazali (d. 1111). The significance of these two figures for their respective civilizations can hardly be overstated. It is virtually impossible to pick up any major work on the renaissance of the twelfth century dealing with law, logic, ethics, philosophy, reason, and conscience, as well as the

---

[91] See B. Lewis, "Some Observations on the Significance of Heresy in the History of Islam," *Studia Islamica* 1 (1953): 43–63.

[92] On the Islamic concept of acquisition, see Chap. 3 n91 and the references there.

founding of the universities, that does not give a major (and positive) role to the teachings and writings of Abelard.[93] Likewise, no account of Islamic philosophy and theology of this period is complete that omits the writings and influence of al-Ghazali.[94]

For Abelard, unaided reason can sort out the contradictions of human and divine writ. In his view there is a great need to sort out these contradictions of faith and belief and to arrive at a more solid foundation, one based on reason and logic. He wrote his great works, such as *Sic et non*, it has been suggested, "because he gloried in finding subjects of high importance for the exercise of reason."[95] Consequently, Abelard communicated to his students "his confidence in reason"[96] and its ability to resolve the contradictions of faith and doctrine. While he had enemies who accused him of heresy and un-Christian motives, he rose to the occasion by defending the uses of reason. In a letter to Heloise he affirmed, "I do not wish to be a philosopher if it means conflicting with Paul nor to be an Aristotelian if it cuts me off from Christ."[97] For him it was evident that there is a unity of truth, that "truth in search of itself has no enemies,"[98] and he set out various defenses of his position. In the *Dialectica,* for example, he wrote:

But if they grant that an art militates against faith, without any doubt they are admitting that it is not knowledge. For knowledge is a comprehension of the truth of things; wisdom, in which faith consists, is a species of it. This (sc. wisdom) is the discernment of what is honorable or useful. But truth cannot be opposed to truth. Truth cannot be opposed to truth or good to good in the way that false can be found set against false or evil against evil; all things that are good are harmonious and

---

[93] E.g., Rashdall, *The Universities of Europe;* Haskins, *The Renaissance of the Twelfth Century;* Grabmann, *Geschichte der Scholastische Methode;* William Kneale and Mary Kneale, *The Development of Logic* (Oxford: Oxford University Press, 1962); David Knowles, *The Evolution of Medieval Thought;* M.-D. Chenu, *Nature, Man, and Society in the Twelfth Century;* Berman, *Law and Revolution.* Lynn Thorndike is typical: "There is no more familiar, and possibly no more important figure in the history of Latin learning who flourished during the twelfth and thirteenth centuries than Peter Abelard"; *History of Magic and Experimental Science* (New York: Columbia University Press, 1923), 2: 4.

[94] See D. B. MacDonald, *The Development of Muslim Theology, Jurisprudence, and Constitutional Theory* (New York: Charles Scribners, 1903); Majid Fakhry, *A History of Islamic Philosophy,* 2d ed. (New York: Columbia University Press, 1983); F. E. Peters, *Aristotle and the Arabs;* Watt, *Islamic Philosophy and Theology;* Louis Gardet and M. M. Anawati, *Introduction à la Théologie Musalmane,* 2d ed. (Paris: J. Vrin, 1970); Muhsin Mahdi, "Islamic Theology and Philosophy," *Encyclopedia Britannica* 9 (1974): 1012–25; and Harry Wolfson, *The Philosophy of Kalam* (Cambridge, Mass.: Harvard University Press, 1976), among others.

[95] Kneale and Kneale, *The Development of Logic,* p. 202.

[96] Ibid., p. 203.

[97] As cited in Michael Haren, *Medieval Thought: The Western Intellectual Tradition from Antiquity to the Thirteenth Century* (New York: St. Martin's, 1985), p. 106.

[98] Kneale and Kneale, *The Development of Logic,* p. 203.

congruent. All knowledge is good, even knowledge of evil, and cannot be lacking in the just man. For the just man to be on guard against evil it is necessary for him to have known in advance what evil is; he could not avoid it unless he knew what it was. . . . On these grounds therefore we prove that all knowledge, which is from God alone and proceeds from his gift, is good. Consequently it must be allowed too that the study of all knowledge is good . . . but study of that learning is especially to be undertaken in which greater truth is seen to be present. This however is dialectic, for to it all discernment of truth or falsehood is subject, in such a way that as the leader of the whole realm of learning it has all philosophy in its princely rule.[99]

In Abelard's view, all knowledge is good, including the knowledge of that which is evil, and there should be no boundaries around the free acquisition of knowledge. Such knowledge, moreover, is a gift of God.

For al-Ghazali, in contrast, the pretensions of the philosopher are far in excess of what they can demonstrate. As a result of his deep probings of the tools of logic and philosophy, their virtue and their limitations, he came to set impossibly high standards for the acquisition of knowledge. This meant that only knowledge that could be logically demonstrated was acceptable; anything less was to be discarded. There is in this epistemological conservatism a premonition of an overbearing Humean skepticism. For al-Ghazali was embarked on a search for absolute certainty in knowledge, "for that knowledge in which the object is disclosed in a fashion that no doubt remains . . . that no possibility of error or illusion accompanies it."[100] In this quest for infallible knowledge Ghazali found no hope in philosophy, much less in theology. As a result he castigated those who "wallow in systematic theology," for those who pursue that course, who delve into arcane logical and philosophical analyses are in grave religious danger. It is "the simple folk" who are "far from this danger . . . and the rest of those common folk who have not waded into research and inquiry nor wallowed in systematic theology as if it were an absolute standard of reference."[101] It is for this reason, Ghazali continues, for example, the prevention of loss of religious faith and the acceptance of false religious doctrine that "the Fathers proscribed research and inquiry and the wading into systematic theology and the examination of these matters."[102] All are forewarned that "for a certainty . . . everyone who forsakes the pure faith in God and His messenger and his Book and wades into research has become entangled in this danger

[99] As cited in Haren, *Medieval Thought*, p. 106.
[100] William M. Watt, ed. and trans., *The Faith and Practice of al-Ghazali* (London: Allen and Unwin, 1953), pp. 21–2.
[101] *Book of Fear and Hope*, tr. William McKane (Leiden: E. J. Brill, 1962), p. 68.
[102] Ibid.

[of going to Gehenna and punishment]."[103] There are snares and intellectual traps at every turn, so that "anyone who alights on a tenet which he has caught from these researchers through the display of the wares of their intellect," whether or not proofs of these ideas are presented or not, if the believer doubts it, then "he is corrupt in his religion, and if he trusts in it, he is thinking himself secure from the stratagems of God, being self-deceived by his deficient intellect."[104] In short, no person who wades into research can be found freed from these dangers, and only pursuit of the knowledge of God through spiritual mediation can achieve the desired goal.

While these contrasts between Abelard and al-Ghazali reveal the different metaphysical commitments to reason and rationality championed by the two civilizations, there are still other levels of difference that suggest different sociological consequences for freedom of inquiry and the pursuit of knowledge. As we saw earlier, the papal revolution (built on the recovery of the Roman civil law and its transformation through the fashioning of the new canon law) created a variety of new social and institutional arrangements that had far-reaching consequences. These consequences can be summarized in the following manner.

1. The recovery of the Roman civil law precipitated the construction of a new system of law based on the assumption of universalistic application. That is, the law was thought to apply uniformly within its jurisdiction, but since it was built in conformity to reason and natural law, it was in principle a universally applicable law that transcended the boundaries of community, ethnic group, and religion. This was even more noticeable in the formation of the Law Merchant, which specifically sought to develop a set of universal legal rules and principles to govern trade and transactions between parties of different countries and political systems.[105]

2. Since the building of the new systems of law (urban, merchant, royal, manorial, and so forth) required the application and use of reason as well as conscience, these metaphysical capacities were imputed to men and they became permanent constituents of Western legal systems. Ultimately they were to gain their greatest significance in the common law world of England and the United States, where lay juries were to become a permanent element of the adjudication process. In such a context the presumption of man's capacity to reason and to establish legal facts became a basic constituent of man as well as a fundamental assumption of political and legal process. By

---

[103] Ibid., p. 70.
[104] Ibid.
[105] See Berman, *Law and Revolution*, chap. 11, as well as W. A. Bewes, *The Romance of the Law Merchant*, reprint (London: Sweet and Max Franklin, 1969).

attributing such capacities to all adult citizens, one might say that such systems bestowed a surprising amount of trust on Everyman. Even within the medieval context, however, reason and conscience were salient elements in legal action and sociocultural process.

3. In acknowledging natural law, as well as recognizing reason and conscience as inalienable elements of man's constitution, the legal revolution of the twelfth and thirteenth centuries established new standards for eliminating unjust laws, whether they be customary, royal, or ecclesiastical. This was a major breakthrough in the construction of objective and universal standards for judging the justice and equity of social relations and probably served as a model of such external standards used to evaluate other human constructions in ethics, science, and politics.

4. Having established external standards for evaluating the reasonableness of law and legal principles, the European medievals brought into existence a hierarchy of legal authority and jurisdictions. At the apex of this edifice were natural law and natural reason, to which all others had to conform. Below that was divine law, and below that all the secular legal authorities headed by kings, princes, down to cities, towns, and corporate bodies within them. Although the relative ascendancies of these jurisdictions were always open to challenge, there was an implicit hierarchy, with enacted legislation prevailing over custom. At the heart of this development one finds the idea of legitimate domains, that is, jurisdictions and the implied constitutional limits imposed by the rule of law and enactment.

5. The theory of jurisdiction rested on the fundamental idea of treating collectives of individuals as whole bodies, corporations that had legitimate interests to so organize. They also received a bundle of legal rights – to own property, to have legal representation, to sue and be sued, and so forth. Hence the legal revolution created a whole new realm of legal and social actors, corporations that ranged from charitable and fraternal groups, to universities, communities, cities, and nation states. Each of these entities was empowered to enact its own ordinances as well as to adjudicate internal disputes.

6. The recognition of collectives as unitary legal actors results in the establishment of two levels of representation on the basis of "what touches all should be approved by all." The first was the local and internal level whereby decisions within the enterprise were made on the basis of majority vote or the will of "the greater and sounder part." Second, the principle of "what touches all should be considered and approved by all" implied that all such entities had the right to be represented before kings and princes meeting in assembly or in court. As this principle of due process and representation was put into effect, it created a new concept of political consent,

whereby rulers had to receive the approval of the governed, especially before levying taxes.

7. The theory of corporate existence, as understood by Roman civil law, distinguished between the property, goods, debts, liabilities, and assets of the corporation and those of individual members. A debt owed by the corporation was not owed by the members individually. Likewise, ownership of property by the corporation was not equivalent to jurisdiction by the head of the corporation, and those empowered to adjudicate within the corporation were not the owners of the property. Most importantly, the allegiance of the individual members was said to be to the corporation, not to other members of the corporation personally. These ideas served to create a foundation for a public versus a private sphere of action and commitment.

8. All of these distinctions reinforce the fundamental principle of the separation of powers, above all, the separation of the religious and the temporal orders, something not possible within Islamic legal theory.

Having set out these revolutionary developments, which took place in the twelfth and thirteenth centuries, we should be careful to note that not all of these ideas were equally implemented. While medieval constitutionalism was a reality, we know that it had several defects, two in particular. First, short of the threat of revolutionary overthrow of the ruler, there were no adequate mechanisms to deter rulers who trampled on the social and political rights of the citizenry.[106] With the arrival of the modern nation state, this problem became acute, leading to various political revolutions. Second, while the constitutionalism of the twelfth and thirteenth centuries (and thereafter) rested upon the implicit idea of the rule of law and the constraints of natural law within the confines of corporate structures, the latter idea was too vague to allow the striking down of unconstitutional rules and regulations, whether in the context of communities, corporations such as the church, or the city and nation states. The crafting of the American Constitution, therefore, remains a great landmark in the history of political rights and due process.

And finally, it might be said, as Freud did, that it is one thing to have a slight flirtation with an idea and quite another thing to be wedded to it and to win the idea a permanent place in the stock of received wisdom. Thus the canonists and the Romanists were the great architects of a host of new legal conceptions bordering, for example, on the idea of impersonal service. Yet, as Max Weber also recognized, it was one thing to have such an idea and quite another to fashion a social system in which powerful "psychological

---

[106] See Charles McIlwain, *Constitutionalism, Ancient and Modern*, rev. ed. (Ithaca, N.Y.: Cornell University Press, 1947).

sanctions," perhaps originating in religious commitment, "held the believer
to it."[107] Nevertheless, we have enough evidence to conclude that a vast
intellectual and legal revolution occurred in the twelfth and thirteenth centuries in the West, and it did its part in transforming medieval society so that it
was a most receptive ground for the rise and growth of modern science. The
same cannot be said of Arabic-Islamic civilization nor of China.

[107] Weber, *The Protestant Ethic and the Spirit of Capitalism* (New York: Scribners, 1958), p.
98.

5

Colleges, universities, and sciences

We are now in a position to develop a more detailed comparative view of the institutional arrangements and cultural climates of the civilizations of the West and Islam insofar as they supported or inhibited the development of modern science. We have seen that, despite the prohibitions surrounding the study of the ancient or foreign sciences in Islam, great progress was made. Indeed, for nearly five hundred years the natural sciences reached their highest state of development in the world among those who used the Arabic language in the Middle East.[1] Given the high state of scientific development – mathematical, computational, theoretical, and experimental – in a language clearly foreign to Europe, it is evident that the cultural advantages of the Arab Middle East far exceeded those of Europe until considerably after the thirteenth century. For that reason one would have anticipated that the Arab world would have made the great leap to modern science long before the Europeans. Such an expectation would also follow from a long-standing theorem in the sociology of science. That is, given any base of cultural objects with a specifiable number of separate units that can be combined and recombined in different configurations, the number of new combinations and permutations (inventions and discoveries) is a mathematical function of the existing base. The larger the existing base, the more new scientific and intellectual innovations that should be expected. This idea was developed by the American sociologist William F. Ogburn into the thesis of the inevitability of simultaneous, independent, and multiple discovery. In the 1920s he and Dorothy Thomas culled the history of science and technology and

---

[1] For a useful overview of the Arabic heritage in the mathematical sciences, see E. S. Kennedy, "The Arabic Heritage in the Exact Sciences," *al-Abhath* 23 (1970): 327–44. For an Arabic–Chinese science comparison, see Chap. 2, section entitled "The Achievements of Arabic Astronomy," as well as Chap. 7, "The Problem of Chinese Science."

found 148 such simultaneous, independent, and multiple discoveries, including the calculus, non-Euclidean geometries, and the law of the conservation of energy.[2]

In 1961 this thesis was greatly extended by Robert Merton and Elinor Barber, whose study of significant scientific discoveries yielded 264 cases of multiple and independent discovery. Among these there were 179 doublets, 51 triplets, 17 quadruplets, 6 quintuplets, and 8 sextuplets.[3] From this perspective, once the cultural base is prepared, new inventions and discoveries will be found by independent and multiple investigators. Thus, given the scientific advantages of Arabic-Islamic civilization noted above (and in Chapter 2), it would have been reasonable to expect Arabic science to give birth to modern science. Furthermore, we also know that the Arabs worked out the first successful non-Ptolemaic models in astronomy and that the very models worked out by the Marâgha school of the thirteenth and fourteenth centuries (especially by al-Tusi and Ibn al-Shatir) were eventually used by Copernicus. Thus the breakthrough to modern astronomy was, paradoxically, mathematically worked out by the Arabs, yet unobtainable for them because of their failure to make a metaphysical leap, that is, to place the same mathematical models in the new heliocentric framework Copernicus dared to suggest. Arabic science did not achieve this accomplishment and, in fact, went into a state of decline.

To understand and explain this anomalous outcome, we must examine the contrasting institutional arrangements of the two civilizations as they developed in the twelfth and thirteenth centuries. Although one may argue that the development of mathematical astronomy hinged on technical issues, the preceding considerations cast a different light on the subject. Consequently, we must direct our attention, on the one hand, toward those developments that empowered men to believe they possessed legitimate powers of reason that would reveal the hidden (and antiscriptural) design of the universe and, on the other, to the institutional breakthroughs that occurred, thereby opening up spheres of intellectual discourse and participation. In the last chapter, we reviewed the many intellectual breakthroughs that opened up

---

[2] The list is reprinted in W. F. Ogburn, *Social Change* (New York: Delta Books, 1966), pp. 90ff.

[3] Robert K. Merton, "Singletons and Multiples in Science," reprinted in *The Sociology of Science: Theoretical and Empirical Investigations,* ed. Norman Storer (Chicago: University of Chicago Press, 1973), chap. 16, at p. 364. It might also be observed that it is not necessary for the success of this thesis that all multiple discoveries be exact identities, anymore than it is necessary to say that, because Fords, Chevrolets, Audis, Volkswagens, etc., are not identical, the modern automobile was reinvented each time a new type of car was invented. Scientific principles have stronger and weaker forms of representation: all that is required is the notion of rough functional equivalence.

new realms of reason and discourse, as well as fraternal association in the West. The central question, from the point of view of the comparative sociology of science, is not whether one group of people or another made a technical discovery that surpassed those of the Greeks, Arabs, Indians, or Chinese, but whether breakthroughs occurred that opened new possibilities for free inquiry, intellectual criticism, and uninhibited discourse. To examine this question we must look at the comparative design of the social institutions of the two civilizations, European and Islamic.

Madrasas: Islamic colleges

As we saw earlier, the principal institution of higher learning in Arabic-Islamic civilization was the madrasa, or college. It was basically a place of study that evolved through various phases, producing in the end the institutional form generally referred to as the college. Dating at least to the ninth century, madrasas were founded as charitable trusts under the law of waqf.[4] As such they were pious endowments, with the endowment being used to maintain the educational property and to pay the salary of the professor, the trust administrator, and later to support students. With this arrangement, tuition was free for students.[5] The famous philosopher and religionist al-Ghazali attended a madrasa in Tus in the 1070s, and there he and his brother received not only instruction but food and lodging free of charge.[6] There were other forms of schooling open to the public, but the madrasa was the central Islamic institution of higher education. For example, the masjid schools were similar to the madrasas, and though they first evolved the complex of attached living quarters for students (the *masjid-khan* complex), the masjids were pious endowments that, once created, became legally separate from their founders – unlike the madrasas. Both the madrasa and the masjid-khan were schools of law, but in the madrasa the founder could appoint himself as head and if, he wished, appoint his descendants in perpetuity.[7] The governing legal factor was, however, that the law of waqf specified that nothing inimical to the tenets of Islam could be undertaken and,

---

4 See Henry Catton, "The Law of Waqf," in *The Law in the Middle East*, ed. M. Khadduri and H. Liebesny (Washington, D.C.: The Middle East Institute, 1955), pp. 203–35; and Makdisi, *The Rise of Colleges* (Edinburgh: Edinburgh University Press, 1981), chap. 3.
5 Makdisi, "On the Origin and Development of the College in Islam and the West," in *Islam and the Medieval West*, ed. Khalil I. Semaan (Albany, N.Y.: SUNY Press, 1980), pp. 32f and 38.
6 W. Montgomery Watt, *Islamic Philosophy and Theology: An Extended Survey* (Edinburgh: Edinburgh University Press, 1985), p. 86.
7 Makdisi, "On the Origin and Development of the College," p. 28.

therefore, nothing could be taught within the pious endowments that was in the least contrary to the spirit of Islam.

I would also mention the institution called the *majlis*. This term apparently derived from the word that designated the place where a learned man sat to give his instruction. It came to be used very broadly to refer to any place where learned discussions took place habitually or which was set aside for such a purpose. For example, Islamic hospitals often had a majlis, that is a classroom, or simply a room set apart for learned discussion or the recitation of medical texts.[8]

It has been suggested by many students of Arabic-Islamic civilization that the movement sponsoring the widespread construction of madrasas in the Middle East must be seen as a joint endeavor to encourage the study of Islamic law and to preserve Islamic traditionalism.[9] We saw in Chapter 2 that Islamic civilization maintained a sharp distinction between the Islamic sciences and the foreign or ancient sciences. It was generally believed that Greek philosophical inquiry, and especially its deeper attachments to certain metaphysical notions, was antithetical to the beliefs of Islam.[10] The specific points on which it was declared to be so were the question of whether or not the world was created or had existed eternally, whether God perceived particulars or only universals, and whether or not there is bodily resurrection, which the philosophers denied and the Muslims affirmed. The Quran is sprinkled with passages asserting that God created the earth, the stars, the sun, and the moon: "Your Lord is Allah, who in six days created the heavens and the earth and then ascended His throne."[11] "Allah is He who created the heavens and the earth and caused water to come down from the clouds, and brought forth therewith fruits for your sustenance."[12] Consequently, Islamic metaphysics and cosmology were sharply at variance with many of the tenets of Greek philosophy. Although Arabic translators of the Greek philosophical corpus could be selective in what they chose to translate – for example, they could favor the Platonic mystical tradition that many Muslim philosophers found congenial and omit other works, such as the *Timaeus* – the fact remained that Greek philosophy rested on metaphysical assump-

---

[8] Ibid., pp. 10–12.

[9] George Makdisi, "Muslim Institutions of Learning in Eleventh-Century Baghdad," *Bulletin of the School of Oriental and African Studies,* 24 (1961): 1–55; and see Makdisi's reexamination of the thesis, in *The Rise of Colleges,* pp. 301–5. Also see Hodgson, *The Venture of Islam* 2: 323–34 and especially pp. 438ff.

[10] See Ignaz Goldziher, "The Attitude of Orthodox Islam Toward the Ancient Sciences," in *Studies in Islam,* ed. Merlin Swartz (New York: Oxford University Press, 1981), pp. 185–215.

[11] Sura 7: 5.

[12] Sura 14: 32.

tions which remained at variance with the Quranic tradition. These ranged from assumptions about the nature of creation, to the nature and uses of logical argumentation, to the rationality of man. The upshot was the exclusion of the sciences of the ancients from the curriculum of the schools of higher learning.[13] The bedrock of instruction was Islamic law (fiqh), along with Quranic studies, Arabic, grammar, the sciences of tradition (hadith), and enough arithmetic to equip legists and qadis to divide up inheritances.

As it happened, not a few law professors became proficient in certain aspects of the foreign sciences and, perhaps surprisingly, books containing the wisdom of the natural sciences were allowed in the libraries of the madrasas as well as many mosques. The prohibition regarding the natural sciences chiefly applied to teaching them. Nevertheless, there were very strong psychological impediments, as Ignaz Goldziher has shown,[14] to the study and pursuit of the foreign sciences, and there were times when they were completely prohibited and book burnings occurred. Injunctions against the study and teaching of logic and the foreign sciences, moreover, were not confined to the eastern caliphate but extended to North Africa and Spain. According to George Makdisi, the Almohad ruler of North Africa and Spain, al-Mansur (reign: 1184–99), "was intent upon putting an end to literature on logic and philosophy in the lands under his sway. He ordered that books on these subjects be burned, and he prohibited their study whether in public or in private, threatening capital punishment for those found studying them."[15] Still, Makdisi insists, in normal times "there was nothing to stop the subsidized student from studying the foreign sciences unaided, or learning in secret from masters teaching in the privacy of their homes, or in the *waqf* institutions, outside of the regular curriculum."[16] It was possible for professors who had mastered significant portions of both the Islamic sciences and the foreign sciences to surreptitiously teach the foreign sciences under the ruse of teaching hadith (the traditions of the Prophet). Thus the scholar Sadr ad-Din b. al-Wakil (d. 1316) is said to have taught "medicine, philosophy, Kalam, and other fields belonging to the 'ancient sciences,'"[17] under the guise of hadith. Just how effective, rigorous, and thorough such instruction could be remains problematic. The teaching of so-called prophetic medicine, based on collections of sayings on medical topics, attributed to the Prophet, came to be a rival of Galenic medicine. For

[13] Makdisi, *The Rise of Colleges*, p. 78.
[14] Goldziher, "The Attitude of Orthodox Islam."
[15] Ibid., p. 137.
[16] Ibid., p. 78.
[17] Ibid.

some historians of Islamic medicine, it was little more than quackery in religious disguise.[18]

Thus, loopholes in the madrasa system account for the fact that Arabic science achieved great heights in medicine, optics, mathematics, and astronomy, although it did not make the breakthrough to modern science. The philosophical and natural sciences were intentionally excluded from the madrasas, but they were often present sub rosa.

Each madrasa was focused on the teaching of law and legal studies, and each was devoted exclusively to a single school of law – Shafi'i, Hanafi, Hanbali, or Maliki. It was quite rare for more than one school of law (madhhab) to be taught in a single madrasa,[19] nor was any other form of law, for example, Greek, Roman, or customary, taught.

We should also note the peculiarities of the method of instruction and the awarding of certification and degrees. While the core of instruction in the madrasas was focused on the study of law and the Islamic sciences, the range of subjects taught and their sequence was often haphazard. Ideally, the subjects taught should follow a prescribed sequence. Instruction should begin with the science of hadith, that is, the process of collecting and transmitting sayings attributed to the Prophet. This included study of the biographies of the famous men who compiled these monumental collections of the sayings of the Prophet. Next the initiate would study the two complementary roots, religion and law, the so-called principles of religion (*usul ad-din*) and the roots of law (usul al-fiqh). Following that, one should learn (memorize) the special legal tradition to which one belonged (madhhab). After that, the initiate should be introduced to the divergences of law (*khalif*) within his school, as well as those between different schools of law. Lastly, within the religious sciences, one should learn dialectic (*jadal*), the art of disputation.[20] The so-called auxiliary sciences, however, had both wide and narrow definitions. For some students they included all the tools of grammar, lexicology, morphology, metrics, and rhyme, as well as prosody and even tribal history and Arab genealogy.[21] For others, time was better spent on studying Quranic exegesis and analyzing hadiths, or fiqh, and less time was spent on

---

[18] See J. Christopher Bürgel, "Secular and Religious Features of Medieval Arabic Medicine," in *Asian Medical Systems: A Comparative Study*, ed. Charles Leslie (Berkeley and Los Angeles: University of California Press, 1976), pp. 44–62 at pp. 46f; and E. G. Browne, *Arabian Medicine*, reprint (Cambridge: Cambridge University Press, 1962), pp. 12–13.

[19] Makdisi, *The Rise of Colleges*, p. 304, and passim. Also see Carl F. Petry, *The Civilian Elite of Cairo in the Later Middle Ages* (Princeton, N.J.: Princeton University Press, 1981), pp. 331f, for what appear to be the first madrasa in Cairo (the Salikiya, ca. 1241) with all four schools of law represented in it.

[20] Ibid., p. 79.

[21] Ibid.

argumentation and dialectic. One scholar might become extremely proficient in Quranic exegesis, Arabic grammar, and phonology; another, fully proficient in Quranic exegesis, hadith, prosody, tribal history, and so forth. There was no standard curriculum. What a student learned depended upon what his master knew and how many (and which) other masters he decided to study under. Those biographies of Arabic scholars that have survived suggest that there were men who developed broad intellectual skills, including proficiency in logic, grammar, and the Islamic sciences, as well as a passing knowledge of the ancient sciences. The impression one is left with is that the truly superior scholars were those who were fully proficient in all of the Islamic sciences and who had more than a superficial knowledge of the ancient sciences. The problem was, however, that anyone who gained a reputation for excellence in the Greek sciences became an easy target of the traditionalists who could simply issue a fatwa – a legal ruling – condemning that person and his studies. There was little chance of escaping the jeremiads of the fundamentalists and traditionalists, who were either ignorant of the foreign sciences or, despite knowledge of them, condemned their pursuit. The charge that a scholar in one of the madrasas was teaching the ancient sciences would be taken very seriously as a religious and legal matter, since it directly violated the founding assumptions of the madrasa under the law of waqf. In brief, without the philosophical guidance of the structure of thought and research imbedded in the Greek corpus, "there was no set curriculum that all had to follow" in the madrasa.[22] What was stipulated was four years to complete the study of law; the student's studies were terminated when he produced a written *ta'liqa* report pointing out some questions of law drawn from his master's lectures and from his readings. Termination meant that the student had reached a new level of proficiency and was now to enter into a close companionship with an established master for an indefinite period of time.

The method of certification in the madrasa was based on the ijaza, the permission or authorization to transmit the material learned, principally books. This model of certification seems to have grown out of the practice of collecting isnads, chains of transmitters validating the authenticity of the teaching transmitted.[23] Accordingly, the granting of an ijaza was generally tied to the transmission of a particular book or set of sayings by the Prophet and his companions. Students could go from one learned master to another, learning his books and collecting ijazas. There was no quota of ijazas to be

---

[22] Ibid., p. 84.
[23] Makdisi, *The Rise of Colleges*, pp. 140ff; and cf. Pederson, *The Arabic Book* (Princeton, N.J.: Princeton University Press, 1984), pp. 31–4.

obtained nor any specified sequence of them. Scholars dictated their books aloud and students copied them down. Once the copying had been corrected, usually through an oral recitation by the student, the student received an authorization, written in the scholar's own hand, to transmit the book to others. Ijazas were absolutely tied to the authorizing individual, not to the school. They could be granted only by the scholar himself, not the school, the sultan, emir, or anyone else.[24] Likewise, a master could refuse to grant the authorization if he so chose. In this system of educational certification, the model stressed the ideal of authentic, even certain knowledge transmitted from one scholar to another, authenticated both by the chain of transmitters and the reputation of the particular scholar. Of course, when a scholar published a new work, there was no review process, and the students accepted it as the work of a great master of the material in question. As noted above, publication took the form of dictating the book directly to students.

In the case of law, however, it appears that the ijaza was of a broader nature, that is, the student was authorized to issue legal opinions (fatwas) and to teach law. According to Professor Makdisi, "the authorization to teach law and issue legal opinions was given after an examination had taken place."[25] The exam was oral and "took place on particular books that had been studied by the candidate." As a student, the jurisconsult al-Shirazi "was authorized to teach law and issue fatwas by two professors; one of them had examined him on several questions in several fields," while other professors examined him on particular books, the result of which authorized him to teach those also.[26] This pattern, reported in several cases by Makdisi, suggests that there may have been an incipient movement toward a more general license to teach represented by the licentia docendi in Europe, though no general certificate of competence issued by a collective faculty actually developed autonomously within Islamic higher education. Among other reasons for this failure to develop uniform standards was the fact that there was no faculty: there was simply a gathering of scholars, each teaching his own unique collection of learned works memorized from the past. It is curious, however, that the case discussed by Makdisi, in which legal scholars are reported to have been examined by two or more professors, dates from the fifteenth century when influences deriving from the European universities might have been present. It remains the case that no equivalent of the bachelor's degree or the licentia docendi emerged in the medieval Islamic madrasa.

[24] See Makdisi, *The Rise of Colleges*, p. 271.
[25] Ibid., p. 151.
[26] Ibid.

In addition to copying and memorizing legal materials, the schools also developed a method of disputation that required students to develop the ability to find "disputed questions." Makdisi suggests that this is the origin of the scholastic method in Islam (and possibly the West).[27] The reasons for developing this method, however, suggest quite a different function for it than was developed by Abelard and the European users of dialectic. In Islamic culture and civilization the method grew out of the need to achieve an orthodox consensus regarding religious doctrine. Legal scholars, as all believers, were enjoined "to encourage good and forbid evil." Moreover, this duty was taken so seriously that an official position was created, the market inspector (muhtasib), whose task it was to surveil the marketplace in search of all forms of behavior at variance with Islamic teachings.[28] Toward that end, it "was . . . incumbent upon a doctor of law who opposed a given doctrine to raise his voice against it, lest he be considered to have accepted it tacitly. Silence had positive value; the system had no place for abstention."[29]

This seems an unlikely source, however, of the dialectical method used by Europeans who sought to arrive at new harmonies of doctrine, not just a discarding of suspect doctrines. The European method also proceeded to find new principles of law (something ruled out in Islamic law after al-Shafiʿi) and by so doing transformed the European system, as we saw in Chapter 4. In Europe the legal system was placed on a new theoretical foundation through the use of logic and dialectic. This resulted in the elimination of contradictions among legal sources – that is, the Bible, church fathers, and Roman law – and the establishment of new legal standards, such as conformity with natural law, conscience, and those standards of reasonableness that flow from past customary usage and ideas of justice.

In the Islamic case, on the other hand, the incorporation of the method of disputation (taʿliqa) led only to the solidification of legal doctrine, not new organizing concepts and principles. Moreover, there could never be a subordination of the teachings of the Quran to the light of reason or to natural law. The Quran was the speech of God, which of necessity had to be authentic. This logic furthered the insistence that the Quran was authentically presented only in the Arabic language.

At the same time the method of dialectic in the madrasas made potential adversaries of masters and students, for the student was enjoined to find difficulties in the professor's teaching. This method of teaching was based on the taʿliqa report, a written set of notes and commentary that the student

---

[27] Ibid., chap. 3, especially pp. 105ff.
[28] See the article on "Hisba," *EI*[2] 3: 485–9.
[29] Makdisi, *The Rise of Colleges*, p. 106.

took down from the master jurisconsult along with other readings. Once assembled, this material was studied and memorized and then submitted by the student "to the master for examination and quizzing with a view to being promoted to the class of *ifta'*" those who may issue legal opinions.[30] When the student had successfully mastered such a body of material and defended it before a master, his studies were terminated and he graduated to the class of *suhba,* the group of companions or assistants who were totally devoted to the aid and assistance of their master. Unlike the European technique of dialect, which also relied on the posing of questions, the Islamic method did not aim at resolving (*resolutio*) the disputed questions. Thus Ibn ʿAqil was a master jurisconsult and theologian who wrote a large mutli-volume work for the instruction of beginners.[31] He described his method as follows: "In writing this work I followed a method whereby first I presented in logical order the theses, then the arguments, then the objections, then the replies to the objections, then the pseudo-arguments [of the opponents for the countertheses], then the replies [in rebuttal] of these pseudo-arguments – all of this *for the purpose of teaching beginners the method of disputation.*"[32]

What is missing here is that element aimed at resolving the contradictions and paradoxes of the legal questions. That is, the European method of discussing disputed questions focused not on showing the frailties of someone else's legal opinion (logic) or method of discussing something, but on mastering issues through their solution. In the European universities legal and theological disputes had a subtly different thrust: "At the beginning of each question authorities who oppose, or seem to oppose, each other are set in array, and then the teacher shows his mastery by producing distinctions of meaning that suffice to solve the problem and dispose of all the difficulties."[33] Furthermore, the Europeans, both in law and theology, proceeded with the notion that the system was still growing and evolving. This was especially so in ecclesiastical (canon) law. Thus, "unlike the Roman law, the canon law did not constitute a closed *corpus,* but continued to grow."[34]

[30] Ibid., p. 114.
[31] Ibn ʿAqil, it should be noted, was a member of the Muʿtazila movement, the "Islamic rationalists," and was forced to recant certain views because of attacks on him by the orthodox. See *Ibn Qadama's Censure of Speculative Theology,* ed. and trans. George Makdisi (London: Luzac, 1962); and above, Chap. 3.
[32] Makdisi, *The Rise of Colleges,* pp. 256 and 117.
[33] William Kneale and Martha Kneale, *The Development of Logic* (Oxford: Oxford University Press, 1962), p. 206.
[34] Haskins as cited in Stephan Kuttner, "The Revival of Jurisprudence," in *Renaissance and Renewal in the Twelfth Century,* ed. Robert L. Benson and Giles Constable (Cambridge, Mass.: Harvard University Press, 1982), p. 306.

Abelard, for example, "approached his task with the belief that it was possible to make discoveries."[35] In law the method of disputation concerned actual principles of law (not questions of fact), and it served as a training ground of great importance, for it was by this means that there "was developed the courage to draw audacious analogies, to handle far flung principles of equity, to fill the lacunae of the law by intuition and imagination."[36] For this reason the dispute of questions in the European context was a major source of the dynamism and development of law, not, as in the Islamic case, a search for potentially heretical or plainly illogical conclusions. The Islamic model of disputation was designed to find fault with the reasoning or method of one's master (or opponent), not to establish new harmonies, much less to build up a new system of law.

The level of antagonism that the Islamic system could generate is illustrated by al-Ghazali's second ta'liqa report prepared under al-Juwaini. After reading the report, his master is reported to have exclaimed, "You have buried me alive! Could you not have waited until I was dead!"[37] Thus the spirit of these exercises seems to have been directed toward finding fault with a particular master and his doctrines, not with the problems of the law as a general system. The event of the student's graduation was consequently looked upon with some apprehension. On the one hand, the student could fail to perform well as a jurisconsult, but of greater danger was the possibility that he might "become an adversary in the arena of disputations and issue legal opinions contesting those of the master." To counter or delay this possibility, the master could hire such a student to repeat the lessons of the master as a "repetitor."[38] Often, however, disputations between parties resulted in the use of invective and insults and precipitated protracted quarrels, altercations, violence, and death.[39] The social implications of such contentiousness led authorities to ban disputations and the use of logic and philosophy.[40]

As training designed to make students of law proficient as muftis and jurisconsults qualified to issue legal opinions, no doubt the system of disputations worked well in preserving the status quo and in constantly weeding out suspect opinions. It was even essential to Islam, according to Pro-

---

[35] Kneale and Kneale, *The Development of Logic*, p. 204.
[36] Hermann Kantorowicz, "The Quaestiones Disputatae of the Glossators," *Tijdschrift voor Rechtgeschiedenis/Solidus Revue d'Histoire du droit* 16 (1939): 1–67, at pp. 5–6.
[37] Makdisi, *The Rise of Colleges*, p. 127.
[38] Ibid.
[39] Ibid., pp. 133–7.
[40] Ibid., pp. 137–9.

fessor Makdisi, because the "method was part and parcel of the Islamic orthodox process for determining orthodoxy."[41] Where it failed was in the creation of a set of objective and universal standards of law, against which all other laws and principles could be judged. Since the legal principles of Islamic law had been given once and for all, in the Quran and the Sunna, and in the principles of fiqh worked out by al-Shafi'i, the only task left was to use logic, in the narrow sense, to uncover faulty reasoning and thus preserve the doctrinal status quo. Moreover, no effort was made to integrate, to systematize and codify the divergences of the four schools of law. Note was taken, in instruction, of some divergences within and between schools of law, but no effort was made to reconcile the differences. The presupposition of the unity of the schools was tacit; no actual unified legal system was produced based on a set of universal legal rules and principles. "Islamic law does not claim universal validity; it is binding for the Muslim to its full extent in the territory of the Islamic state, to a slightly lesser extent in enemy territory, and for the non-Muslim only to a limited extent in Islamic territory."[42] The four schools, moreover, agreed to disagree, thereby preserving the traditional and personalistic character of Islamic law.

Universities and the West

The history of the rise of universities in the West in the Middle Ages is generally well known.[43] The uniqueness of this development – in comparison to the history of institutions of higher learning in other civilizations of the world – and its sociological implications for the pursuit of science have not yet, however, been appreciated. As late as 1971, Joseph Ben-David could write that "the European university was originally not different from the arrangements for higher learning of other traditional societies such as India,

[41] Makdisi, "The Scholastic Method in Medieval Education: An Inquiry into Its Origin in Law and Theology," *Speculum* 49 (1974): 649.
[42] Joseph Schacht, *Introduction to Islamic Law* (Oxford: Oxford University Press, 1964), p. 199.
[43] The standard works on this include Hastings Rashdall, *The Universities of Europe in the Middle Ages*, new ed., 3 vols., ed. A. B. Emden and F. M. Powicke (Oxford: Oxford University Press, 1936); A. B. Cobban, *The Medieval Universities* (London: Methuen, 1975); Stephen d'Irsay, *Histoire des universités, française et etrangères*, 3 vols. (Paris: J. Vrin, 1933). These have now been supplemented by specialized studies including the following: Gordon Leff, *Paris and Oxford in the Thirteenth and Fourteenth Centuries* (New York: John Wiley, 1968); Nancy Siraisi, *Arts and Sciences at Padua* (Toronto: The Pontifical Institute, 1973); Mary McLaughlin, *Intellectual Freedom and Its Limits in the Twelfth and Thirteenth Centuries*, reprint (New York: Arno Press, 1977); *Rebirth, Reform, and Resilience: Universities in Transition, 1350–1770*, ed. James Kittelson and Pamela Transue (Columbus: Ohio State University Press, 1984); and the essays in *Science in the Middle Ages*, ed. David C. Lindberg (Chicago: University of Chicago Press, 1978).

China, and Islam."[44] If our discussion thus far has shed any light on this subject, this claim flies in the face of historical and legal scholarship, above all, in the European-Islamic context.[45] It is even more unlikely in the case of China, as we shall see in Chapter 7.

The uniqueness of the European university can be seen on three levels: legal and social organization, curriculum, and philosophical and metaphysical commitments. From a structural and legal point of view, the madrasa and the university were at polar opposites. Whereas the madrasa was a pious endowment under the law of religious and charitable foundations, and as such was neither an entity legally independent of religious precepts or commitment nor an autonomous corporation, the universities of Europe were legally autonomous corporate enterprises that had many legal rights and privileges. These included the capacity to make their own internal rules and regulations and the right to own and sell property, to have legal representation in various forums, to make contracts, and of course to sue and be sued.[46] While many European universities grew out of cathedral schools and religious orders, this was not a prerequisite nor was it invariably the case. The University of Bologna, for example, was a lay institution and only because of its unparalleled study of Roman law were several of its students and masters later to become popes and church officials.[47] Unlike their Islamic counterparts, the universities of Europe were not established as pious endowments, but rather as autonomous legal entities with characters as "whole bodies" of students and scholars who were empowered to carry on their affairs as they saw fit. As scholarly guilds, "they sprang into existence . . . without express authorization of king, pope, prince, or prelate. They were spontaneous products of that instinct of association which swept like a great wave over the towns of Europe in the course of the eleventh and twelfth centuries."[48] Of course, this did not mean that they ignored the religious precepts of Christianity, but it did mean that other principles –

---

[44] Ben-David, *The Scientist's Role in Society* (Englewood Cliffs, N.J.: Prentice-Hall, 1971), p. 47.

[45] Although Ben-David discusses the fact that students and scholars formed corporations, he seems unaware of the broader political and legal significance of this concept as I have set it out. He says, for example, that "this corporate device . . . was not entirely unique to Europe" (p. 47) when there is no evidence for its existence in the most significant civilization adjacent to the West, namely, Arabic-Islamic civilization. Likewise, I have not been able to find evidence of its existence in Chinese law. Hence the corporate legal form as we know it is indeed unique to Roman law and its European modification brought about by the revolution in canon law of the twelfth and thirteenth centuries.

[46] Pearl Kibre, *Scholarly Privileges in the Middle Ages* (Cambridge, Mass.: Medieval Academy of America, 1974); and Gordon Leff, *Paris and Oxford*, pp. 70–4.

[47] D'Isray, *Histoire* 1: 89; Rashdall, *Universities of Europe* 1: 108ff.

[48] Rashdall, *Universities of Europe* 1: 15.

those of the requirements of learning, of law, especially the ideals of natural law, reason, and conscience – had equally strong bearing on what they decided (collectively and individually) and on what was studied. The course of study was not dictated solely by religious scruples or theological aversions, though religious purists and fundamentalists did attempt to impose such restrictions.

The fact of the incorporated status of European universities is (and was) of paramount importance insofar as it affected the future course of learning and organization within institutions of higher learning. As noted earlier, where a collective of individuals is treated legally as a corporate entity, it is granted a set of rights and prerogatives that are lacking without that status – which of course all madrasas lacked. One of the most salient features the Roman law recognized was that a corporate body is not uniquely tied to its current membership or directives: when the present members die or otherwise relinquish their control over the corporation, the corporation continues to live and to make decisions about its present and future activities. In other words, a corporation is a social (as well as a legal) enterprise that transcends the current membership, board of trustees, directors, or other fiduciaries. It also transcends their particular conception of the mission and goal of the enterprise. It is a collective actor whose will is always subject to the vote of the current membership, and it may therefore elect to pursue a course of action – even a course of study – that was neither imagined nor intended by the founders.

By way of contrast, every Islamic college (madrasa) was a direct and restricted product of the explicit legal instrument of its founder. Its provisions had to be carried out exactly as stated in the original founding document. The founding instrument, issued personally by the founder (not by a legal consultant and without the benefit of legal formulae) was reviewed by a qadi (judge) to insure its conformity to Islamic law and then placed on file.[49] In every respect, therefore, such an instrument establishing a pious endowment was a personalistic document, and if the founder failed to incorporate essential features, it was confined to such existence in perpetuity. On the other hand, if the founder failed to appoint a trustee to administer the trust, the qadi would appoint one, "for no trust was allowed to fail for want of a trustee."[50] But once the trust was set up, no changes were allowed either by the founder or its trustee. In effect, the endowment was rigidly restricted to the precise formulation of the founder's legal instrument and no future

[49] Makdisi, *The Rise of Colleges,* pp. 35–6.
[50] Ibid., and Schacht, *Introduction to Islamic Law,* p. 209, on the individualistic nature of Islamic law, including the law of waqf.

changes were permitted. It would follow, likewise, that property, once made part of the waqf, could not be alienated, since it was a permanent gift to the service of God. It is reported, nevertheless, that unscrupulous rulers did sometimes expropriate waqf property in order to recycle its construction materials, which were often in short supply, as well as for more mercenary reasons.[51] The religious and legal principles, nevertheless, forbade such expropriation.

For these reasons it is easy to see why the university as an organizational form was a far more viable instrument of social organization and action than the madrasa, and why the Islamic colleges were narrow and rigid insofar as growth and development was concerned. It has also been pointed out that during the medieval period, the word madrasa technically meant "a school of law": "There was no *madrasa* without *fiqh*."[52] Professor Makdisi sums up the situation as follows: "Comparing the two situations, we see in one case strict adherence to the stipulations of the founder; in the other, there is flexibility and a great amount of leeway." Similarly,

the trustees [of a corporate trust] can perpetuate themselves, being replaced as they come to the age of retirement, or otherwise end their trusteeship. The lack of change in the former [the unincorporated charitable trust] stunts growth and development which, on the other hand, becomes possible for the latter. This is one of the reasons why untold numbers of colleges in Islam and the West came into existence and then disappeared.[53]

Given this view of the matter and the important role that European law played in the emergence of universities as autonomous corporate bodies, it would be surprising to find the same organizational form in Islam, China, or elsewhere.[54]

The structural and organizational differences between universities and Islamic colleges, however, do not end here. Another significant difference inheres in their methods of certification. The model of certification in the Islamic college was the ijaza, the permission to transmit certified knowledge,

---

51 Makdisi, *The Rise of Colleges*, pp. 40ff.
52 Gary Leiser, "Medical Education in Islamic Lands from the Seventh to the Fourteenth Centuries," *Journal of the History of Medicine and Allied Sciences* 38 (1983): 48–75; Leiser reminds us that "the verb *darasa*, derived from *drs*, simply means 'to study.' In the medieval era this form of the verb was never used; only the second, *darrasa*, which means 'to teach fiqh.' A *dars* was a 'lesson in fiqh,' and a madrasa was a 'place' where *fiqh* was taught," p. 56. This follows Makdisi's earlier work.
53 Makdisi, *The Rise of Colleges*, p. 235f.
54 It should be made clear that today Western colleges and, especially, American colleges are also corporations, based on modifications of the model of the medieval university. Makdisi reviewed this issue, including the famous Dartmouth College case (1819) in which it was affirmed by the Supreme Court that Dartmouth is a corporation under the Constitution of the United States. See *The Rise of Colleges*, pp. 229–37.

granted solely by the individual master. Here we also find an impediment to the development of collective, generalized, and impersonal standards for evaluating scholarship. The Islamic model in the madrasa rested on the transmission of knowledge certified by the chain of transmitters and the personal authorization of the master scholar. There was no collective certification or validation that would have been the product of the collective action of a group of scholars. Nor were there bachelor's or doctoral degrees to be awarded upon completion of a course of study. In the European universities, by way of contrast, the license to teach (the licentia docendi) was a certification of competence granted only by the head of the university, the chancellor, after the candidate had been examined by the faculty and the chancellor. Such an examination could not take place in the madrasa, for, among other reasons, "there was no faculty."[55] There were only individual masters, each variously authorized to transmit writings previously received through authorized transmission, though also capable of dictating his own works.

There was, however, another, positive side to the heavy stress in the madrasas on the study of received works and their memorization. The stress placed on the memorization and recitation of ancient texts also resulted in the close study and memorization (outside of the madrasa) of large portions of works like Ptolemy's *Almagest* and Euclid's *Elements*. It seems fair to say that the memorization and constant reexamination of such works could have a salutary effect, as there is much to be learned from a complete mastery of scientific (and mathematical) classics whose principles are more or less up-to-date (as those two works were).[56]

On the other hand, it might be asked whether such memorization and mastery are equivalent to learning paradigms, in Thomas Kuhn's sense of that word.[57] His view is that paradigms represent something on the order of tacit knowledge, but it is also knowledge that enables the scientist to see one problem as being like another in terms of some set of external and abstract criteria. There are, according to this view, certain typical ways of treating, for example, falling bodies, the movements of pendulums, and so forth. The question is, therefore, whether these conceptual aids are learned through the step-by-step examination and memorization of a classical text, or whether they are learned by actively solving practice problems posed by the instructor. Such exercises serve to teach those typical ways of treating scientific

[55] Makdisi, *The Rise of Colleges*, p. 271.
[56] For the role of memorization and repetition in the madrasas, see *The Rise of Colleges*, pp. 99–103.
[57] *The Structure of Scientific Revolutions*, 2d enlarged ed. (Chicago: University of Chicago Press, 1970).

and mathematical problems that otherwise are only implicit in a scientific text.

Finally, when the model of learned scholarship is that of the scholar who has memorized the great texts (and prides himself on being able to recall to memory all the important works of his field), it seems likely that the major effort of learning will be directed toward memorizing the classics and not toward correcting their weaknesses and advancing beyond the current state of the art. Indeed, many students of Islamic culture and civilization have blamed its stagnation on the rigidification of thought that resulted from slavish memorization of texts and extreme fidelity to traditional authorities. Indeed, the excessive stress on memorizing and the concomitant lack of understanding of what had been learned have been discussed as a defect of education in Arab societies down to the early twentieth century.[58]

There are two ways to evaluate the relative merits of the methods of instruction in the two sets of institutions of higher learning – Islamic and European. Some medieval historians of European science feel that we know too little about this to reach a conclusion: "Did the student memorize some or most texts, which may have been prohibitively expensive? Did they solve problems? Was the abacus used in practical arithmetic? Were Arabic numerals employed for computations? On these and other vital matters we are largely ignorant."[59] In the Arabic-Islamic case, however, the mode of instruction of the medieval period seems to have come down nearly intact to the early nineteenth century, and accounts of the process of learning seem to confirm the pattern of students siting at the feet of their masters and memorizing large amounts of material.[60] From the various accounts of Islamic education, there appears to have been a much greater stress on memorization and repetition, though a distinction did exist between memorizing something and actually mastering it.[61] In the case of specifically scientific training, there is less systematic information about the process of learning, and what we do have comes from autobiographical accounts, usually of those outstanding scholars who made permanent marks on the history of Islamic scholarship. Such reports tend to give us a picture of the exceptions rather than the rule. The question is not whether exceptional individuals can suc-

---

[58] See Dale F. Eickelman, "The Art of Memory: Islamic Education and Its Reproduction," *Comparative Studies in Society and History* 20, no. 4 (1978): 485–512. This is probably related to the Islamic tradition, still very much alive, of highly esteeming those who have memorized the entire Quran, and who bear the honorific title of "Hafiz."

[59] Edward Grant, "Science and the Medieval University," in *Rebirth, Reform, and Resilience: Universities in Transition, 1300–1700*, pp. 68–102 at p. 77.

[60] For a penetrating look at traditional Islamic education in Eastern Islam and Iran, see Roy Mottahedeh, *The Mantle of the Prophet* (New York: Simon and Schuster, 1985).

[61] See Makdisi, *The Rise of Colleges*, pp. 99–105.

ceed despite adversity, but whether a new social system, fully supportive of scientific investigation, developed, and what the institutional configuration needs to be in order (1) to make the breakthrough to modern science, and (2) to make the pursuit of science a continuous and honorable occupation free from moral and political taint. In the case of scientific training, in medicine and the like, there was also a strong encouragement to memorize scientific works, whether they be the *Almagest* or works by Euclid, Galen or the great Arabic physicians. But this is far from the whole story.

The second approach is to consider the character and methods of examinations. This seems to me the solution to the problem of evaluating the two educational patterns, at least from this distance in time. In the one case – that of Europe – there was a degree system, in which a core curriculum was imposed and a process of collective evaluation took place to determine whether mastery of the core curriculum had been achieved. In the other, a personalized and arbitrary system of self-selected courses (and masters), with the attendant individualized collection of ijazas (permissions), was perpetuated. Clearly, of the two systems, the Islamic model had a much higher reverence for traditional knowledge passed down through the centuries, represented by the institution of the ijaza, which was modeled after the collection of isnads, chains of transmitters of religious sayings. As Professor Makdisi puts it, "The authorization to teach was tied primarily to the book. It guaranteed the transmission of authoritative knowledge."[62] Conversely, the Arabic-Islamic system devalued impersonal standards bereft of the personal chains and the master. While the practice of assigning students to a master at the European universities existed, the students did not collect permissions from him but were examined by the whole faculty. Indeed, the English universities instituted examination procedures whereby readers, for both exams and theses, included evaluators from other universities.[63] In short, the Europeans were keen on establishing a degree system whereby they could judge the relative merits of candidates according to uniform standards of achievement. This is reflected in the standardized sequence of courses, the periodic examinations, and the general certification of competence by the masters collectively as a faculty, with the final awarding of the degree by the chancellor. A. B. Cobban describes the degree-granting process of the medieval European universities in the following manner.

[62] Ibid., p. 140. Cf. George Vajda, "Idjaza," *EI²* 3: 1021.
[63] See James Weisheipl, "Curriculum of the Faculty of Arts at Oxford," *Medieval Studies* 26 (1964): 143–85, at pp. 145ff for assigning students to masters; and Rashdall, *Universities of Europe* 3: 141ff for the examination procedure at Oxford; as well as Leff, *Oxford and Paris*, pp. 137–80.

The regime was hard, the struggle exacting, and the students were expected to display a dedication equal to the precious endeavor on which they were engaged. Although the student was not obliged to sit written examinations for the attainment of a degree, he was nevertheless severely taxed at every point in his undergraduate career. One could say that the degree was awarded on the basis of a total and continuous assessment of the student's performance. . . . The undergraduate had to attend the prescribed number of lectures in each course, and had to have a detailed knowledge and understanding of stipulated texts. Moreover, the student had to prove himself an academic practitioner by the mastery of a series of complex exercises specified for different stages in the course. Finally, he was subject to lengthy probing oral examination.[64]

This suggests a strong desire on the part of the European masters to erect objective standards by which all students could be evaluated, based on a more or less standard set of courses. All students were required to study not just the seven liberal arts, but the natural books of Aristotle, as well as moral philosophy and metaphysics.[65] In comparison to the personally tailored, even haphazard, pattern of learning in the madrasa, this system was highly structured. It was a system designed to produce individuals capable of making fine intellectual distinctions and of carrying out logical analyses within a broad common framework that included the natural sciences, logic, philosophy, and metaphysics. And it reached its goal through "prolonged practical oral assessment."[66]

In making this comparison, we should remember that the Islamic college was devoted to Islamic law and the auxiliary Islamic sciences. It was designed to produce individuals capable of issuing legal decisions and of keeping "intact what had been handed down."[67] Even in this area of nonscience instruction, however, there were no required courses or books, other than the Quran. Judged by medieval European standards, instruction was highly personalized, often whimsical in its selection of topics for study, and lacking in uniformity. Science education took place outside this formal setting, but scholars presume (relying on many accounts regarding the teaching of medicine, for example)[68] that it followed a similar personalistic pattern based

---

[64] A. B. Cobban, *The Medical Universities* (London: Methuen, 1975), p. 209.
[65] Grant, "Science and the Medieval University," pp. 71–9; and Weisheipl, "Curriculum of the Faculty of Arts at Oxford," pp. 143–85.
[66] Cobban, *Medieval Universities*, p. 210.
[67] Makdisi, *The Rise of Colleges*, p. 105.
[68] Goitein, "The Medical Profession," in *A Mediterranean Society*, 2 vols. (Berkeley and Los Angeles: University of California Press, 1971) 2: 247–8; Michael Dols, "Introduction," *Medieval Islamic Medicine: Ibn Ridwan's Treatise "On the Prevention of Bodily Ills in Egypt"* (Berkeley and Los Angeles: University of California Press, 1984), pp. 3–73; Gary Leiser, "Medical Education," pp. 48–74; and Bürgel, "Secular and Religious Features," pp. 48ff.

on the inclination of the master. However, since the masters of the foreign
sciences outside the madrasas had the works of Aristotle, Euclid, Ptolemy,
and Galen as a standard, one may suppose that training in science had more
structure and higher standards of expectation than was the case in legal
and theological matters. For example, in medical education it is said that
"the closest thing to universally required reading was the Sixteen Books of
Galen."[69] But it is clear that those standards represented by Greek philosophy and science were never adopted within the Islamic colleges, and for that
reason (along with the reasons for their prohibition in the first place) the development of natural sciences finally aborted after the thirteenth and fourteenth centuries in the Islamic world. It should also be pointed out that
many students in the madrasa pursued their studies as apprentices for long
periods of time. Makdisi, for example, reports several individuals who spent
one and a half to two decades apprenticed to their masters.[70] The lack of a
clearly defined terminus and a standard curriculum turned them into perpetual students, a most inefficient use of human talent. To some degree this prolonged period of study and apprenticeship parallels the years and decades often spent by traditional Chinese students preparing for the state-sponsored
examinations in China.

The absence of a fixed curriculum and a terminal degree also heightened
competition between those constantly engaged in disputations. Since there
was no final degree, no final demarcation of a job well done, the constant
public disputing for the purpose of displaying one's own brilliance and
upstaging others created perpetual rivalries and kept the fully competent
fuqaha (law professor) always on guard, waiting for the next newcomer to
upstage him, with the resultant public disgrace or, worse, outpouring of
insult and invective. "It was through disputation that excellence in a field of
knowledge was established. To be 'top man' in one field, one had to prove
that he was 'unbeatable' in the field, and the best way to prove it was to
engage in public disputation."[71]

Finally, to return to the European setting, there is additional evidence of a
push among Europeans for uniform and objective standards of teaching in
the universities. They were bothered by the fact that, in their eyes, there
were "few stable criteria by which one university could evaluate the standard
attained by a graduate of another [university] without carrying out its own
investigation."[72] As a result, an effort to standardize, or to equalize teaching
opportunities, was attempted by creating a standard license – the *ius ubique*

[69] Leiser, "Medical Education," p. 62.
[70] *The Rise of Colleges*, pp. 96ff.
[71] Makdisi, "The Scholastic Method," p. 65 n45.
[72] Cobban, *Medieval Universities*, p. 31.

docendi – issued by all universities with a *studium generale* that guaranteed the recipient universal rights to teach. But this effort largely failed. It was implemented in part to give the graduates of newly founded universities the same aura of respect enjoyed by those from Paris and Oxford. The pope's resort to equalization by papal bull, however, did not suffice. Accordingly, "the principle of the mutual recognition of degrees and the teaching license broke down even in the case of major universities like Oxford and Paris" in the fourteenth century;[73] the universities refused to accept the licenses of graduates of other schools without first imposing new examinations. The important point, nevertheless, is that the Europeans were not willing to accept the potentially idiosyncratic (or deficient) educational standards followed in other localities. They desired (and moved to create) a uniform set of evaluative standards.

In short, these developments suggest that there was a concerted effort among European academics to create objective, impersonal, and universal standards of intellectual achievement, and the first step in that direction was the imposition of oral exams administered by the various faculties as a whole, or by examiners selected from their ranks. At Oxford, for example, "when the candidate presented himself before the chancellor for the license in arts, he had to swear that he had read certain books, and nine regent masters (besides 'his own' master who presented him) were required to testify or 'depose to their knowledge' of his sufficiency (de scientia), and five others to their 'belief' therein (de credulitate)."[74] The second step was the imposition of standard degrees and a uniform license to teach.

Islamic protoscientific institutions

We have seen that what was taught in the Islamic colleges was quite different from what was taught in the European universities. As a general rule, and as a matter of legal and religious precept, the teaching of the natural or foreign sciences was prohibited in the Islamic colleges. Although some fuqaha, law professors, had actually mastered significant portions of the foreign sciences, especially arithmetic and logic, their knowledge of this material was always carefully guarded and almost never publicly shared. Any teaching of it was done in the privacy of the professor's home. It was knowledge carefully and discretely gathered both on one's own and by apprenticing oneself to a learned master of the foreign sciences for private instruction. Nevertheless, there were other protoscientific institutions of learning in Islam.

---

[73] Ibid.; and Rashdall's account, *Universities* 1: 8–14.
[74] Rashdall, *The Universities of Europe* 3: 142.

Islamic hospitals

To some degree the teaching of medicine escaped the prohibition against the teaching of the natural sciences in quasi-public institutions. This occurred when medical texts were taught and discussed in a majlis, a discussion room attached to a hospital, although even here the practice seems to have been private instruction in the home of the physicians.[75] There are also some rare reports of medical madrasas being established.[76] A closer look at medicine and how it was taught in medieval Middle Eastern culture provides a some-what clearer picture of the nature of scientific education during this period.

It has been pointed out by many students of the history of Arabic medicine that there was a high regard for the social role of the physician during this period, and an equally high regard for the utility of the ancient sciences.[77] Detractors of it, however, were also present.[78] On the whole in *Middle Eastern* culture – Christian, Jewish, and Islamic – the physician occupied a high social position. Prominent physicians served as court physicians, in high positions in governmental administration, or as community leaders.[79] In intellectual life they have been described as "the torchbearers of secular erudition, the professional expounders of philosophy and the sciences, disciples of the Greeks, heirs to a universal tradition, a spiritual brotherhood, which transcended the barriers of religion, language, and countries."[80] Far more than other learned men, physicians made it their

---

[75] Leiser, "Medical Education," p. 54f; Bürgel, "Secular and Religious Features," pp. 48–9; Goitein, *A Mediterranean Society* 2: 248; and cf. N. A. Ziadeh, *Urban Life in Syria under the Early Mamluks* (Westport, Conn.: Greenwood Press, 1970), p. 161.

[76] The original sources for these reports seem to provide few details prior to the fourteenth century of such schools, leaving many scholars doubtful of the existence of actual medical colleges; on the absence of details see Ziadeh, *Urban Life in Syria*, p. 155; and Leiser, "Medical Education," p. 56. In the one case found by Leiser, 'Abd al-Rahmin b. 'Ali, who in 1225 is said to have "endowed his house" in Damascus as a medical madrasa, he was probably a specialist in law (a religious science), "as were some of his successors," making the claim of his founding a medical school rather uncharacteristic; pp. 57–8. But see Michael Dols, "Introduction," *Medieval Islamic Medicine*, pp. 26f, where he agrees that it was the physician-jurist who did such teaching if madrasas for medical education existed.

[77] Goitein, *A Mediterranean Society* 2: 241.

[78] See Franz Rosenthal, "The Defense of Medicine in the Medieval Islamic World," *Bulletin of the History of Medicine* 43 (1969): 519–32. For examples of medical controversies which arose, see Joseph Schacht and Max Meyerhof, *The Medico-Philosophical Controversy Between Ibn Butlan and Ibn Ridwan* (Cairo: Cairo University Faculty of Arts, 1937, Publication no. 13); as well as Ibn Ridwan's discourse on health in Egypt, in Dols, ed., *Medieval Islamic Medicine*.

[79] Goitein, *A Mediterranean Society* 2: 242.

[80] Goitein, "The Medical Profession in the Light of the Cairo Geniza Documents," *Hebrew Union College Annual* 34 (1963), p. 177, as cited in Franz Rosenthal, "The Physician in Medieval Muslim Society," *Bulletin of the History of Medicine* 52, no. 4, 477.

business to be thoroughly versed in philosophy, logic, and the natural sciences. Almost all the major philosophers up until the twelfth century earned their living through the practice of medicine.[81] As a result they were perhaps the major social group working toward the assimilation of Greek philosophy and the natural sciences into Islamic culture and civilization.[82]

There appear to have been three paths to becoming a physician. The first was to have the good fortune to be born into a medical family, whose male heads were eager to pass on their knowledge. A second path was that of self-teaching through the study and memorization of the considerable wealth of medical and other books available in Arabic-Islamic civilization during the twelfth and thirteenth centuries. This was the path taken by the famous physician Ibn Ridwan (998–ca. 1069) because of his poor economic situation as a young man.[83] He developed a strong defense for such a course, believing that the direct study of the ancients, especially Galen, was the best method. The third alternative was to learn from a local physician who held classes either at home or possibly in the majlis of the local hospital. There medical texts were read, discussed, and sometimes disseminated to all participants.[84] Public lectures on medicine seem to have been rare, so that private instruction seems to have been the norm.[85] In general, the memorization of medical texts was at the heart of the system. It was common for students to read aloud from texts which the teacher corrected if necessary. "Students often memorized major works in the common belief that knowing a text by heart must precede its understanding."[86] The physician ʿAbd al-Latif al-Baghdadi (d. 1231) advised his students: "When you read a book make every effort to learn it by heart and master its meaning. Imagine the book to have disappeared and that you can dispense with it, unaffected by its loss."[87] But physicians, like other scholars, also dictated their works to students, and these lecture notes were often made into handbooks for students.[88]

It is with regard to its hospitals, however, that Islamic medical practice advanced beyond other cultures. Although there are earlier models of significant elements found in the Islamic hospital, it appears to have taken on a

---

[81] S. Pines, "Philosophy," in *The Cambridge History of Islam*, vol. 2, edited by P. M. Holt (New York: Cambridge University Press, 1970), p. 784.
[82] Cf. Max Meyerhof, "Science and Medicine," in *The Legacy of Islam*, 1st ed. (Oxford: Oxford University Press, 1931); and F. Rosenthal, *The Classical Heritage in Islam* (Berkeley and Los Angeles: University of California Press, 1975), pp. 183ff.
[83] Leiser, "Medical Education," p. 51f.
[84] Ibid., p. 53; and cf. Bürgel, "Secular and Religious Features."
[85] Goitein, *A Mediterranean Society* 2: 248; Bürgel, "Secular and Religious Features," p. 48.
[86] Dols, *Medieval Islamic Medicine*, p. 30.
[87] As cited in Makdisi, *The Rise of Colleges*, p. 103.
[88] Leiser, "Medical Education," p. 60.

new identity in the Islamic world. The hospital was a place of treatment not only for the physically ill but also for the insane.[89] In addition, the larger and more famous hospitals had special rooms set aside for patients suffering from similar ailments. For example, the three notable hospitals of Baghdad, Damascus, and Cairo – respectively, the ʿAbudi (founded in 987), the Nuri (1154), and the Mansuri (1284) – had rooms equipped for such specialists as the physiologist, the oculist, orthopedist, surgeon, phlebotomist, and cupper.[90] Between the thirteenth and the fifteenth centuries six hospitals were founded in Damascus. The famous Nuri hospital had ten different classes of staff members, including three physicians, a pharmacist, an oculist, a superintendent, and a chief administrator of the waqf.[91] In those hospitals that boasted all of the above features, plus a library and discussion rooms for instruction, all the makings of the modern teaching hospital were present.[92] Yet here again the legal impediments constituted by the law of waqf made the hospitals rigid institutions with precarious futures, because they were founded under the religious law of trusts that permitted no deviation from the original founder's stipulations. Suggestions have also been made that the hospitals were largely for the poor and the incurable.[93] Nevertheless, idealized exhortations of master physicians to medical students enjoined them to "constantly attend the hospitals and sick-houses; pay unremitting attention to the conditions and circumstances of their inmates, in the company [of] the most acute professors of medicine."[94]

As with other areas of learning, medical students went from one master to another.[95] It appears that obtaining a position in a hospital was difficult, apparently achieved by only the most prominent physicians.[96] Unlike in the learning of law, there was a great deal of concern about a common standard of evaluation for determining the competence of physicians. Although some

---

[89] Cf. Bürgel, "Secular and Religious Features," pp. 49, 52; and Dols, "The Origins of the Islamic Hospital: Myth and Reality," *Bulletin of the History of Medicine* 61 (1987): 367–90, at p. 388. Also see the article on the prototype Middle Eastern hospital, "Gondeshapur," *EI²* 2: 1119–20.

[90] Bürgel, "Secular and Religious Features," p. 49. And see Carl Petry, *The Civilian Elite of Cairo,* p. 332, for a description of the al-Mansuri.

[91] Nicola Ziadeh, *Damascus under the Mamluks* (Norman, Okla.: University of Oklahoma Press, 1964), p. 56.

[92] Cf. A. Sayili, "The Emergence of the Proto-Type of Modern Hospital in Medieval Islam," *Studies in the History of Medicine* 4 (1980): 112–18.

[93] Dols, "The Origins of the Islamic Hospital," p. 31.

[94] E. G. Browne, *Arabian Medicine,* reprint (New York: Cambridge University Press, 1962), p. 56, citing al-Majusi; so identified by Martin Levey in *Early Islamic Pharmacology* (Leiden: Brill, 1973), p. 172.

[95] Goitein, *A Mediterranean Society* 2: 248.

[96] Ibid., p. 249f.

historians believe that teaching physicians "used to examine their pupils and to give some sort of certificate to the successful ones,"[97] others argue that "there is no evidence to suggest that systematic examinations were given at the end of the course of study or that diplomas were granted."[98] Here too the system relied upon the ijaza, the authorization to transmit a text. Given the many avenues by which a person could acquire medical knowledge – from relatives, from different physicians, including those specializing in so-called Prophetic medicine,[99] as well as through self-teaching – there seems little likelihood that anything approximating a standard license or certification for medical practice existed. As with learning in general, there was no standard or prescribed method of learning medicine, above all for those who proposed to become autodidacts. The closest thing approximating a curriculum was the "Sixteen Books of Galen."[100] Outstanding physicians expected their students to learn Greek philosophy and logic as well as medicine and, consequently, "dozens of books on medicine were read besides those of the Greek."[101] In such a situation lacking diplomas, degrees, a standard curriculum, and an organized body of professionals to enforce minimum standards, it is not surprising to learn that charlatanism was widespread.[102] The classic statement of the nature of this medical malpractice was written by Razi (864–ca. 925) in the early tenth century:

The tricks of these people are numerous and it would be difficult to mention all of them in a treatise such as this. They are insolvent and believe they can inflict pain on the public for absolutely no reason at all. There are among them those who claim that they can cure epilepsy by making a cross-shaped incision on the middle of the head. Then they produce things they have brought with them which the patient is led to believe were extracted from the incision. Some of them pretend to extract from the nose a venomous snake. They put a toothpick or piece of iron in the nose of the unfortunate patient and rub it until blood begins to flow. Then the quack picks up from there something he had already prepared, like this animal, which he claims to have extracted from the veins of the liver. Some pretend to remove cataracts from the eyes. They scrape the eyes with a piece of iron. Then they place a fine coating on it which they extract as if it were the cataract. Some pretend to suck water out of the ear. They put a tube in it and then put something into the tube from their mouths

[97] J. Christoph Bürgel, "Secular and Religious Features," p. 49.
[98] Dols, *Medieval Islamic Medicine*, p. 32.
[99] See Bürgel, "Secular and Religious Features," pp. 54–61; as well as Browne, *Arabian Medicine*, pp. 12–13.
[100] Leiser, "Medical Education," p. 62; and cf. Dols, *Medieval Islamic Medicine*, pp. 9ff.
[101] Ibid., p. 63.
[102] Ibid., p. 66; Bürgel, "Secular and Religious Features," p. 50.

which they suck out. Some insert worms generated in cheese into the ear or into the roots of the teeth and then extract them.[103]

And this is only half of Razi's catalogue of medical malpractice.

What is sociologically interesting, however, is the fact that many notable physicians were attentive to the problems of quackery and incompetence and sought to combat them through the creation of examination procedures. By the late thirteenth century, an Arabic handbook for examining physicians was produced with the title "The Examination of All Intelligent Physicians."[104] The author of the work, Ibn ʿAbd al-Jabbar al-Sulami (d. 1207), claims to have written the book for the express purpose of allowing anyone to examine a physician, since it was divided into sections, each with its own questions and answers. "The ten sections tested the physician on the pulse, urine, fever and crises, symptoms, medication, therapeutics, the eyes, surgery, bone setting, and the principles of medicine."[105] Unfortunately, this impressive move to create standard examinations for physicians (doubtless inspired by earlier works, including Galen's) could not be institutionalized because there was no recognition in Islamic law of autonomous legal entities – professional guilds, corporations, universities, or even legally autonomous cities – that could enact a set of ordinances to apply uniformly to all practitioners. "The so-called corporations of physicians, surgeons, and oculists are so designated only because chiefs called *raʾises* were appointed by the state to maintain standards of teaching, practice, and discipline in the professions. There is no indication that these functionaries represented guild solidarities."[106] As we have seen, Islamic law had no concept of a corporation, or a collective treatment of actors, as a single legal entity. Similarly, Professor Goitein says, we "look in vain for a term designating a guild in the Muslim handbooks of market supervision which have come to us from the twelfth century. . . . There was no such term because guilds in a strict sense of the word had not yet come into being."[107] This is so because such a system of legal rules creating a professional organization (which would protect the public) would also grant legal privileges to a separate group within the Muslim community – something unauthorized by the Shariʿa and contrary to the spirit of Islam. In the last analysis, the system of medical education rested on the issuing of ijazas to students who learned medical texts (though

---

103 As cited by Gary Leiser, "Medical Education," pp. 66–7. This is Leiser's revised translation of the passage originally printed in Elgood, *Medical History of Persia and the Eastern Caliphate* (Cambridge: Cambridge University Press, 1951), pp. 251–3.

104 Leiser, "Medical Education," p. 70.

105 Ibid.

106 Ira Lapidus, *Muslim Cities of the Later Middle Ages* (Cambridge, Mass.: Harvard University Press, 1967), p. 96.

107 Goitein, *A Mediterranean Society* 2: 82f.

that was not absolutely required) and on the ministrations of the market police – the muhtasibs – who were authorized to issue a license of good conduct to a practicing physician.[108] Goitein suggests that in order "to work independently as a physician one needed a license granted not by a university or scientific corporation, which did not exist, but by a prominent physician, who was authorized by the government, normally it seems by the chief of the market police."[109] But it is not clear what the license could consist of. There are various reports of local rulers insisting on forcing physicians to take examinations, usually administered by a notable physician who had an elevated office, perhaps "chief of medicine," though historians are uncertain about what such a title implied.[110] Professor Bürgel points out that if the muhtasib was authorized to examine and license physicians to practice medicine, then judged by the muhtasib's handbook, "the acknowledged level of medical practice must have been deplorably low."[111] In short, we are left with "the impression that the entire [examination] procedure was arbitrary. And because charlatanism continued to be a major problem, examinations seem to have exerted a very limited influence in enforcing high standards."[112]

The role of the muhtasib should be considered carefully, for though he is often described as the market inspector,[113] he was in effect a roving enforcer of religious and moral standards and was allowed to administer optional (*ta'zir*) punishments that pertained to all cases in which the legal facts did not constitute a *hadd* crime (a crime against God in which punishment was mandatory), but in which there was a strong presumption of wrongdoing. In other words, the market inspector administered "out of court" justice.[114] The office of the muhtasib was a direct result of the Islamic conception of *hisba*, that is, of the duty of every Muslim to promote the good and forbid evil. This collective duty was given to the market inspector and thereby gave

---

[108] Ibid., p. 247; Bürgel, "Secular and Religious Features," p. 50.

[109] Goitein, *A Mediterranean Society* 2: 247.

[110] Leiser, "Medical Education," p. 72, refers to LeClerc's comment that he did not know what the phrase meant.

[111] Bürgel, "Secular and Religious Features," p. 50.

[112] Ibid., p. 71.

[113] See Ziadeh, *Damascus*, pp. 89–91, for a good overview of the muhtasib's functions. The most extensive discussion (in English) is Sami Hamarneh, "Origin and Functions of the Hisbah System in Islam and Its Impact on the Health Professions," *Sudhoff's Archiv für Geschichte der Medizin und der Naturwissenschaften* 48 (1964): 157–73. Hamarneh uses most of the same sources as Max Meyerhof, "La surveillance des professions médicales et para-médicales chez les Arabs," *Bulletin de l'Institut d'Egypt*, 26 (1944): 119–34; reprinted in Meyerhof, *Studies in Medieval Arabic Medicine*, ed. Penelope Johnstone (London: Variorum Reprints, 1984), chap. 11.

[114] See Coulson, *Conflicts and Tensions in Islamic Jurisprudence* (Chicago: University of Chicago Press, 1969), p. 84; and Coulson, "The State and the Individual in Islamic Law," *International and Comparative Law Quarterly* 6 (1957): 49–60.

him a great deal of freedom to admonish wrongdoers, since a great many acts and activities unspecified in the Quran were nevertheless thought to be antithetical to the spirit of Islam and could, therefore, be disapproved of or punished by the market inspector.[115] Thus, if the chief of medicine in a city were elevated to this post, he was a scientific professional functioning as a moral and religious official, one tied to the government but without any grounding in a strictly professional association. In other words, since there was no professional association of physicians, there could be no collective, nonreligious, and scientific consensus about such matters. The legal constraints of Islamic law prevented physicians from becoming a professionally autonomous group with its own legal rules and regulations that would set standards for medical practice as a whole. Consequently, the clear separation of medical and scientific matters from religious matters could not be achieved because law in Islam means the Command of God: it means "what one must do to pass the reckoning on the day of judgment." The result was a sort of regulation that was fundamentally religious.[116]

Nevertheless, from our point of view, it is impressive that the Arab physicians – Muslims, Christians, and Jews – recognized the problem of creating standards of competence and moved toward creating a set of objective and impersonal standards in the form of medical handbooks by which anybody – physician or layman – might examine an individual for his medical competence. Even more impressive were the heights of scientific sophistication achieved by Arabic physicians in the thirteenth century, especially in Damascus and Cairo. Perhaps the best illustration of this is found in the work of Ibn al-Quff (1233–86) and the Cairene, Ibn al-Nafis (1210–88). Both of these men had worked at the hospitals in Damascus and Cairo, Ibn al-Nafis having been a senior apprentice to al-Quff at the Nuri hospital in Damascus. Ibn al-Quff was apparently the first Arab physician to call for a standard set of weights and measures in medicine and pharmacy. He is also known to have excelled in anatomical descriptions of the body, particularly of the heart and the blood system. He "described accurately and with much care what we now call the capillary system, which connects arteries with veins for the completion of blood circulation. The phenomenon was fully explained four hundred years later in the monumental work of the Italian

---

[115] On *Hisba*, see *EI²* 3: 485–9; and Dols, *Medieval Islamic Medicine*, p. 33 and n170.

[116] These considerations make it unlikely that either China or the medical practice of the Islamic Middle East was the source of "medical examinations" in the West, as claimed by Joseph Needham: "China and the Origin of Qualifying Examinations in Medicine," in *Clerks and Craftsmen in China and the West* (New York: Cambridge University Press, 1970), pp. 379–95.

anatomist, Marcello Malpighi (1628–1694), with the aid of the micro-scope."[117] In an anatomical description of the heart, he noted:

The heart has four outlets of which two are on the right side. The one branching from the Vena Cava, carries the blood. In the orifice of this blood vessel – which is thicker than any of the other openings – there are three valves which close from the outside in. The second blood vessel is connected with the arterial vein and through it nourishments from the lungs come. I, heretofore, know of no one ever describing these valves.[118]

It was four centuries later before Europeans fully explained these structures and functions. Ibn al-Quff also described in great detail the stages of growth of embryos. After providing a general characterization of a human fetus for the first six to seven days and for thirteen to sixteen days, he says that it

gradually is transformed into a clot, and in 28 to 30 days into a small "chunk of meat." In 38 to 40 days the head appears separate from the shoulders and limbs; the brain and heart are formed before other organs and are followed by the liver. The fetus takes its food from the mother in order to grow and to replenish what it discards or loses. The author spoke of three membranes covering and protecting the fetus, of which the first connects arteries and veins with those in the mother's womb through the umbilical cord. The veins pass food for the nourishment of the fetus, which arteries transmit air. . . . By the end of seven months all organs are com-plete.[119]

Ibn al-Nafis, Ibn al-Quff's apprentice, is noteworthy for his discovery and description of the pulmonary system, a feat repeated by William Harvey in the seventeenth century.

---

[117] Sami Hamarneh, "Arabic Medicine and Its Impact on Teaching and Practice of the Healing Arts in the West," *Oriente e Occidente* 13 (1971): 395–426, at p. 420. And see Hanarneh, "Thirteenth-Century Physician Interprets Connection Between Arteries and Veins," *Sud-hoff's Archiv für Geschichte der Medizin und der Naturwissenschaften* 46 (1962): 17–26.

[118] As cited in "Arabic Medicine and Its Impact," p. 420.

[119] Sami Hamarneh, "The Physician and the Health Professions in Medieval Islam," *Bulletin of the New York Academy of Sciences* 47 (1971): 1088–1110, at pp. 1102–3. On the surface this anatomical description suggests a deep knowledge of internal organs that seems to require the performance of dissection, something that experts in Arabic medicine believe was not generally practiced (see below). Careful thought about this example might suggest that a great deal can be known about fetuses through all of the misfortunes of miscarriage and abortion. Miscarried and aborted fetuses as well as umbilical cords do not require dissec-tions to observe. The fetus at that period is also virtually transparent, thereby revealing its internal organs. Likewise, the afterbirth ("three membranes protecting the fetus") is also part of births and miscarriages. Nor should we discount knowledge of fetuses obtained in other cultures (Greek, Indian, or Chinese), to which Ibn al-Quff may have had access. In addition, Christian and Jewish physicians may have had fewer inhibitions regarding Islamic law and gained additional knowledge to which Ibn Quff may have been privy. In any event, this is a question for specialists in the history of Arabic medicine to resolve.

Despite all these medical advances – and there were many others, including some in ophthalmology that were not superseded for hundreds of years – nothing was to come of these efforts, since medicine and the health professions were to become, as some say, a casualty of the orthodox reaction which "ousted philosophy from most of the countries of Islam," so that "in the course of time medicine too, fell into disrepute, until it reached total eclipse which was not overcome until modern times."[120] Despite its great sophistication, above all, its superiority in comparison to the West prior to the twelfth and thirteenth centuries, Arabic-Islamic medicine was not able to achieve the breakthrough that would have allowed physicians to pursue their calling and practice their art in a progressive spirit, leading to new discoveries and medical treatments.[121] Instead, E. G. Browne observed, when he met with a council of Iranian physicians in Tehran in 1887, not one had any knowledge of modern medicine.[122]

The problems and impediments that blocked the breakthrough to the modern practice of medicine can be summed up in the following manner. The first impediment was the fact that medical education was largely confined to the Islamic hospitals and not taught in the Islamic schools of higher education, namely, the madrasas. Like the madrasas, the hospitals were religious endowments and therefore constrained by the religious law. This meant that they were not autonomous legal entities equivalent to corporations, but were founded under religious auspices and the pledge not to undertake anything inimical to the spirit and letter of Islamic religious law. As religious entities, they lacked the potential for change and redirection that the corporate structures of the universities of the West made possible.

Second, experts on the history of Arabic medicine assert that the dissection of human bodies was strictly forbidden by religious law.[123] As Professor Bürgel put it, "Our sources do not contain the slightest indication of anybody having dared to trespass this custom. Yuhana Ibn Masawaih, a great physician of the earlier period (d. 857) who was a Christian and a free-thinking rationalist in demeanor, dissected apes." Yet Bürgel knew of "no parallel to this exception" in later Islamic medical prac-

---

120 Goitein, *A Mediterranean Society* 2: 241.
121 There appears to be a consensus among students of Arabic medicine that after the twelfth century, the general standards of medicine (despite some outstanding physicians in the thirteenth century) steadily declined. Bürgel calls it "the most deplorable decay imaginable," in "Secular and Religious Features," p. 53. Cf. Hamarneh, "The Physician," p. 1097; and Levey, *Early Islamic Pharmacology*, p. 170, where he suggest that Islamic pharmacology started to decline in quality in the twelfth century.
122 Browne, *Arabian Medicine*, p. 93.
123 Bürgel, "Secular and Religious Features," p. 54.

tice.[124] At the same time, the Hippocratic oath prohibited the performance of abortions.

Third, among the more pious Islamic believers, the idea was widespread that illness was connected with evildoing on the part of the sick person and was a punishment visited by God. These ideas were further related to the orthodox views against natural causality. The belief in the omnipotence of God denies natural causes, so "that no such thing as natural causality existed. The apparent relation between cause and effect was a delusion of the senses, and all actions and phenomena were immediately caused by the prime cause which was God."[125] Since these ideas were espoused by the religiously devout and orthodox, medical practitioners schooled in Greek philosophy represented a strong challenge to the orthodox point of view rooted in the religious law. In addition, since physicians could not join together in an autonomous self-regulating guild or corporation, their scientific standards had no way of becoming firmly established in the institutional structure of Islamic civilization. They remained under the supervision of the office of the market inspector, an inherently religious office.

The observatory

The second protoscientific institution in Islam is the observatory. For a time, the astronomical school of Marâgha in western Iran actually thrived and made singular contributions to the development of the modern science of astronomy. It did this by assembling a staff of astronomers and instrument makers who constructed an observatory of unprecedented size and scope of activity. For these reasons a consideration of this project is in order, for though the project failed to survive, it did serve as a model that shaped future conceptions of the observatory.

Founded in 1259 just south of Tabriz, the Marâgha observatory represented new heights in the science of astronomy within Islam and the world. Perhaps because it was founded under the direction of the religious scholar and astronomer Nasir al-Din al-Tusi (d. 1274), Marâgha appears to have been founded under the umbrella of the law of waqf, that is, the religious law of endowments. This is remarkable, for though astronomy was implicitly acknowledged to be a "handmaiden of religion,"[126] it was associated with astrology, and the latter's claim to predict the future ran directly counter to the teachings of Islam. That is, from a strictly religious point of view,

---

[124] Ibid.
[125] Ibid., p. 55.
[126] Sayili, *The Observatory in Islam* (Ankara: Turkish Historical Society Series 7, no. 38, 1960), pp. 27, 127.

only God knows the future, and those who claim such knowledge are usurpers of God's unique and inimitable qualities. For that reason, several other observatories were destroyed because of their alleged association with astrology.[127] Nevertheless, the Marâgha observatory was built with great precision over several years and for the purpose of making exacting astronomical observations, which, in the nature of the case (to correct deficiencies in existing zij tables),[128] could not be completed in less than thirty years. Not only was the observatory built on a grander scale than any before, it was staffed with astronomers, instrument makers, and mathematicians and contained a huge library reported to number four hundred thousand volumes.[129] It also possessed several unique instruments, including terrestrial and celestial spheres, a large armillary sphere, and climatic maps of the earth.

Still another innovation of the Marâgha observatory was that it gave instruction in the natural sciences. It is said that almost a hundred of al-Tusi's students were instructed in astronomy and the natural sciences there, and that the funds of the local ruler were used for this purpose.[130] Such an arrangement suggests that the teaching of the natural sciences was officially recognized and supported by the local rulers, if not the local religious scholars, the 'ulama'.

The fact that the observatory was founded under the law of waqf would suggest that the institution had suitable legal protection and could be expected to enjoy a long life, indeed, perpetual existence. But such was not the case. By 1304–5 the observatory had ceased to function; it had enjoyed a lifetime of barely forty-five years, or, according to Aydin Sayili's most liberal estimate, fifty-five to sixty years.[131] In the middle of the fourteenth century, a visitor to the site had only ruins to view. In short, the Marâgha observatory, a brilliant but short-lived experiment in scientific research, soon passed away.[132] Yet it was the astronomers associated with this observa-

---

[127] Ibid., pp. 171, 202, 292, 314.

[128] Zij tables are handbooks of astronomical observations containing matrix listing of the periodic positions of the planets as seen from various points on the earth, some of which contain ten thousand entries. The Arabs constructed many such tables. See E. S. Kennedy, "A Survey of Islamic Astronomical Tables," *Transactions of the American Philosophical Society*, new series, 46, pt. 2 (1956): 123–77; and now, David King, "On the Astronomical Tables of the Islamic Middle Ages," *Colloquia Copernicana* 3 (1975): 36–56; and also "The Astronomy of the Mamluks," *Isis* 74, no. 274 (1983): 531–55, at p. 541.

[129] Sayili, *The Observatory*, p. 194.

[130] Ibid., p. 219.

[131] Ibid., p. 213.

[132] There were many other observatories founded in the Islamic world, both before and after the Marâgha observatory. But none rivaled the achievements of the Marâgha observatory, though the observations recorded at Samarqand in the fifteenth century probably were more accurate. See Sayili for these other observatories.

tory, al-Tusi, al-'Urdi, and al-Shirazi, who contributed solid new additions to the planetary models later perfected by Ibn al-Shatir, the timekeeper of Damascus. It was these models that were duplicated by Copernicus (see Figures 2–5, Chapter 2).

Given the virtues of the religious trust (waqf) as an institutional form in Islam, it is not surprising that the Marâgha observatory came to an untimely end, for every step necessary to advance the cause of science was in fact an "heretical innovation" (bid'a)[133] from the point of view of the religious scholars (the 'ulama'). In other words, in Islam a religious trust was the only legal instrument and organizational form available for such a purpose but it could be created only on condition that nothing contrary to the letter and spirit of the law be undertaken. It was under the law of waqf that many different religious and charitable acts were undertaken in medieval Islam. These included the building of bridges, hostels, orphanages, and so on.[134] The teaching of philosophy and the natural sciences, however, because they were considered foreign and hence antithetical to the spirit of Islam, could not be officially protected under the law of waqf. The fact that the Mongol rulers of that period were heterodox believers probably had much to do with why the observatory was constructed in the first place. According to Aydin Sayili, the record of the observatory's founding "clearly shows that the main purpose for the foundation of the Marâgha observatory was an astrological one."[135] This suggests that the founders were indeed daring in this undertaking and that it was only a matter of time before a reactionary opposition would be mounted by the religious scholars.[136]

It is clear that a number of legal, cultural, and organizational reforms would have had to materialize in order for the observatory to evolve into a modern scientific institution endowed with legally protected intellectual autonomy. I have already noted that according to the law of religious endowments, only those activities that were dedicated to religious edification could be supported. Strictly speaking, the teaching of the foreign or ancient sciences was not part of that mission in the eyes of the orthodox, even if, under the protection of the Mongol ruler Hulagu, the director of the observatory, al-Tusi, managed to have it chartered under the law of waqf. Sayili's account

---

[133] Bernard Lewis, "Some Observations on the Significance of Heresy in the History of Islam," *Studia Islamica* 1 (1953): 52ff.

[134] Makdisi, *The Rise of Colleges*, pp. 38ff.

[135] Sayili, *The Observatory*, p. 202.

[136] The need for royal patronage and protection by natural scientists, and the consequent decline of scientific activity when it was withdrawn, is emphasized by Willy Hartner, "Quand et comment s'est arrêté l'essor de la culture scientifique dans l'Islam?" pp. 319–38 in *Classicisme et déclin culturel dans l'histoire de l'Islam*, ed. R. Brunschvig and G. E. von Grunebaum, pp. 332–3.

of the founding of the observatory under the law of waqf suggests that this was, in fact, an innovative use of that legal instrument.

If the founders and trustees of the Marâgha observatory had acted as if the trust were a corporation in the Western sense of a legally autonomous entity capable of charting its own course, the observatory would clearly have been an extralegal entity vis-à-vis the Shari'a. A close scrutiny of its activities would have resulted in their being labeled criminal, since from a strictly Islamic point of view, those actions that violate Islamic principles are both offensive to God and criminal.[137] In the end, it was only by a radical stretch of legal conceptions that the activities planned for the Marâgha observatory could be encompassed under the law of religious endowments. On the other hand, there was no other legal status for such collective endeavors, and the idea of a separate, self-governing, and autonomous institution with its own rules and regulations, independent of the religious law, was simply unthinkable.

The observatory operated in an environment in which there was widespread suspicion toward and condemnation of the open pursuit of the natural sciences. Even if, as apparently happened, the law of waqf was used as a legal fiction to support un-Islamic activities for a time, those activities could not be disguised for long. As we saw in Chapter 2, the religiously orthodox and especially the partisans of the traditions of the Prophet were often violently opposed to the study of logic, philosophy, and the natural sciences. Consequently, the open pursuit of the natural sciences could be maintained only for as long as the local ruler could keep the 'ulama', the local religiously minded scholars (both those associated with colleges and those not so associated), placated. When strong-willed and dynamic leaders arose from among such pious folk, they could become effective leaders in a community.

Thus, among the early Mamluks in greater Syria of the thirteenth and fourteenth centuries (during the time Ibn al-Shatir was working in Damascus), there were forceful religious personalities who, by their examples of piety and force of character, intimidated rulers and thereby enforced religious scruples. For example, there was a certain religious master of the hadith sciences, Ibn as-Salah, Shaykh of Dar al-Hadith al-Ashraifiyya (d. 1243), about whom it was said that he "never permitted anybody in Damascus to study 'logic and philosophy.'" Likewise, the same source says that "kings obeyed him."[138] While the powers of such a man are necessarily

---

[137] For the Islamic distinctions between criminal and other kinds of offenses, see Matthew Lippman, Sean McConville, and Mordachai Yerushalmi, *Islamic Criminal Law and Procedure* (New York: Praeger, 1988), pp. 37ff; as well as Schacht, *Introduction to Islamic Law*, pp. 76ff, 175–87.

[138] Ziadeh, *Damascus*, p. 249, n42.

limited, the ʿulamaʾas a group were indeed a significant force to be reckoned with. "Even without holding office, many of the ʿulamaʾ, especially men of strong personality and character, could force their point of view on the rulers, because they had the support of the public, who respected them for their learning, sincerity, and enthusiasm."[139]

The life of the great Islamic philosopher and traditionalist Ibn Taymiyya (d. 1328), another contemporary of Ibn al-Shatir in Damascus, provides an additional example of such a leader and his influence on the community. Ibn Taymiyya was publicly active in the name of the Faith and went with his friends to local shops where they emptied out and broke wine bottles. At the same time they administered the taʿzir punishments, the optimal admonitions due all violators of the spirit of Islam. He also visited the military troops preparing to repulse the Mongol invaders. The Mongols too were Muslims and, theoretically, should not be harmed by other Muslims. But Taymiyya "spoke to the men about unity and victory, saw [to it] that the amirs and other people swore to be faithful, and took trouble to explain the legality of fighting the Mongols although they were Muslims like themselves."[140] In short, the ʿulamaʾ who opposed the study of the foreign sciences were a force to be reckoned with. Given the absence of any institutional sources of autonomy, above all, legally sanctioned autonomy, observatories and other places dedicated to the study of the natural sciences could become objects of violent attack and had no legitimate defense insofar as the religious law was concerned.[141]

In addition, if a research program designed to explain the disparities between observation and theory then current in astronomy were to be successful, the Muslim astronomers would have to adopt the heliocentric system. As in the West, it is fair to assume that the religiously orthodox would have had little tolerance for such a scientific and religious revolution. While the orthodox had not embraced the authority of Aristotle in support of the geocentric system, it was taken to be obvious that the earth was the center of the universe, and the moon's revolutions around the earth were obvious enough to have been mentioned in the Quran. Recent study of traditional Islamic cosmology indicates that as late as the nineteenth century, traditionalists took pains to assert that the earth did not move and that any form of heliocentrism was false as shown by the traditional Islamic teachings. Thus

[139] Ibid., p. 184.
[140] Ibid., p. 185.
[141] This pattern of violent reproach toward students of the natural sciences continued long into modern times. Evidence of the persecution and killing of those learned in the natural sciences in the eighteenth century is reported in H. A. R. Gibb and Harold Bowen, *Islamic Society and the West*, reprint (London: Oxford University Press, 1965), 1/2: 139–64.

Figure 6. In *The Tales of the Prophets of al-Kisa'i* (ca. 1200, ed. and trans. William Thackston [Boston: Twyane Publishers, 1978]), the earth is depicted as held by the hand of an angel, who stands on a rock, in turn supported by a bull, who stands on a fish. This mythological story was rendered into artistic form in the seventeenth century by the Persian painter Abul Hassan (ca. 1620). It is further embellished by placing a Mughul ruler, Jahangir, son of Akbar the Great, on top of the world slaying an enemy. (Reproduced by kind permission of the Trustees of the Chester Beatty Library, Dublin.)

Anton Heinen reports on a late – probably late nineteenth-century – alteration of a fifteenth-century manuscript on Islamic cosmology in which the modern commentator, anxious about the implications of a certain passage, inserted the line "They hold it [the earth] firm, and it is not moved as a ship on the ocean"[142] (Figure 6). When the manuscript in question (as-Suyuti's book on traditional Islamic cosmology) was written in the late fifteenth century, Copernicanism was unknown among the Arabs. In addition, as-Suyuti (1445–1505) even then alluded to the fact that the foreign sciences were considered forbidden.[143]

Furthermore, the fifteenth and sixteenth centuries witnessed a revitalization of traditional Islamic cosmological doctrine, which is attested by the fact that as-Suyuti's book was widely emulated and copied throughout the Middle East.[144] This book, like its imitators, was based solely on traditional Islamic sources and Scriptures. It contained no references to the work of the Marâgha school nor any work falling within the Greek-inspired astronomical tradition of the *Almagest*.

In the face of these intellectual, legal, and practical difficulties, the wisdom of the Islamic cultural elite led them to create the office of timekeeper, the muwaqqit, attached to the mosque.[145] That is, the custom of creating in local mosques the office of timekeeper, occupied by a competent astronomer, became widespread after this time. There is at present very little in Western languages about the office and role of the muwaqqit. According to David King, the existence of the muwaqqit is not reflected in documents prior to the mid-thirteenth century when the Mamluks,[146] the former slaves in Egypt, emerged to rule that country and greater Syria.[147] Given the emergence of the office of muwaqqit, we can only surmise that the Islamic religious scholars, perceiving a religious need (that is, to establish the correct times of day) and the intellectual threat of a foreign science becoming indispensable and autonomous within the life of the Muslim community, co-

---

[142] Anton M. Heinen, *Islamic Cosmology: A Study of As-Suyuti's "Al-Hay'a as-saniya fi l-hay'a as-sunniya,"* with critical ed., trans., and commentary (Beirut: Franz Steiner Verlag, 1982), p. 15.

[143] Ibid., p. 13.

[144] Ibid., pp. 7ff.

[145] Sayili, *The Observatory,* mentions the office of muwaqqit on many occasions in his study, suggesting that it was "a firmly established institution" (p. 127, but also pp. 241–3, 315, and 361–2); however, no overview of the institution, its founding, duties, or limits, is provided.

[146] David King, "The Astronomy of the Mamluks," 534f. But we must await the appearance of David King's study, "On the Role of the Muwaqqit in Medieval Islamic Society," forthcoming in the *Proceedings of the Second International Symposium on the History of Arabic Science* (Aleppo).

[147] A good introduction to the social, political, and administrative life of the Mamluks is Ziadeh, *Urban Life in Syria.*

opted the threat by creating the religious office of the official timekeeper – located in the mosque, not the madrasa or the observatory – and thereby forestalled an internal crisis. This is remarkably parallel to the way the problem of licensing and regulating physicians was handled, whereby the task was given to the muhtasib, an essentially religious and moral official, not to the community of professionals composed of physicians.[148] In effect, this is probably another illustration of A. I. Sabra's concept of the "naturalization" or "Islamicizing" of the natural sciences in Islam by assimilating them to the intellectual outlook of the Islamic worldview.[149] As Professor Sabra put it, "The final results of all this is an instrumentalist and religiously oriented view of all secular and permitted knowledge. This is the view that accompanied the limited admission of logic and mathematics and medicine into the madrasa and the conditional admission of the astronomer into the mosque."[150] What is more, Sabra argues, what we have here "is not a general utilitarian interpretation of science, but a special view which confines scientific research to very narrow, and essentially unprogressive areas."[151]

Given these problems of adapting to and accommodating the intellectual and institutional requirements of modern science in the Arabic-Islamic world, it is time to consider the nature of scientific education and its place in European society and social institutions.

Western universities and the place of science

During the last two decades, historians of medieval science in Europe have staked a firm claim "that the medieval university provided to all an education that was essentially based on science."[152] This is a far different assessment than that suggested by Joseph Ben-David's allusions to "the peripherality of science in the medieval university."[153] Indeed, Edward Grant asserts that the medieval university "laid far greater emphasis on science than does its mod-

---

148 Dols, *Medieval Islamic Medicine*, p. 34 n172, alludes to the view of Felix Klein-Frank that "such inspection by the muhtasib meant social and professional disdain of the medical professions."
149 A. I. Sabra, "The Appropriation and Subsequent Naturalization of Greek Science in Medieval Islam: A Preliminary Assessment," *History of Science* 25 (1987): 236–8.
150 Ibid., p. 240.
151 Ibid., p. 241.
152 Edward Grant, "Science and the Medieval University," in *Rebirth, Reform, and Resilience: Universities in Transition, 1300–1700*, pp. 68–102, at p. 68.
153 *The Scientist's Role in Society*, pp. 50ff; so also in all his earlier papers, especially "The Scientific Role: The Conditions of Its Establishment in Europe," *Minerva* 4, no. 1 (1965): 15–54. It seems likely that this was one of the major reasons why Thomas Kuhn, in his original review of the book, referred to the first half of it as a disaster. See Kuhn, "The Growth of Science: Reflections on Ben-David's 'Scientist's Role,'" *Minerva* 10, no. 1 (1972): 167.

ern counterpart and descendent."[154] In other words, European universities accomplished what the Islamic colleges could not – that is, the establishment of a curriculum centered on the teaching of science – and they placed an even greater emphasis on scientific subjects than contemporary universities.

In the context of our present study it is essential to recognize that the natural sciences became "the foundation and core of a medieval university education" precisely because of the unprecedented translation activity of the twelfth and thirteenth centuries.[155] Through this process the very best of the accumulated scientific wisdom of the Greek and Arabic traditions was brought into Europe. This monumental translation feat,[156] located in Spain, Sicily, and northern Italy, made available to the West, in scarcely a hundred years, the corpus of Aristotle and his commentators, along with other seminal Greek and Arabic works. Once incorporated into the university curriculum, these works dominated scientific thought for the next four hundred years.[157] Among these newly translated writings we find Euclid's *Elements*, Ptolemy's *Almagest*, Ibn al-Haytham's *Optics*, the algebra of al-Khwarizmi, the medical writings of Galen, those of Hippocrates, and of course the medical *Canon* of Ibn Sina (Avicenna). At the same time, unlike the situation in Arabic-Islamic civilization, these works were directly incorporated into the curriculum of the universities, not taught secretly, surreptitiously, or only in the privacy of an individual's home. Nor did they have to be taught in the guise of religious traditions derived from the Scriptures. As Professor Grant put it, this development, occasioned by the new translations, "laid the true foundations for the continuous development of science to the present."[158] Moreover, by 1200, "two of the three greatest universities of Christendom, Oxford and Paris, were already in existence with curricula based on the new science."[159]

As noted earlier, the natural science works of Aristotle became a major focus of the curriculum of the university, forming one division within the three philosophies – natural philosophy, moral philosophy, and metaphysics. These books included the *Physics, On the Heavens, On Generation and Corruption, On the Soul, Meteorology, The Small Works on Natural Things*

---

[154] Grant, "Science and the Medieval University," p. 70.
[155] Ibid., p. 68.
[156] This story has been told in many places since Charles Haskins, *Studies in the History of Medieval Science* (Cambridge, Mass.: Harvard University Press, 1928); now see Edward Grant ed., *A Source Book of Medieval Science* (Cambridge, Mass.: Harvard University Press, 1974) (hereafter cited as *A Source Book*); and David C. Lindberg, "The Transmission of Greek and Arabic Learning to the West," in *Science in the Middle Ages*, pp. 52–90.
[157] Grant, "Science and the Medieval University," p. 69.
[158] Ibid.
[159] Ibid., p. 70.

(*Parva Nauralia*), as well as biological works such as *The History of Animals, The Parts of Animals,* and *The Generation of Animals.* It is with these books, Professor Grant observes, that we find "the treatises that formed the comprehensive foundation for the medieval conception of the physical world and its operation."[160] Since all of these works were required parts of the medieval university curriculum, it is possible to speak of a shared learning experience that was "essentially scientific."[161]

These were studied along with the seven liberal arts, that is, the *trivium* (composed of grammar, rhetoric, and logic) and the *quadrivium* (arithmetic, geometry, astronomy, and music).[162] It should be noted, however, that the old seven liberal arts had undergone a radical transformation in the twelfth and thirteenth centuries. The result was that the quadrivium in large part had been absorbed into natural philosophy, as well as enlarged in scope.[163] As Gordon Leff put it, by 1215 "dialectic and philosophy had virtually replaced the other liberal arts."[164]

While in the thirteenth and fourteenth centuries the sequence of study had not evolved into a fixity, there was a preferred sequence. The ideal was to complete the trivium and quadrivium before beginning the study of the natural sciences, while moral philosophy and the natural sciences were to be mastered before beginning metaphysics.[165] One can hardly overlook the fact that the quadrivium, above all with its expanded content after the great translation movement, under any other name would constitute the exact sciences. While not all universities put the same stress on the teaching of the exact sciences, they were clearly part of the curriculum and came to reflect the scientific importance of Euclid's *Elements* and at least the early books of Ptolemy's *Almagest*. This was the most important work in astronomy up until Copernicus and was translated twice in the twelfth century, once in about 1160 and again in 1175 by Gerard of Cremona.[166] Moreover, the Europeans were quick to introduce new books into the curriculum that made for briefer summaries and easier expositions of the principles of natural science for the use of students. Perhaps the most notable example of this

---

[160] Ibid., p. 78.

[161] Ibid.

[162] For the early evolution of these studies from Greece, see Paul Abelson, *The Seven Liberal Arts,* reprint (New York: Russell and Russell, 1965); as well as more recent studies. For Rashdall's account of the curriculum at Oxford, see *The Universities of Europe* 3: 153–60.

[163] Among others see Grant, "Science and the Medieval University," pp. 93–4 n9; Nancy Siraisi, *Arts and Sciences at Padua,* pp. 109ff (re *scientia naturalis* and *metaphysica*).

[164] Leff, *Paris and Oxford,* p. 119.

[165] James Weisheipl, "The Curriculum of the Faculty of Arts at Oxford," pp. 143–85, p. 148.

[166] Olaf Pedersen, "Astronomy," in *Science in the Middle Ages,* ed. David C. Lindberg (Chicago: University of Chicago Press, 1978), p. 313.

is a work by John of Sacrobosco (ca. 1274–56), *On the Sphere*. It contained both the astronomical and cosmological thought of the medievals and was based on the writings of Aristotle and Ptolemy.[167] It is of some importance that this popular book of Sacrobosco refers to the universe as "the machine of the world," which rather dramatically suggests the orderliness and apparent determinism of the world order. It is a cosmological view that might have been privately entertained by some Arab astronomers, but it was a view that ran directly counter to the occasionalist views of the legists and theologians who dominated orthodox thinking and the Islamic colleges. It should also be noted that, while astrological thinking was often popular in the ruling courts of Islam, the claim of astrologers to be able to predict the future led to reprisals as well as the destruction of several observatories. In contrast, the medieval European universities often taught astrology, and this discipline served as "a sort of provisional metaphysics" that was later modified and displaced in the thirteenth century when Aristotle's works, especially the *Metaphysics* and the *Physics*, were fully adopted.[168]

Alongside the new Aristotle in the curriculum, there also emerged in the twelfth and thirteenth centuries a body of scientific knowledge that has been called the *corpus astronomicus*.[169] This body of astronomical materials included standard texts, scientific instruments, and collections of data, that is, tables of astronomical observations, which allowed the determination of local time as well as the prediction of astronomical events such as eclipses and conjunctions of heavenly bodies. It was this body of materials that laid the foundation for mathematical astronomy and set the scientific puzzles that Copernicus was to grapple with three hundred years later.

In brief, there is every indication that in the twelfth and thirteenth centuries Europeans enthusiastically embraced the fundamental works on science that came to them from Greek and Arabic sources. Most important, however, they institutionalized the study of these materials by making them the central core of the university curriculum. With a system of examinations in place and the main body of a new scientific curriculum spelled out, the West took a decisive (and probably irreversible) step toward the inculcation of a scientific worldview that extolled the powers of reason and painted the universe – human, animal, inanimate – as a rationally ordered system. In-

---

[167] See Grant, *A Source Book*, pp. 442ff.

[168] Richard Lemay, *Abu Ma'shar and Latin Aristotelianism in the Twelfth Century* (Beirut: American University of Beirut, Oriental Series no. 38, 1962), p. 8.

[169] Pedersen, "Astronomy," pp. 315ff; and John North, "The Medieval Background to Copernicus," in *Vistas in Astronomy*, vol. 17, ed. Arthur Beer and K. Aa. Strand (New York: Pergamon Press, 1975), pp. 3–16; especially 8ff.

deed, it was described as a world machine that could be known with precision.[170]

On the other hand, it should come as no surprise that this conceptually rich and methodologically powerful body of secular learning posed a threat to Christian teachings and that it aroused considerable opposition in the thirteenth century when it was introduced into the universities.[171] From the point of view of the clash between philosophy and religion, we see the ubiquitous conflict of sacred and secular worldviews. The same tensions that had arisen earlier in Arabic-Islamic civilization upon the arrival of Greek philosophy were now manifest in the Christian West. In the writings of Aristotle there were philosophical ideas that stood fundamentally opposed to Christianity. From an Aristotelian point of view, the world was eternal – it had always existed and would always exist. As we saw earlier, this ran counter to the Islamic view that the world was created by God. Since the Muslims had always argued that the Quran was the final revelation of God, thereby perfecting the Judeo-Christian revelations, they also accepted a modified version of the six-day creation story of the Old Testament.[172] Once Aristotle and his teachings had been translated into Arabic and Arabs began to find out what his teachings were, a powerful encounter occurred. The upshot was the creation of Islamic philosophy (falsafa). This movement in turn spawned Muslim theology (kalam), that is, the science of religion or disputation that strove to establish the principles of religion using the techniques of philosophy. The earliest and most significant of these theological movements was the Muʿtazilites, the so-called Islamic rationalists.[173] After early successes, the Muʿtazilites were defeated so that even theology was frequently condemned by the orthodox. When the madrasa movement was launched in the eleventh century, both philosophy and theology as well as the natural sciences were excluded. In short, to defend itself against the incursion of an alien philosophy containing fundamentally incompatible

---

[170] Benjamin Nelson stressed the ubiquity and importance of this idea in the Middle Ages in several of his papers; see "Certitude and the books of Scripture, Nature, and Conscience" in *On the Roads to Modernity*, pp. 121–52; as well as pp. 154, 160, 162, 190, and 197 n6.

[171] Grant, "Science and Theology in the Middle Ages," in *God and Nature: Historical Essays on the Encounter Between Christianity and Science*, ed. David C. Lindberg and Ronald L. Numbers (Berkeley and Los Angeles: University of California Press, 1986), pp. 52–3.

[172] As we noted earlier, Sura 7:54 as well as Sura 32:4 of the Quran refer to God's creation of the world in six days. More difficult passages such as Sura 41:9–12 (containing references to eight days) are also confined within the six-day chronology by Islamic exegetes.

[173] See F. E. Peters, *Aristotle and the Arabs* (New York: New York University Press, 1968), especially chap. 6, as well as Peters, *Allah's Commonwealth* (New York: Simon and Schuster, 1973), pp. 428–69; Watt, *The Formative Period of Islamic Thought* (Edinburgh: Edinburgh University Press, 1973), chaps. 7 and 8; and my account above in Chap. 3, the section entitled "Reason and Conscience."

metaphysical presuppositions, Islamic civilization banned Greek philosophy and science from its institutions of higher learning.

In the West the situation was different. Because most of Aristotle's works had been lost in the West up until the translation activities of the twelfth and thirteenth centuries, Christianity had been spared this frontal assault of Greek philosophy. On the other hand, it might be said that Christianity at its birth was aided by the handmaiden of Greek philosophy, since so many of the early Christian converts had written in and spoke Greek. In effect, when Christianity emerged into the world, it did so in an essentially Hellenistic environment that powerfully shaped its language, metaphysics, and metaphors.[174] Until the time of the Reformation, this powerful Greek influence on Christian dogma and theology was scarcely noticed. Who was to notice that the very concept of catechism or the doctrine of one substance (*Homoousios*) was of Greek and not Semitic provenance? Nevertheless, Christianity did keep its distance from certain deeply rooted metaphysical conceptions which it challenged only later.

With the arrival of the new Aristotle in the twelfth and thirteenth centuries, however, the confrontation between Aristotle and Christian thought could no longer be forestalled. It should be remembered, moreover, that in the eleventh and twelfth centuries, before the arrival of the new Aristotle, educated Europeans at the major centers of learning, and especially at the School of Chartres, had wholeheartedly embraced Platonic thought, especially the *Timaeus*. This had led to an intellectual revival and conceptual revolution that some scholars have said laid the foundation for modern science.[175] Within that worldview, the universe is a perfectly ordered and well-regulated cosmos in which natural causes operate. Within it man is fully endowed with reason, which enables him to map out the cosmic order, as difficult as that task is recognized to be.

When it asserted the eternity of the world, the new Aristotle posed great problems for the Christian West. "Left unchallenged, Aristotle's eternal world would have undermined, if not destroyed, one of the central themes of Christianity,"[176] namely, the idea that the world had been created by God in his infinite goodness, and that the end of that world was in preparation according to God's own plan. If Aristotelian cosmology were correct, this view would have to be repudiated.

---

[174] See Edwin Hatch, *The Influence of Greek Ideas on Christianity*, reprint with foreword, new notes, and bibliography by Frederick C. Grant (Gloucester, Mass.: Peter Smith, 1970).

[175] Tina Stiefel, *The Intellectual Revolution in Twelfth-Century Europe* (New York: St. Martin's, 1985); and see my discussion above in Chap. 3, the section entitled "Reason, Man, and Nature in Europe."

[176] Grant, "Cosmology," in *Science in the Middle Ages*, pp. 265–302, at p. 269.

Unlike the Muslims, the Europeans took "the invited guest" into their house and gave it a place of honor, namely, at the center of the university curriculum. The result was, as noted above, the transformation of the seven liberal arts into "the vast new domain of philosophy outlined by the works of Aristotle."[177] This transformation of the organization of the liberal arts curriculum at the University of Paris, for example, transformed the masters of the arts into a "faculty of rational, natural, and moral philosophy."[178] The net effect of this educational innovation was to grant philosophy (including natural philosophy) an autonomy and independence within the university previously unavailable in the West, and not achieved in Islamic or Chinese institutions of higher learning until the twentieth century.

Given the tensions inherent in this encounter, some Europeans reacted with alarm. This took the form of various pronouncements (in Paris) against Aristotle and even the condemnation of those who taught certain philosophical theses. But this reaction, profound as it was, was too little and much too late. By the middle of the thirteenth century the natural books of Aristotle had found their way into the curricula of the universities. In fact, a statute of 1255, enacted by the whole Faculty of Arts at Paris, prescribed the reading of the natural books of Aristotle as well as the time to be spent lecturing on them.[179] By this time, moreover, a large literature of disputed questions in the natural sciences had grown up. It was a product of the lectures that the masters gave, and it contained compilations of their summaries of major questions, as well as original treatises.[180] This literature reflects the concerted way in which a large set of questions in the natural sciences – physics, astronomy, cosmology, mechanics, and so forth – were assiduously discussed and debated. These included such questions as the constitution, nature, and conditions of the transformation of nature; whether the world is

---

[177] McLaughlin, *Intellectual Freedom and Its Limits*, p. 63.
[178] Ibid., p. 40. Additional insights into the transformation of the arts curriculum can be found in Pearl Kibre, "Arts and Medicine in the Universities of the Later Middle Ages," in *The Universities in the Late Middle Ages*, ed. Jacques Paquet and J. Ijsewign (Louvain: Louvain University Press, 1979), pp. 213–27; and Guy Beaujouan, "The Transformation of the Quadrivium" in *Renaissance and Renewal in the Twelfth Century*, pp. 463–87.
[179] Grant, *A Source Book*, pp. 43–4.
[180] The *questio* literature is illustrated in Grant, *A Source Book*, pp. 199–204, and has been discussed by many recent historians of medieval science; e.g., Weisheipl, "The Curriculum of the Faculty of Arts"; John Murdoch, "From Social to Intellectual Factors: An Aspect of the Unitary Character of Late Medieval Learning," in *The Cultural Context of Medieval Learning*, ed. John Murdoch and Edith Sylla (Boston: Reidel, 1975), pp. 271–338; Grant, "Cosmology," pp. 266ff; Grant, "The Condemnation of 1277, God's Absolute Power, and Physical Thought in the Late Middle Ages," *Viator* 10 (1979): 211–44; Grant, "Science and the Medieval University," in *Rebirth, Reform, and Resilience*, especially pp. 80ff; and McLaughlin, *Intellectual Freedom and Its Limits*, among others.

singular or plural; whether the earth turns on its axis or is stationary; "whether every effecting thing is the cause of that which it is effecting"; whether things can happen by chance; whether a vacuum is possible; whether the natural state of an object is stationary or in motion; whether luminous celestial bodies are hot; whether the sea has tides; and so on for virtually every known field of inquiry.[181] It is hard to imagine a more concentrated diet of scientific questions pertaining to the nature, composition, mechanisms, and patterns of the natural world. Perhaps even more difficult to imagine today is the mandatory discussion of all these questions by the whole student body in the arts curriculum.

Given this framework and agenda, fully established in the universities of Europe by the mid-thirteenth century, it is also difficult to imagine that all such naturalistic and philosophical discussion could have been extirpated from the minds of academics and their students when they went out into the world. In other words, this university training must have deeply ingrained the scientific point of view in the minds of European intellectuals in the twelfth and thirteenth centuries. More importantly, this set of naturalistic questions (and their pursuit) is of the same order as those continuously posed by scientific thinkers down to the present. To take a plain example from the more recent history of astronomy, "Why does the sky darken at night?" While this formulation of the problem is intended to appeal to the curious layman, in fact this problem did puzzle first-rate astronomers for centuries, from at least the time of Newton, and was only resolved in the twentieth century.[182] And it was through questions formulated in such a manner by the European medievals that we see the formation of the scientific agenda which was to come into its own with the scientific revolution of the sixteenth and seventeenth centuries.

Even more important were the legal and institutional structures that had been reconstructed due to the work of the Romanists and the canonists from the late eleventh century. As we saw earlier, these jurists had been busy creating the new institutional arrangements that were to govern early modern life henceforth. In the first instance, this took the form of recognizing groups of men as "whole bodies," that is, as corporations which had autonomy and jurisdiction. They had legally defined autonomy that permitted them to run their own affairs, independent of the influence of outside forces. The universities were the foremost of these corporate groups at the time. Of

---

[181] See Grant, *A Source Book*, pp. 199–200; and Grant, "Science and the Medieval University," pp. 82ff.

[182] This homely but fascinating example is discussed by the physicist and science writer Alan Lightman in *A Modern Day Yankee in a Connecticut Court and Other Essays on Science* (New York: Viking, 1986), pp. 128–32.

course, the Bishop of Paris, Stephen Tempier – who issued the condemnation of 1277 – might (as he did) attempt to separate philosophy from theology and try to prevent the arts faculty at the University of Paris from
discussing theological questions. But such an effort to separate matters of
faith from ethics and naturalistic inquiry could hardly prevent naturalistic
inquiries from being carried on.[183] Aside from that, the efficacy of such a
maneuver is doubtful, internally and externally vis-à-vis the University of
Paris. The structure of the academic enterprise internally was meant to be a
consensual order. That is, corporations and *studia generale* were meant to
make decisions regarding internal matters on the principle of "what affects
all must be approved by all"; but more importantly, decisions were to be
made by a majority or by "the greater and sounder part" (see above, "Universities and the West"). It was perfectly possible for a faculty to vote to
restrict or abolish certain studies, but it would have to come to a group
resolution.[184]

Conversely, we have seen that in the Islamic colleges there was no faculty
empowered to vote on such a course of action. The whole educational enterprise there was totally individualistic, even personalistic, and there was no
legal freedom to alter the course of study of the colleges endowed under the
religious law of trusts. They were neither consensual groups nor legally
autonomous entities in the Western sense. In Europe it seems doubtful that
even a papal bull would have been more effective than the edict of Paris
since, once again, some kind of local faculty act of approval would have to be
forthcoming if the question concerned the precise subject matter that the
masters were to teach and students were to study.

In addition, the canonists (the ecclesiastical lawyers) had been at work
creating the new canon law, summarized by Gratian's monumental *Concordia discordantium canorum* (c. 1140), in which it had been established that
man has reason and conscience and that these are standards according to
which even the Scriptures must be evaluated insofar as legal pronouncements
are concerned. Gratian had already shown in specific instances, for example,
regarding the question of whether or not priests should be permitted to read
profane literature,[185] that even where there are religious authorities opposed
to such an action, one must use reason to weigh the arguments and evidence.

---

[183] The list of condemned doctrines is printed in Grant, *A Source Book*, pp. 45–50.
[184] It is perhaps only speculation, but Professor Grant suggests that "it is highly probable that
the arts masters generally disapproved of the condemnation" of 1277; "The Condemnation
of 1277," p. 213 n5.
[185] See Berman's discussion of this, in *Law and Revolution* (Cambridge, Mass.: Harvard
University Press, 1983), pp. 147f; an English translation of this is in A. O. Norton, *Readings in the History of Education*, reprint (New York: AMS, 1971), pp. 58–75.

He solved the contradiction between the positions by declaring that "anyone (and not just priests) ought to learn profane knowledge not just for pleasure but for instruction, in order that what is found therein may be turned to the use of sacred learning."[186] Thus natural reason had been put into effect to resolve a disputed question and, in the process, had elevated reason and its search for truth to a lofty height.

The very idea of the search for truth had likewise become part of the credo, the ethos, of academics of this time. While Peter Abelard had died around 1144, his maxim that "truth in search of itself has no enemies" was both alive and sustained by other sources, including Aristotle's own teachings. In taking the pursuit of philosophy as a serious enterprise, the view was widespread that logic and dialectic were not just speculative endeavors but had given birth to a universal method for the study and treatment of all areas of inquiry. Logical studies, the masters had concluded, "supplied a universal method, a means of access to the principles of all methods; dialectic was, as Peter of Spain said, 'the art of arts and the science of sciences.'"[187] Accordingly, theologians and masters of arts, "defining their function as the pursuit of truth, claim[ed] freedom on this ground."[188] Finding inspiration in Aristotle's *Ethics* and other sources, they proclaimed that "those whose function is the contemplation of truth are raised above kings and princes."[189] To be a friend of truth became a serious calling for these philosophers and theologians. "Few phrases are repeated more frequently by the masters" of the thirteenth and fourteenth centuries, writes Mary McLaughlin,

and in contexts more significant for intellectual freedom, than the words, "friend of truth," by which the philosopher's duty is described. When John of Maligny challenges the authority of the chancellor, when John of Jandun exalted the opinions of Averroes, when Durand of Saint-Pourcain rejected the doctrine of Aristotle and Thomas Aquinas, they did so as "friends of truth." John of Meun sang the praise of William of Saint-Amour, "for that he the truth upheld" against the friars, the enemies of truth and free-speaking. It was as the "friend of truth" that Nicholas of Autrecourt rose to launch on Aristotle's philosophy an assault he intended to be devastating.[190]

In short, the metaphysics of exalting reason and rationalism in the pursuit of truth was deeply embedded in the vocabulary and discourse of the Europeans of the time, and they were not prepared to give it up, even if it meant imagining the impossible for the sake of theological harmony.

---

[186] Berman, *Law and Revolution*, p. 147.
[187] McLaughlin, *Intellectual Freedom and Its Limits*, p. 52.
[188] Ibid., p. 306.
[189] Ibid.
[190] Ibid., p. 308.

When the condemnation of 1277 occurred, the doctrine of the eternity of the world and the thesis of the impossibility of a plurality of worlds were condemned (among other philosophical presuppositions), but the demand that the integrity of philosophical pursuit be maintained persisted unabated.[191] Indeed, as many medievalists have suggested, the condemnation actually had the effect of encouraging theologians to imagine non-Aristotelian possibilities that otherwise would not have been entertained.[192] Professor Grant suggests that even among those sympathetic to the condemnation, the effect was to more sharply articulate physical arguments and metaphysical possibilities that otherwise would not have been worked out. These intellectual exercises, in effect, thought experiments, often took shape within the context of an eleventh-century distinction between the absolute power of God and the ordained power of God.[193] With the former, God was omnipotent and could do whatever he chose, while according to the latter, God had created an actual plan and order for his creation that established its own limitations. "From this crucial distinction, it followed that once God had decided the natural order of our world from among the innumerable, initial possibilities, he would not tamper with the plan by substituting from the store of unused possibilities."[194] Accordingly, one finds John Buridan granting that God could have created any number of possible worlds beyond the present one, but we should not take the existence of these worlds seriously, "unless we have independent evidence for so believing, evidence derived from one or all of the ordinary sources, that is, from the senses, experience, natural reason" or Scripture.[195] Similarly, Nicole d'Oresme (ca. 1325–82) imagined a non-Aristotelian world that allowed him to consider whether the directionality of objects within it would conform to Aristotle's beliefs or not, and he concluded in the negative. By such means, d'Oresme was led to entertain consistent but conceptually novel cosmological possibilities while retaining the conviction that "there never has been nor will there be more than one corporeal world."[196]

---

[191] See note 180 above.

[192] The thesis that the condemnation of 1277 liberated medieval thought and thus allowed modern science to develop was first argued by Pierre Duhem. See the discussion by Benjamin Nelson in *On the Roads to Modernity*, pp. 127–8; and Grant, "The Condemnation of 1277," pp. 216–18. Nelson and Grant both disagree with Duhem that the condemnation actually marked the birthday of modern science; Nelson, *On the Roads to Modernity*, p. 127; Grant, "The Condemnation of 1277," p. 217.

[193] Grant, "The Condemnation of 1277," p. 215; and see William J. Courtenay, "Nominalism in Late Medieval Religion," in *The Pursuit of Holiness in Late Medieval and Renaissance Religion*, ed. Charles Trinkaus and Heiko Oberman (Leiden: E. J. Brill, 1974), pp. 26–59.

[194] Grant, "The Condemnation of 1277," p. 215.

[195] Ibid.

[196] Ibid., p. 223.

In effect, the condemnation of 1277, especially as regards the question of the absolute power of God to make as many worlds as he pleases, encouraged some of the most gifted minds of the era to construct new and interesting solutions to problems in the physics of Aristotle; "and though the speculative responses did not replace or cause the overthrow of the Aristotelian world view, they did challenge some of its fundamental principles and force their attention on 'medieval thinkers.' They made many aware that things might be quite otherwise than were dreamt of in Aristotle's philosophy."[197] Rather than discouraging the pursuit of philosophy (and cosmology) during the period of its existence (it was annulled in 1325), the condemnation served to reinforce the struggle to maintain the autonomy and independence of philosophy.[198] In a word, the study of the natural sciences was deeply entrenched in the curriculum of the medieval universities. Although such study (and the questions it raised) provoked a condemnatory response from the authorities at the University of Paris, the pursuit of natural inquiries weathered the storm. Indeed, it became even stronger and more articulate than it had been at the outset. The study of the natural sciences and pursuit of philosophical truth, in effect, had become institutionalized in the universities, the central institutions of higher learning in the West. Nothing was to disturb that state of affairs thereafter. What did become a matter of controversy was the bifurcation of the empirical and the mathematical ideals of scientific methodology, reflected in the different outlooks and practice of science on the Continent and in England, especially in the seventeenth century.[199] But that is another story.

Medicine and the universities: the quest for universalistic norms

Before I draw this chapter to a close, I would like to examine the reception of the new medical knowledge brought forth in the translations from Greek and Arabic. For it is here that one finds the precocious development of a new profession in the modern sense, and another realization of the European tendency to establish generalized norms of intellectual thought and social practice.

As had happened with the previously lost or unknown philosophical writings of the Greeks and Arabs, so too the newly translated medical

---

[197] Ibid., p. 241.
[198] This is also a theme of McLaughlin's *Intellectual Freedom and Its Limits.*
[199] This is the theme, mentioned in Chap. 1, addressed by Thomas Kuhn in "Mathematical vs. Experimental Traditions in the Development of Physical Science," *Journal of Interdisciplinary History* 7, no. 1 (1976): 1–31.

writings were taken directly into the heart of the new European universities. At Padua, Paris, Bologna, Montpellier, Oxford, and elsewhere, new medical faculties, so-called higher faculties, sprang up.[200] In short order these centers of higher education made it mandatory for all medical students to first study the arts before entering upon medical studies. After following the prescribed course of study, including internships under senior physicians, students were examined and then awarded diplomas that entitled them to practice medicine. Concern about medical malpractice on the part of doctors of medicine in college faculties led to the enactment of statutes and other regulations all over Europe that served to restrict the practice of medicine to those who had been properly certified either by a local board of medical specialists or civil public authorities, or by a university faculty. These enactments did not fully control all medical practitioners, nor did they eliminate all malpractice and unwarranted medical claims as judged by the standards of the day. But the fact is that the European medievals, beginning in the twelfth century, developed a "variety of ways of evaluating and attesting to the competence" of medical practitioners, and these legitimating procedures "went far beyond the forms of legal recognition of medical practitioners in the Roman world."[201] The effects of this development were evident in cities all over Europe, but especially in Bologna, Paris, Padua, Montpellier, and Oxford. The net effect of this was to turn medicine into the first secular profession in Europe. For some historians of medieval medicine this was a process centered on the institutionalizing of medical education, thereby taking it out of the hands of amateurs and laymen.[202] For everywhere the possession of a university degree in medicine was recognized as entitlement to practice medicine.

While some in the West may see this as a power struggle focused on the monopoly of educational resources, it in fact served to set standards of medical education and practice, standards that were not set in the Arabic-Islamic context. As we saw earlier, in the Islamic setting there was no degree or diploma that could be issued to medical students upon completion of

---

[200] See Charles Haskins, *Studies in Medieval Science;* Vern Bullough, *The Development of Medicine as a Profession* (New York: Hafner, 1966); Charles Talbot, "Medicine," in *Science in the Middle Ages,* pp. 391–428; Paul O. Kristeller, "The School of Salerno: Its Development and Its Contribution to the History of Learning," in *Studies in Renaissance Thought and Letters* (Rome: Edizioni di Storia e Letturatura, 1956), pp. 495–551; Nancy Siraisi, *Arts and Sciences at Padua;* Siraisi, *Medieval and Early Renaissance Medicine: An Introduction to Knowledge and Practice* (Chicago: University of Chicago Press, 1990), chap. 3; Luke Demaitre, "Theory and Practice in Medical Education at the University of Montpellier in the Thirteenth and Fourteenth Centuries," *Journal of the History of Medicine* 30 (1975): 103–23; among others.

[201] Nancy Siraisi, *Medieval and Early Renaissance Medicine,* p. 17.

[202] E.g., Bullough, *The Development of Medicine.*

their studies, since there was no faculty that could issue such a certification. Not only was there no faculty in the sense of a collegially organized body of masters, but the physicians in general were excluded from the Islamic colleges, and their domain of instruction was either private tutoring or, on a more limited basis, instruction through the reading of texts in the classrooms attached to hospitals. But these latter were in no sense degree-granting institutions. Since there were no faculties and there was no examination system administered by a collective body of masters or practitioners, the standard acknowledgment of studies completed was the ijaza, the authorization to transmit a text. Furthermore, anyone might, as the famous Ibn Ridwan of Cairo did, teach himself medicine and then practice the art and, by dint of rhetoric and persuasive writing, convince others of his medical authority. In the meantime, there was no effective way either to eliminate charlatanism or to establish a standard curriculum, since everything depended upon individual inclination and personal contacts. It was up to the putative master physician to guide his students as he saw fit, through the literature he knew; and while he would read texts with his students, and even take up difficult questions, there was nothing equivalent to an examination held by a group of medical practitioners. Though the market inspector (muhtasib) in large cities was entrusted with granting certificates of good conduct to physicians to allow them to practice, there is no indication of a general practice of licensing through the administration of an examination. There are three known cases, for example, in which official action was taken to restrict medical malpractice in the Abbasid period:

The first occurred in the early ninth century when a bogus list of medicines was given to a group of pharmacists who were asked to provide them. Those pharmacists who brought in prescriptions were banished, and those who said the list made no sense were retained. The second instance took place when the Caliph al-Mutqtadir learned that a man treated by a doctor had died and so forbade anyone to practice medicine before passing an examination. Sinan b. Thabit b. Qurra then administered an examination to all doctors in Baghdad. Two centuries later, ibn al-Tilmidh, after becoming chief physician in the Abbasid capital, quizzed a man who had little scientific understanding of medicine but much practical experience.

From this Gary Leiser concludes that "the decisions to give examinations were arbitrary and that the examinations themselves were poorly organized."[203] More importantly, these events and the conditions that provoked them did not result in the development of a generalized system of examinations and certification. Similarly, the market inspector could not rely on

---

[203] Gary Leiser, "Medical Education," pp. 67–8; based on Ibn Abi Usaibi'a's famous biographical history of medicine.

possession of a college diploma or other certificate of educational achievement, since none existed.

Finally, given the nature of Islamic law, there could be no legislation such as was passed in the cities of Europe, whereby those without medical degrees were officially enjoined by the legal system to desist from practicing medicine. For example, the "Parisian medical faculty from its inception claimed the right to regulate medicine and its allied groups within the confines of the city."[204] While enforcement was difficult, Parisian doctors underscored their exclusive right to practice medicine by bringing cases of unauthorized medical practice to the attention of the courts. This resulted in further acts of regulation by the ecclesiastical and secular authorities (including the king). The result was a system of "careful supervision of each of the clearly defined divisions of medicine."[205] In Bologna, similar regulations were put into effect. "The college of Doctors of Medicine at Bologna in 1370, again in 1395, and in 1401, specifically forbade anyone from practicing medicine or surgery without their permission."[206] This took the form of requiring outsiders to provide suitable credentials of their qualifications, or forcing them to study medicine at Bologna for at least three years. "Any claim to have studied [medicine] outside of Bologna had to be attested by three trustworthy witnesses; moreover, the applicant would have to be examined, licensed, and approved" by physicians of the college.[207] All over Europe, the twelfth, thirteenth, and fourteenth centuries saw a variety of groups spring into existence whose main goal was the regulation of medical education and practice. Many localities "formed organizations of various kinds that regulated the admission of members through examination and other requirements; some medical corporations also obtained the legal right to approve or otherwise regulate medical practitioners in their regions."[208] In other places civil authorities licensed medical practitioners, as did some kings, but for various reasons, the process of evaluation and certification tended to revert back into the hands of those with university training, or those otherwise attested, and locally autonomous groups of practitioners.

In short, the Europeans were impelled to establish uniform standards of medical education and practice and to institutionalize the means by which this could be accomplished. This was possible because legally distinctive spheres of regulation and autonomy existed. The first level of autonomy was that maintained by university faculties of medicine, which could establish

[204] Bullough, *The Development of Medicine*, p. 99.
[205] Ibid., p. 108.
[206] Ibid., p. 106.
[207] Ibid.
[208] Siraisi, *Medieval and Early Renaissance Medicine*, p. 18.

their own rules of education and training vis-à-vis the college curriculum. Secondly, given the fundamentally corporate nature of the legal system, the benefits of a medical education (as also a legal education, or simply a baccalaureate) granted the holder exclusive rights. These rights in turn could be (and were) recognized by the legal establishment and were enforced to the degree that the secular domains had jurisdiction, usually only within urban territories. In contrast, such exclusive rights were not a possibility in Islamic law, because Islamic ideology prevented the creation of exclusive groups with their own legal autonomy. All groups at all times within the Muslim umma (community) had to be regulated by the religious law that recognized no such spheres of professional autonomy. As we shall see in Chapter 7, there was a similar exclusion of autonomous professional associations by traditional Chinese law and a failure to establish in China a recognized medical profession that could be legally (and autonomously) regulated.

6

◁══════════════════════════════▷

Cultural climates and the ethos
of science

We turn now to a consideration of the larger context of Arabic-Islamic civilization and to the structural impediments that prevented the breakthrough to modern science there. In the preceding chapters I have described the nature of Arabic science and some of its major accomplishments over the course of its several-hundred-year hegemony. I have also described the institutional structures within which scientific and intellectual life was carried on during the heights of Arabic-Islamic civilization. I shall now accentuate the dimensions of the cultural breakthroughs – legal, institutional, and intellectual – that took place in the European West and resulted in the institutionalization of science there, in contrast to the cultural and structural impediments that inhibited the rise of modern science in the world of Islam.

A major access point to the problem of the rise of modern science can be found in the multiple institutional arrangements that create and sustain the role of the scientist. In directing our attention to that problem, we should consider the broader institutional arrangements that entail the scientific role-set. From one point of view, those values and commitments that constitute the scientist's role-set can be called the ethos of science, as I noted in Chapter 1. As Robert Merton originally expressed the idea, the norms of the ethos of science "are expressed in the form of prescriptions, proscriptions, and permissions."[1] And these are centered on the values of universalism, communalism, organized skepticism, and disinterestedness.

Although sociologists of science in the past have attempted to view the role of the scientists as that of a narrowly defined cultural actor, I have suggested just the opposite: the scientist is and always has been a purveyor of

---

[1] Robert K. Merton, *The Sociology of Science: Theoretical and Empirical Investigations* (Chicago: University of Chicago Press, 1973), p. 269.

knowledge affecting the widest reaches of thought and even metaphysics – though scientists today would deny involvement in any such thing. In part this disclaimer is a defensive self-effacement that may derive from the narrowness of contemporary scientific specialization and problem solving. On the other hand, it comes from a neglect of the overall role of science in society and the ways in which the scientific vision shapes all of our perceptions of physical, cultural, and psychological reality. Nevertheless, the breakthrough to modern science centered on the freedom of nonecclesiastical elites to describe (and explain) the known world – freely, publicly, and completely – in terms radically at variance with the received religious wisdom. Although Copernicus and other astronomers, especially Galileo and Kepler, were intent to establish the science of astronomy, there is no denying the fact that their scientific claims radically altered the Judeo-Christian worldview. It is true, too, of course, that the new world system of astronomy profoundly shattered and offended the Islamic cosmological vision. As we saw earlier, Muslims were still attempting to defend the traditional geocentric worldview in the late nineteenth century.

From such a point of view, the breakthrough to modern science was a breakthrough that destroyed the received worldview and established a new and legally protected institutional location within which intellectual inquiries of the most far-reaching cultural and intellectual consequences could be carried on without hindrance. Of course, this did not mean that no one, above all, no political or ecclesiastical authority, would challenge the new findings of science; it meant rather that institutionally and even legally the philosopher cum scientist had the right to exercise his reason and to express his thoughts in public forums and that this activity was presumed to be legitimate, even if it transgressed traditional assumptions. No such presumption was available in Islamic law. A religious figure (a legist) could, simply de jure, issue a legal ruling, a fatwa, declaring that some thinker had violated the religious law and that therefore "his blood could be shed," just as was done recently in the case of the writer Salman Rushdie, when the Ayatollah Khomeini declared that Rushdie and all those involved with publication of his book "are sentenced to death."[2]

As I suggested earlier, if we follow Benjamin Nelson's insight, it is not a question of whether the Arabs, Chinese, or Indians surpassed the Greeks in optics, chemistry, medicine, astronomy, or mathematics, but whether there was "a comprehensive breakthrough in the moralities of thought and in the logics of decision" that opened up "wider universalities of discourse and

---

[2] See Daniel Pipes, *The Rushdie Affair: The Novel, the Ayatollah, and the West* (New York: Birch Lane Press, 1990), p. 27.

participation."[3] Furthermore, from this point of view, the norms of science described by Robert Merton give concrete detail to the nature of the institutional breakthroughs that signify the institutionalization of the role of the scientist.

The arresting of Arabic science

As we have seen, during the golden era of Arabic-Islamic civilization vast sums of money were expended on learning and the acquisition of knowledge.[4] As one surveys the domains of knowledge in which Islamic civilization left its imprint, above all in the natural sciences (astronomy, mathematics, medicine, pharmacology, optics, and so forth), the achievements of this civilization assume singular proportions. The arts of decoration and architecture were also not lacking in originality and superb representation. There was, in short, no lack of talent, dedication, and inventive genius. Since the mainstream institutions of higher learning, the madrasas, systematically excluded the study of philosophy and the natural sciences, it is all the more impressive that hundreds of scholars traveled great distances to acquire philosophical and scientific knowledge from individual masters and in the process achieved great heights of learning. In the natural sciences this achievement is most aptly illustrated by the development of astronomical planetary models of the universe that were mathematically equivalent to those of Copernicus, which were to appear some two hundred years later.[5]

In brief, up until the thirteenth and fourteenth centuries Arabic science was so promising and developed as to be called the most advanced in the world. In the case of astronomy, it is obvious that this supremacy existed until the mid-sixteenth century, when the astronomical models of Ibn al-Shatir and the Marâgha school were superseded by the new astronomical system of Copernicus. Even in mathematics, as E. S. Kennedy has pointed out, great mathematical inventiveness is to be found in Islamic civilization as late as the fifteenth century.[6]

There is a tendency to explain the arresting of Arabic science by reference to geopolitical developments, that is, the invasion of Eastern Islam (in the thirteenth century) by the Mongols and the reconquest of Spain (beginning

[3] Nelson, *On the Roads to Modernity*, p. 99.
[4] See Chap. 2, note 1 for a definition of this era.
[5] Ibn al-Shatir's models are dated about 1350 while Copernicus's great work, *De revolutionibus*, was only unveiled for the world on his deathbed in 1543, though his heliocentric writings date to about 1511 and were known in astronomical circles, especially in Germany but also in Rome. For the precise definition of this equivalence, see below, "Internal Factors."
[6] E. S. Kennedy, "The Exact Sciences in Timurid Iran," in *The Cambridge History of Iran* (Cambridge: Cambridge University Press, 1986), 6: 568–80.

in the eleventh century). This explanation, however, neglects the actual course of scientific development in Arabic-Islamic civilization, and in the case of Spain, it greatly distorts the significance of Spain to the civilization of Islam. It also fails to consider the actual nature of Arabic-Islamic cultural institutions and their ultimate impact on freedom of thought and inquiry.

To put Spain in proper perspective, an analogy may be useful. If the Russians should reclaim (and resettle) the state of Alaska, that in itself would signify a political and military weakness on the part of the United States, but the (hypothetical) future development of the United States, culturally, economically, and so forth, could hardly be blamed on the Russian reconquest of Alaska. Similarly, the reconquest of Spain, beginning around 1045, signified the internal political weakness of the Islamic regime on the Iberian peninsula. Historians agree that it was internal subversion that brought about the decline of Islamic political hegemony in Spain.[7] This was coupled with growing Muslim intolerance, for both Jews and Christians, and a similar negative reaction on the part of Christians. This intolerance had forced many Christians and Jews in Spain and Maghrib to migrate east.[8] The significance of the Islamic culture of Spain for Islamic civilization (and the West) is quite another matter.

If Spain had persisted as an Islamic land into the later centuries – say, until the time of Napoleon – it would have retained all the ideological, legal, and institutional defects of Islamic civilization. A Spain dominated by Islamic law would have been unable to found new universities based on the European model of legally autonomous corporate governance, for, as we have seen, corporations do not exist in Islamic law. Furthermore, the Islamic model of education rested on the absolute primacy of fiqh, of legal studies, and the standard of preserving the great traditions of the past.[9] This was

---

[7] See Ira Lapidus, *A History of Islamic Societies* (New York: Cambridge University Press, 1988), pp. 382ff; Watt, *A History of Islamic Spain* (Edinburgh: Edinburgh University Press, 1965), pp. 86–91; and Thomas Glick, *Islamic and Christian Spain in the Early Middle Ages* (Princeton, N.J.: Princeton University Press, 1979), pp. 46–50.

[8] Ibid., p. 49. Actually, the Jews appear to have fled eastward and the Christians migrated north to Christian areas.

[9] According to Hodgson, "Education was commonly conceived as the teaching of fixed and memorizable statements and formulas which could be adequately learned without any process of thinking as such. A statement was either true or false, and the sum of all true statements was knowledge. One might add to the accessible sum of true statements to be found in one's heritage, but one did not expect to throw new light on old true statements, modifying or outdating them. Hence knowledge that mattered, *'ilm*, as against the facts of 'common knowledge' which even illiterates picked up as a matter of course, was implicitly conceived as a static and finite sum of statements, even though not all the potentially valuable statements might be actually known to anyone at a given time." *The Venture of Islam* 2: 458. This view, it seems to me, applies best to the official and institutional forms of educational practice, less to the falasufs, the philosophers following Aristotle's model.

symbolically reflected in the ijaza, the personal authorization to transmit knowledge from the past given by a learned man, a tradition quite different from the West's group-administered certification (through examination) of demonstrated learning.

In the actual event, the founding of Spanish universities in the thirteenth century, first in Palencia (1208–9), Valladolid, Salamanca (1227–8), and so on, occurred in long-established Christian areas, and the universities were modeled after the constitutions of Paris and Bologna.[10] The various Arabic histories of Spain indicate that there were no Islamic colleges – madrasas – built in Spain prior to the fourteenth century, when one was founded in Granada ca. 1349.[11] Of course, such considerations do not condone the *reconquista,* much less the persecution of Jews and Muslims by the Inquisition.

In addition, those especially great intellectual figures of Spain, for example, Ibn Bajja, Averroes, and Maimonides, were probably far more significant for the West than they were for Arabic-Islamic civilization. Averroes and Maimonides were persecuted at one time or another by their own countrymen – Averroes by his co-religionists, and Maimonides by zealous Muslims urging conversion to Islam, which forced his emigration to Cairo where he took up his medical practice and wrote his books. Ibn Bajja, on the other hand, died early of poisoning, probably administered by his jealous co-religionists.[12]

Furthermore, Maimonides' great work, *The Guide of the Perplexed,* was rejected by the orthodox Jewish leadership. "In the thirteenth and fourteenth centuries it was violently denounced for being anti-religious and as vehemently defended against the charge."[13] The naturalistic approach to philosophy and sacred subjects was not revived in Jewish culture until

---

[10] Rashdall, *The Universities of Europe in the Middle Ages,* 2: 63–114.

[11] George Makdisi, "Madrasa," *EI²* 5: 1128. The same point is made by L. P. Harvey, *Islamic Spain, 1250–1500* (Chicago: University of Chicago Press, 1990), pp. 190, 230. This probably accounts for the fact that when Anwar Chejne discusses "Sciences and Education" in *Muslim Spain* (Minneapolis: University of Minnesota Press, 1974), chap. 9, he makes no reference to any educational institutions, that is, to actual schools, colleges, or their analogues. Apparently all instruction was in private homes, sometimes in mosques, and in so-called salons or literary clubs (*masjids*), p. 181. Of course, this is consistent with traditional Islamic education, except for the absence of madrasas for the teaching of law and the Islamic sciences.

[12] Moody, "Galileo and Avampace [Ibn Bajja]: Dynamics of the Leaning Tower Experiments," in *Roots of Scientific Thought,* ed. Philip P. Wiener and Aaron Noland (New York: Basic Books, 1957), pp. 176–206 at p. 200. A useful but still sketchy attempt to explain the facts of Ibn Rushd's disgrace caused by the sultan is in Dominique Urvoy, *Ibn Rushd (Averroes)* (London: Routledge, 1991), pp. 34–6.

[13] Shlomo Pines, "Maimonides," in *Encyclopedia of Philosophy* (New York: Macmillan, 1967) 5: 134.

Spinoza.[14] Not least of all, both Averroes and Maimonides were masters of the literary techniques of concealment and unexpected disclosure that had been highly developed in Islamic as in Judaic culture because of the pervasiveness of religious restriction.[15] Averroes, despite his stature as a judge, was banished toward the end of his life, his books burnt, and forced to emigrate to Morocco (in 1195) where he died in 1198.[16] Moreover, the great philosophical debate Averroes joined when he attempted to refute Ghazali's "Refutation of Philosophy" fell on deaf ears in Arabic-Islamic civilization. There was no significant response forthcoming among Islamic philosophers. As W. M. Watt has put it, "Though the work of Averroes was know in the east, its outlook was so foreign to these men that it had nothing to say to them."[17] Instead, it was in the West, paradoxically, that his writings, and above all his commentaries on Aristotle, were taken with the utmost seriousness.[18] Beyond the details of these philosophical debates, it must be remembered that Arabic-Islamic civilization without Spain (effectively after the fall of Seville in 1248, and definitively after 1492) continued on, largely undisturbed (despite the Mongols' invasion) until the incursions of Napoleon into Egypt in 1798.

It is true that Baghdad was sacked by the Mongols in 1258, but here again Islamic civilization's resilience triumphed with the consequence that Islamic culture and institutions were quickly revitalized and that part of the world remained (and remains) deeply Islamic. What is more, it was just such invaders (led by Hulagu) from outside the traditional heartland of Islam who supported the building of the Marâgha observatory and thus fostered the development of the first non-Ptolemaic astronomical models of the universe. One might say that the late thirteenth and early fourteenth centuries, especially in Damascus and Cairo, represented the pinnacle of Arabic scientific

---

[14] For the philosophical and religious issues arising from the *Guide*, see Leo Strauss, *Persecution and the Art of Writing*, reprint (Westport, Conn.: Greenwood Press, 1973); and Paul Johnson, *A History of the Jews* (New York: Harper, 1987), especially pp. 287ff for later European reactions to Maimonides.

[15] See above, Chap. 2, pp. 82–3.

[16] Iskandar and Arnaldez, "Ibn Rushd," *DSB* 12: 2; and Watt, *A History of Islamic Spain*, p. 140.

[17] Ibid., p. 141.

[18] It might also be pointed out that some social historians place the appearance of organized science in Spain in the tenth century (Glick, *Islamic and Christian Spain*, p. 257), and that a century later, between 1025 and 1075, the portion of Christian scientists exceeded that of Muslims (ibid., fig. 6, "Ratio of Christian and Muslim Scientists," p. 260). Unfortunately, Glick's masterful study fails to discuss the institutional bases of scientific inquiry. Instead he transforms the traditional Islamic pattern of personal contacts and influence (and the absence of colleges) into schools of science (pp. 253–61), thereby giving a modern connotation to an ancient and premodern form of learning and social organization.

development in astronomy and medicine. In astronomy the capstone is represented by Damascene Ibn al-Shatir (d. 1375). Although at least fifty-five significant astronomers lived after al-Shatir (under Mamluk rule), not one of them benefited from nor sought to incorporate Shatir's astronomical advances.[19]

In medicine, there is good evidence that Damascus (and Cairo) was the center of a distinguished gathering of first-rate physicians in the thirteenth century.[20] Here one refers to the discoveries of Ibn al-Quff (1233–86) and Ibn al-Nafis (1210–88) mentioned in Chapter 5.

In short, some of the most important scientific developments to be found in Arabic-Islamic civilization occurred either during or after the point in time when external geopolitical factors were supposed to have caused its collapse. Thus, we should consider the most obvious internal factors regarding the development of science, and then we should examine the external and structural factors that are sociological in nature.

Internal factors

If we turn our attention to the achievements of Arabic science, it may be said that its failure to give birth to modern science cannot be explained on the basis of strictly technical and narrowly scientific considerations. As we saw in Chapter 2, many of the planetary models developed by the Marâgha school are mathematically equivalent to those of Copernicus. That is, (1) Copernicus uses the Tusi couple[21] as the Marâgha astronomers did, (2) his planetary models for longitude in the *Commentariolus* are based on those of Ibn al-Shatir, while (3) those for the superior planets in *De revolutionibus* use Marâgha models, and (4) the lunar models of Copernicus and the Marâgha school are identical.[22] Given these facts, it cannot be seriously argued that the Arabs suffered from mathematical deficiencies. Nor can one

---

19 David King, "The Astronomy of the Mamluks," *Isis* 74, no. 274 1(1983): 531–55; and King, "Ibn al-Shatir," *DSB* 12: 357–64.

20 See the references in Chap. 2; and especially, Sami Hamarneh, "Medical Education and Practice in Medieval Islam," in *The History of Medical Education,* edited by C. D. O'Malley (Berkeley and Los Angeles: University of California Press, 1970), pp. 39–71; Hamarneh, "The Physician and the Health Professions in Medieval Islam," *Bulletin of the New York Academy of Sciences* 47, no. 9 (1971): 1088–1110; and Hamarneh, "Arabic Medicine and Its Impact on Teaching and Practice of the Healing Arts in the West," *Oriente e Occidente* 13 (1971), especially pp. 418–22.

21 As we saw earlier, the Tusi couple, an astronomical device invented by the mathematician astronomer Nasir al-Din al-Tusi (d. 1274), shows how a uniform circular motion could be compounded of two circular nested orbits. See Figs. 1–3 in Chap. 2.

22 Noel Swerdlow and Otto Neugebauer, *Mathematical Astronomy in Copernicus's "De revolutionibus"* (New York: Springer Verlag, 1984), p. 46.

claim that they lacked theoretical imagination in astronomy, optics, or physics. In the case of physics, one may recall the originality of Ibn Bajja, whose commentary on Aristotle regarding the dynamics of motion is in a direct path leading to Galileo's theory of free fall.[23] It does not detract from Ibn Bajja to say that the theory was probably first invented by the Christian Neoplatonist Johannes Philoponus in sixth-century Alexandria. No Arabic manuscript of it seems to exist, though Ibn Bajja may have heard of it through various Arab traditions.

In the case of astronomy, the difference between the Copernican and Arabic systems was essentially metaphysical – a choice between geocentric and heliocentric systems, a situation in which there was no empirical data on which to base a definitive choice. Furthermore, the success of that whole endeavor may also be attributed to Arabic successes in perfecting spherical geometry as well as trigonometry. It was the possession of these technical tools that allowed the Arabs to surpass the Chinese astronomers and led to the appointment of Arab astronomers to the Chinese Astronomical Bureau in Peking in the thirteenth century.[24]

Similarly, those who suggest that the failure of Arabic science to yield modern science was due to a failure to develop and use the experimental method are confronted with the fact that the Arabic scientific tradition was richer in experimental techniques than any other, whether European or Asian. As already mentioned, these included at least three separate experimental traditions: in optics, in astronomy, and in medicine. In the case of optics, Ibn al-Haytham's development and use of the idea of experiment must be regarded as one of the most significant in the whole history of science, and this is attested by the great influence his *Optics* had in the West. Although there was a delay in its dissemination after its composition between 1028 and 1038, it continued to be widely influential in the West down to the sixteenth century.[25] In this work Ibn al-Haytham set out to recommence the study of optics in a manner that departed from all earlier writers. Instead of summing up the cumulative wisdom of previous writers on the subject, al-Haytham took a new departure that sought in all cases possible to

---

[23] See Moody, "Galileo and Avempace." The relevant documents from Ibn Bajja, Averroes, and Aquinas to Galileo are translated in *A Source Book of Medieval Science*, ed. Edward Grant (Cambridge, Mass.: Harvard University Press, 1974), secs. 33–56 (hereafter cited as *A Source Book*).

[24] See Aydin Sayili's comments on Arab-Chinese exchanges in *The Observatory in Islam* (Ankara: Turkish Historical Society Series 7, no. 38, 1960), pp. 188–91 and 361–4; as well as Joseph Needham, *Science and Civilisation in China* 3: 186–94, 49, 372ff, and passim (hereafter cited as *SCC*); and Nathan Sivin, "Wang Hsi-Shan," *DSB* 14: 159.

[25] For this dating, see A. I. Sabra, ed. and trans., *The Optics of Ibn al-Haytham* (London: The Warburg Institute, University of London, 1989), 2: xxxiii.

bring mathematics and demonstration to bear on the study of the properties of light and vision. In the process, he used a variety of experimental apparatuses, including specially arranged dark chambers, specially designed apertures for the controlled admission of light, viewing tubes, and so on. Consequently, the concept of an experiment (*i'tibar*) "emerges as an explicit and identifiable methodological tool involving the manipulation of artificially constructed devices."[26]

Although the *Optics* of al-Haytham remained virtually unknown in the Islamic world during the eleventh and twelfth centuries, it was finally rescued by the Persian natural philosopher Kamal al-Din al-Farisi (d. ca. 1320). In his hands we find a spectacular application of the experimental method to the task of explaining the rainbow. To do this, Kamal al-Din contrived a unique experimental situation that simulated the effect of a ray of sunlight striking a drop of water. He achieved this by placing a small sphere filled with water in a darkened room with carefully controlled emission of light. Through his work with this experimental situation he showed that the rainbow is the result of two refractions and one reflection of rays of light inside drops of water.[27]

Before Kamal al-Din obtained a copy of al-Haytham's *Optics*, it was transmitted to the West, where, probably in Spain, it was translated into Latin so that Western writers were familiar with it by the 1250s and 1260s.[28] There, in the Latin West, the work first influenced Roger Bacon (ca. 1220–92) and then all of the major writers on optics, including Robert Grosseteste (ca. 1175–253), Pecham (ca. 1230–92), Witelo (d. after 1275), and Theodoric of Freiburg (ca. 1250–1310).[29] Although the first three chapters of Book I of al-Haytham's *Optics* were not included in the Latin translation and thus his preliminary experimental investigations of light were omitted, the work continued to suggest an experimental approach to the investigation of light and optical phenomena. It remains a remarkable coincidence that Theodoric of Freiburg virtually simultaneously and independently performed (probably about 1304) the same experiment as al-Farisi involving a sphere filled

[26] A. I. Sabra, "Ibn al-Haytham," *DBS* 5: 190; and *The Optics of Ibn al-Haytham*, 2: 14.
[27] A. I. Sabra, "The Scientific Enterprise," in *Islam and the Arab World*, ed. Bernard Lewis (New York: Alfred Knopf, 1976), p. 190; and see Roshdi Rashid, "Kamâl al-Dîn al-Fârisî," *DSB* 7: 212–19.
[28] David C. Lindberg, *Theories of Vision from al-Kindi to Kepler* (Chicago: University of Chicago Press, 1976), pp. 106ff. In *Optics* 2: lxiv, n94, Professor Sabra mentions the fact that al-Mu'taman ibn Hud, who was king of Saragossa from 1081 to 1085, refers to "Alhazen's Problem" regarding certain lemmas in the *Optics*. This suggests that the *Optics* had made its way to Spain in Ibn al-Haytham's own lifetime, though it was not until much later that it was translated and made available to Latin scholars.
[29] See David C. Lindberg, "Lines of Influence in Thirteenth-Century Optics: Bacon, Witelo, and Pecham," *Speculum* 46 (1971): 66–83.

with water. He proposed the same explanation of the rainbow, namely, the passage of light through droplets of water involving two refractions and one reflection of rays of light.[30] It is noteworthy, moreover, that this same experiment was virtually replicated by Descartes in the seventeenth century.[31]

The idea of comparing observations of two states of affairs has long been a practice in astronomy, and according to A. I. Sabra, this was the context from which Ibn al-Haytham drew his notions about experimental proof in mathematical optics.[32] Accordingly, both Bernard Goldstein and George Saliba have referred to astronomical examples in which medieval Jewish and Islamic astronomers compared theory and observations.[33]

A similar pattern of experimentation developed in medicine. For example, al-Razi (d. 925) has been described as a physician noted for his "refusal to accept statements unverified by experiment," his understanding of control experiments, his recording of clinical observations, and his criticism of authorities such as Galen.[34] Although Avicenna (d. 1037) has been criticized for slighting the ideas and work of al-Razi, it remains true that Avicenna's great medical work, the *Canon*, was a prodigious work that held sway over the medical field even in Europe into the sixteenth century. A. C. Crombie has pointed out that the *Canon* contained a set of rules that laid down the conditions for the experimental use and testing of drugs. These rules were in fact "a precise guide for practical experimentation," above all, in the process of discovering and proving the effectiveness of medical substances.[35]

In short, the scientific world of Islam was rich in experimental ideas and these did not go without use in optics, astronomy, and medicine. The prob-

[30] William A. Wallace, "Theodoric of Freiburg," *DSB* 4: 92ff.

[31] A. I. Sabra, *Theories of Light from Descartes to Newton* (London: Oldbourne, 1967), pp. 62–3.

[32] Sabra, *Optics* 2: 14ff.

[33] See Bernard Goldstein, "Theory and Observation in Medieval Astronomy," *Isis* 63 (1972): 39–47; and George Saliba, "Theory and Observation in Islamic Astronomy: The Work of Ibn al-Shatir," *Journal for the History of Astronomy* 18 (1987): 35–43.

[34] See Albert Iskandar, "Ibn Sina," *DSB*, Suppl 15: 498, where he discusses Ibn Sina's graceful overshadowing of al-Razi's progressive spirit of empiricism. And cf. Max Meyerhof, "Thirty-Three Clinical Observations of Rhazes," *Isis* 23 (1933): 321–55. Similarly, in pharmacology there is good evidence of significant advances in the testing and identification of medical materials (*materia medica*) using empirical techniques. The work of Abu 1- 'Abbas an-Nabati (fl. thirteenth century) is an illustration of the fact that great pains were taken by first-rate Arab pharmacologists in testing and describing medical materials and in separating unverified reports from actual tests and observations; see Albert Dietrich, "Islamic Sciences and the Medieval West: Pharmacology," in *Islam and the West*, ed. Khalil Semaan (Albany, N.Y.: SUNY Press, 1980), pp. 50–64.

[35] A. C. Crombie, *Robert Grosseteste and the Origins of the Experimental Science, 1100–1700* (Oxford: At the Clarendon Press, 1953), p. 79.

lem was neither a lack of the development and use of the experimental method nor a lack of mathematical theory. Furthermore, there is little doubt that up until the twelfth and thirteenth centuries, the balance of accumulated knowledge and the presence of well-trained scientific talent significantly favored the Arabs over the Europeans. To mention the most obvious comparison, in the twelfth and thirteenth centuries there was in the West no group of scholars of comparable status in astronomy and cosmological speculation working in a continuous tradition like that composed (in Western Islam) of Ibn Bajja (Avempace, d. 1338), Ibn Tufayl (d. 1185), Averroes (d. 1198), al-Bitruji (fl. 1200), and Maimonides (d. 1204), and in Eastern Islam al-'Urdi (d. 1266), al-Tusi (d. 1274), and Qutb al-Din al-Shirazi (d. 1311). This is not to say that there were no astronomers who did important work in astronomy in the West during this period, but only that the Arabs had already – above all, by the time of Ibn al-Shatir – reached the goal of creating mathematical models that were Copernican in design.[36] Whether or not there were more truly great intellects in the Arabic-Islamic world of the Middle East of that time, Arabic-Islamic civilization clearly had extraordinary intellectual advantages bequeathed through its literary and scientific past, and until that legacy had been transmitted to and assimilated by the West, it was reasonable to expect that its intellectual achievements in the future would far surpass those of the West. But this did not happen.

The problem was not internal and scientific, but sociological and cultural. It hinged on the problem of institution building. If in the long run scientific thought and intellectual creativity in general are to keep themselves alive and advance into new domains of conquest and creativity, multiple spheres of freedom – what we may call *neutral zones*[37] – must exist within which large groups of people can pursue their genius free from the censure of political and religious authorities. In addition, certain metaphysical and philosophical assumptions must accompany this freedom. Insofar as science is concerned, individuals must be conceived to be endowed with reason, the world must be thought to be a rational and consistent whole, and various levels of universal representation, participation, and discourse must be available. It is

---

[36] Detailed accounts of the state of European scientific thought in late antiquity and the High Middle Ages can be found in M. Clagett, *The Science of Mechanics in the Middle Ages* (Madison, Wis.: University of Wisconsin Press, 1959); A. C. Crombie, *Medieval and Early Modern Science*, 2 vols., rev. ed. (New York: Doubleday, 1959); Crombie, *Robert Grosseteste*; Edward Grant, *Physical Science in the Middle Ages*, reprint (New York: Cambridge University Press, 1977); and Grant, *A Source Book*. For a description of the state of European medieval astronomy, see Olaf Pedersen, "Astronomy," in *Science in the Middle Ages*, ed. David C. Lindberg (Chicago: University of Chicago Press, 1978), pp. 303–37.

[37] This is a term I borrow from Benjamin Nelson; see *On the Roads to Modernity*, p. 178.

precisely here that one finds the great weaknesses of Arabic-Islamic civilization as an incubator of modern science.

External factors: cultural and institutional impediments

To focus our attention on these problems of institutional development, I propose the following outline of some impediments that blocked the development of modern science in Arabic-Islamic civilization. In setting out these areas of developmental blockage, no claim is made that they are mutually exclusive. It is useful, moreover, to reconsider these elements in the light of Merton's formulation of the ethos of science. As noted above, if the "prescriptions, proscriptions, and permissions" of the ethos of science are to obtain, they must be imbedded in the institutional apparatus of a society and civilization. If the scientific worldview is to prevail, its elements of universalism, communalism, organized skepticism, and disinterestedness must be given paradigmatic expression in the dominant directive structures of a society. A major clue as to why Arabic science failed to give birth to modern science can therefore be found in the fact that these norms were not institutionalized in the directive structures of Islamic civilization. The arresting of the breakthrough that would have institutionalized the normative ethos of science can be summarized as follows. In doing so, I slightly alter some of Robert Merton's terms so that one can more immediately grasp the wider societal (and civilizational) context that gives legitimacy to all the patterns of conduct.

The failure to develop universalism

The norm of universalism, in Merton's sense, consists in the standardizing of "pre-established impersonal criteria" for judging individual achievements.[38] I would suggest that this impersonalism is parasitic on the larger cultural norms that establish universalism (and personal standards of conduct) for classes of social actors. This is paradigmatically the domain of legal norms, and it is here that we see most dramatically the contrasting images of idealized conduct in the two civilizations. It is here that we see the greatest resistance to the creation of a rationally ordered, hierarchical set of universal legal norms, and therewith the failure to produce universal scientific norms for a scientific community.

If the norm of universalism is to prevail, then all potential participants in

[38] Merton, *The Sociology of Science*, p. 270.

social interaction must be placed on an equal footing. This is done abstractly by creating a set of impersonal standards that apply to all actors regardless of their social status, station in life, or community and ethnic origins. Persons and acts are then judged according to universal standards such as the internal standards of the activity or discipline at hand. Notions of reasonableness also come into play and these have to do with customary practice as understood by the gatekeepers in charge. But in order for these conditions to come into existence, potential participants must be judged as morally neutral or otherwise independent of the taint that is ascribed to membership in a variety of community, ethnic, religious, and similar particularistic groups.

In the case of Arabic-Islamic culture, it proved virtually impossible to achieve this level of moral and ethical neutrality in the realm of thought. And this is so, above all, because of the particularistic nature of Islamic law itself. Consequently, all developments in Islamic law served to reinforce a great variety of particularisms, instead of creating a universal level of discourse.

In addition to being a sacred law, Islamic law is a composite of four major schools of law: the Hanafi, Maliki, Shafi'i, and Hanbali, each named after its personal founder. Over the course of Islamic history there have been hundreds of such personal schools, but most died out, leaving only these four. In the twelfth and thirteenth centuries one still hears echoes of the Zahiri school through the very conservative school founded in Spain by Ibn Hazm, whose influence continued until the eventual expulsion of the Muslims by the Christians in 1492.[39] The four major schools coalesced around powerful individuals whose unique religious and legal gifts allowed them (or one of their followers) to gain a significant following and to develop legal points of view sufficiently distinct as to constitute separate traditions. Accordingly, these religious qua legal identities continued to structure social interaction at all levels ever after.

For example, when the movement to establish colleges (madrasas) throughout the Islamic world was launched in the eleventh century, each college appointed a law professor, the mudarris, who belonged to one of the schools of law, and that meant that the college then became an institution exclusively dedicated to that legal perspective. As we observed in Chapter 4, even when (in the late fourteenth century) more than one school of law was represented in an Islamic madrasa, the professor went from one group of students to another in order to avoid mixing the students and their respective

---

[39] Joseph Schacht, *An Introduction to Islamic Law* (Oxford: Oxford University Press, 1964), is the definitive work on the early schools of law; but also see N. J. Coulson, *A History of Islamic Law* (Edinburgh: Edinburgh University Press, 1964), chap. 3; and Makdisi, *The Rise of Colleges* (Edinburgh: Edinburgh University Press, 1981), pp. 2ff.

systems.[40] The pattern was thereby established such that no effort was made to integrate the legal schools of thought, to overcome the "discord of discordant texts" and to a fashion a single "legal science" into a uniform and universal system of law. Thus, on the most basic level of intellectual discourse, the particularities of one's legal tradition (madhhab) prevented direct dialogue and discussion.

Islamic law, furthermore, retains a very deeply ingrained personalistic bias that manifests itself on many levels. In the first instance, if an individual requires a legal ruling, a fatwa, he may apply to as many legists (qadis, muftis, or similar figures) as he wants until he gets the opinion closest to his wishes.[41] If he fails to get satisfaction within his own school of law, he may turn to one of the others. This situation likewise stems from the lack of the idea of jurisdiction, that is, a delimited domain of legitimacy. No doubt in the West this concept was a product of the investiture controversy and the legal separation of sacred and secular worlds. But the larger point is that without a clear conception of the legitimate domain for social action one can never be sure that one is indeed in a neutral space where the agreed upon principles will prevail. Instead, everything is up for grabs, and may the stronger party win.

Many scholars have also noted the extremely personalistic or individualistic nature of Islamic law as it applies to legal actions taken by the individual. For example, when a person decides to establish a charitable trust (waqf), he draws up the stipulations of the trust, and the provisions he spells out have the force of law.[42] The founder thus declares his intent and purposes and later gives a copy of the document to a qadi who places it on file. Of course the qadi may reject parts (or all) of the provisions if they conflict with the tenets of Islam, but the document is a legal instrument, though it is drawn without standard legal formulae.

On a very different level, we may also note that in Islamic criminal law, such acts as murder and bodily assault are treated as private affairs that permit the victim to retaliate. They are not treated as matters of public interest in which the state has a proprietary concern.[43] Such crimes – *quesas* crimes – are defined as crimes involving retaliation, and this right to retaliation on the part of the injured party is what defines them. The alleged

---

[40] Makdisi, *The Rise of Colleges*, p. 304.
[41] Ibid., p. 277; and Coulson, *Conflicts and Tensions in Islamic Jurisprudence* (Chicago: University of Chicago Press, 1969).
[42] Schacht, as cited in Makdisi, *The Rise of Colleges*, p. 35.
[43] See Schacht, *Introduction to Islamic Law*, and Matthew Lippman, Sean McConville and Mordachai Yerushalmi, *Islamic Criminal Law and Procedure* (New York: Praeger, 1988), chap. 3.

victim and/or his family may retaliate in a completely unpredictable manner. The area in which the central authority maintains a continuing vigilance is that of religion and "crimes against god," the *hudud* crimes. At the same time, the realm of personal injuries is greatly reduced, since "the concept of negligence is unknown to Islamic law."[44] Here again, there is an absence of clearly defined universal standards to which individuals are held accountable.

Instead of establishing a set of universal standards, Islamic law sought to place all acts on an ethical continuum that included the following five categories: (1) obligatory, (2) prohibited, (3) permissible, (4) praiseworthy, and (5) blameworthy (or optional). This left a large area of optional (ta'zir) crimes (and punishments), however, which are in principle violations of the spirit of Islam and the Quran. Consequently, these punishments could be invoked at the discretion of any particular legist or market inspector (muhtasib) as he saw fit. But whenever the legal punishments were put into effect, they were again dispensed according to a highly particularistic set of considerations.

For example, the Shafi'i legist al-Mawardi (d. 1058) wrote the following:

Discretionary punishment is inflicted in cases of offenses for which the Shari'a has not established written [hadd] punishment. . . . It has this in common with hadd penalties: it, too, is a means of punishment which differs with the type of offense. However, discretionary punishments differ from hadd punishment in three respects:

The punishment for respectable persons belonging to the upper classes is less than for low class persons who lead a bad life. . . . Discretionary punishment thus varies according to the status of a person, whereas all men are treated the same way in the application of the hadd punishments. Thus in the case of a man of high standing, it may be enough to turn away from him; with a man of lesser rank it may suffice to speak to him sternly; another may have to be reprimanded sharply in humiliating terms, but without slanderous or injurious implications. Finally, those in the lowest group shall be imprisoned for such a term as may be necessitated by their rank in society and their offense. Some will be held for a day, some for a longer time and some even for an indefinite period. . . . Others will be exiled, if their offenses would tempt other believers to do wrong. . . . Some finally will be flogged, the number of lashes dependent on the gravity of the offense and the behavior of the offender.[45]

In short, Islamic law set the prime example of treating all cases according to the particularities of the case and the individual, and it thereby refrained from establishing a set of uniform and universal principles of fairness and justice. As Joseph Schacht put it, "The aim of Islamic law is to provide

---

[44] Schacht, *Introduction to Islamic Law*, p. 182.
[45] As cited in H. Liebesny, ed., *The Law of the Near and Middle East* (Albany, N.Y.: SUNY Press, 1975), 229.

concrete and material standards, not to impose formal rules on the play of contending interests, which is the aim of secular law."[46] And this stress on the substantive and particularistic aspects of human relations "leads to the somewhat surprising result that considerations of good faith, fairness, justice, truth, and so on play only quite a subordinate part in the system."[47]

In contrast to this, the legists of the West, the Romanists as well as the canonists, sought to achieve a uniform structure of law in which formal and abstract principles served to produce uniform treatment and result. In addition to such ideas as good faith, justice, and so on, they argued that such principles as the reasonableness of the law (or custom), including the incidence of its appearance among different groups of people, its longevity, and matters of equity should be taken into account in the determination of the law. Once defined and promulgated, laws ought to apply universally and equally to all men, kings included, and should not be suspended by local leaders. Similarly, European legal scholars worked out various hierarchies of order and jurisdiction, so that the theoretical bounds for all legal codes could be known; that is, there is a limited domain (sovereignty) for even the divine law, as well as ecclesiastical law (canon law), royal law, manorial law, urban law, and so on. There was an implicit hierarchy of priority regarding statutory and judge-made law. In Islamic law, of course, statutory law was unknown because it was anathema to the Shariʿa.

In a word, one might say that the model of universalism in the West is to be found in its legal system and the model of particularism is found in Islamic civilization in its law. While the animus of Western law was directed toward establishing universal standards that conformed to natural law and natural reason,[48] Islamic culture and its law remained a sacred law, which, despite the legal notion of ijma, that is, consensus of the legal scholars, relied on a personalistic system that dispensed justice through the competing schools of law. This was so because no systematically organized set of laws and principles corresponding to the European canon law (worked out initially by Gratian in the twelfth century) was ever forthcoming in Islamic law. Similarly, the idea of precedent was lacking,[49] and without this idea and that of jurisdiction, that is, limited sovereignty, there could be no uniformity of law in theory or practice (on which, more below). In short, Islamic law in its spirit and its application institutionalized a thoroughgoing particularistic

---

[46] Schacht, *Introduction to Islamic Law,* p. 203.
[47] Ibid.
[48] See Chap. 4; but see also Harold Berman, *Law and Revolution,* pp. 140, 145–7, 196.
[49] Herbert Liebesny, "English Common Law and Islamic Law in the Middle East and South Asia: Religious Influences and Secularization," *Cleveland State Law Review* (1986/6) 34: 19–33.

and personalistic approach to all human encounters. For that reason it proved impossible to establish a neutral zone of scientific inquiry in which a singular set of universal standards – free from the incursions of religious law – could apply without interference.

The failure to develop autonomous corporate bodies

This aspect of the breakthrough to modern science has generally been overlooked. Sociologists (and historians) of science frequently refer to the autonomy of science, but they do so in a completely modern context in which corporate autonomy and legal autonomy are taken for granted. It is a domain hinted at in Emile Durkheim's reference to the "precontractual foundations of contract," but otherwise it has not been discussed in the literature.[50] Similarly, Robert Merton's otherwise useful scheme does not recognize this crucial area. Historians of science have been aware that the autonomy of science was often challenged in the early modern era, and they therefore looked to "scientific societies" and even "invisible colleges" as the incubators of modern science.[51] This view is now in disfavor, especially in the light of reappraisals of the place of science in the medieval university by medieval historians of science. Here again the very existence of scientific societies presupposes a state of legal autonomy that cannot be assumed in Islam or in the West prior to the twelfth and thirteenth centuries.

The failure of legally autonomous corporate bodies to emerge in Arabic-Islamic civilization (prior to the borrowing of Western civil codes in the nineteenth century) is likewise a product of the unique character of Islamic law. Very largely this stems from deep-rooted religious and metaphysical commitments regarding the unitary character of the Muslim community, the umma. From a theological point of view all believers are equally members of God's community, the community of the faithful. Since God has set out the rules for the proper mode of conduct so that the believer will "pass the reckoning on the day of judgment," these rules apply eternally and to every individual soul. Moreover, these laws of God – the Shari'a – are said to be a complete and perfect blueprint for society, and since no provision was made for separating one group of believers from another, it is inconceivable that there should be multiple legal statuses (other than those of kinship, that is,

---

50 Durkheim, *The Division of Labor* (New York: Free Press, 1933), especially pp. 206–16; and see Parsons, *The Structure of Social Action* (New York: Free Press, 1968), 2: 230f.
51 Martha Ornstein, *The Role of Scientific Societies in the Seventeenth Century* (Chicago: University of Chicago Press, 1938); and also Roger Hahn, *The Anatomy of a Scientific Institution: The Paris Academy of Sciences, 1666–1803* (Berkeley and Los Angeles: University of California Press, 1971).

husband and wife, father and son, and so forth) which would confer new or unique legal privileges on one group of Muslims. Furthermore, the whole Muslim community stands in constant judgment by God, and therefore no domain can be separated legally from another and no special benefits or exemptions from holy writ can be granted. Accordingly, the history of Islamic law shows that the idea of autonomous entities, above all, a separation of religious from secular jurisdiction, is foreign to Islamic law. As noted earlier, this separation of the religious from the secular spheres, such as occurred in the European Middle Ages, is one of the most fundamental breakthroughs required for the development of a science of law and for the rise of modern science itself.

Islamic law, moreover, did not develop the idea of a juridic person. As Joseph Schacht put it, "Public powers are, as a rule, reduced to private rights and duties, for instance the right to give a valid safe-conduct, the duty to pay the alms-tax, the rights and duties of the persons who appoint an individual as Imam or Caliph."[52] Islamic law had no provision for legally autonomous groups: corporate personalities such as business corporations, guilds, cities, towns, or universities did not exist in Islamic law. Nor were legally autonomous professions such as lawyers recognized by Islamic law.[53] In fact, Schacht observes:

The whole concept of an institution is missing. The idea of a juridic person was on the point of breaking through but not quite realized in Islamic law, [yet] this did not happen at the point where we should expect it, with regard to the charitable foundation of waqf, but with regard to the separate property of a slave who is being sold not as an individual but together with his business as a running concern.[54]

David Santillana expressed the same view: "Muslim jurists do not know – and that is easy to understand if we think of the political and social differences between the Islamic state and those of the Roman type – [either] the juridical personality of municipalities, [or] . . . that of collectives of persons such as guilds."[55]

In the realm of penal law, Islamic law likewise failed to separate religious rights and duties from those obligations or privileges that might derive from membership in various entities – corporate or collective. As Joseph Schacht put it:

[52] Schacht, "Islamic Religious Law," in *The Legacy of Islam*, 2d ed. (New York: Oxford University Press, 1974), p. 398.

[53] See F. Ziadeh, *Lawyers: The Rule of Law and Liberalism in Egypt* (Stanford, Calif.: Hoover Institution, 1968).

[54] Schacht, "Islamic Religious Law," p. 398.

[55] David Santillana, *Istituzioni di diritto musulmano malichita* 1: 170-1, as cited in S. M. Stern, "The Constitution of the Islamic City," in *The Islamic City*, ed. A. H. Hourani and S. M. Stern (Philadelphia: University of Pennsylvania Press, 1970), p. 49.

Islamic law distinguishes between the rights of God and the rights of human beings. Only the rights of God have the character of a penal law proper, or a law which imposes penal sanctions on the guilty. Even here, in the centre of penal law, the idea of a claim on the part of God predominates, just as if it were a claim on the part of a human plaintiff.[56]

In the area belonging to redress of torts, Islamic law did not distinguish between legal categories of fault, criminal responsibility, and just punishment:

Whatever liability is incurred here, be it retaliation or blood-money or damages, is subject of a private claim, pertaining to the rights of humans. In this field, the idea of criminal guilt is practically non-existent, and where it exists it has been introduced by considerations of religious responsibility. So there is no fixed penalty for any infringement of the rights of a human to the inviolability of his person and property, only exact reparation of the damage caused. This leads to retaliation for homicide and wounds on one hand, and to the absence of fines on the other. There are a few isolated doctrines in some schools of Islamic law which show that the idea of a penal law properly speaking was on the point of emerging in the minds of some Islamic scholars at least, but again, as was the case with the juristic person, it did not succeed in doing so.[57]

In sum, the major source for establishing a domain of autonomy – law – was diametrically opposed to the idea of competing (or complementary) jurisdictions of legal administration, and opposed to the idea of granting particular rights to autonomous classes of individuals. Hence, any form of corporate autonomy – guild, city, university, scientific society, business, or professional corporation – was ruled out by the Islamic conception of sacred law. It thereby blocked the creation of autonomous educational institutions with their rights and privileges such as occurred in Latin Europe in the twelfth and thirteenth centuries.

The persistence of particularism in institutions of higher learning

Another dimension of the personalistic nature of learning in the educational institutions of Islam can be seen in the perpetuation of the ijaza (authorization) system. The persistence of this form of apprenticeship, whereby a student attached himself to a learned teacher and scholar, marks a major turning point in the cultural divide separating modern and ancient science, as Joseph Ben-David has noted (though he is silent on the role of Arabic science and civilization as a contributor to the development of modern

---

[56] Schacht, "Islamic Religious Law," p. 398.
[57] Ibid., p. 399.

science).[58] The point is well taken that even within official institutions of higher education such as the madrasa, no generalized system of teaching and examination developed whereby one received a degree or certificate of achievement in a clearly defined area and which was certified by the institution itself. In tracing the various phases of "the license to teach," George Makdisi argues that the license to transmit hadiths eventually became, in jurisprudence, the "license to teach law and to issue legal opinions."[59] Nevertheless, the system of collecting permissions or licenses to transmit from individual scholars persisted:

The license to teach law and legal methodology, and to issue legal opinions, conferred upon the candidate authority based on his competence in law and legal methodology. This authority and competence resided in the ʿalim (pl. ʿulamaʾ), the learned man of religion, specifically in the jurisconsult, faqih. When the master-jurisconsult, the mudarris, granted the license to teach law and issue legal opinions, he acted in his capacity as the legitimate and competent authority in the field of law. When he granted the license to the candidate he did so in his own name, acting as an individual, not as part of a group of master-jurisconsults acting as a faculty; for there was no faculty. Throughout its history down to modern times, the ijaza remained a personal act of authorization, from the authorizing ʿalim to the newly authorized one.[60]

It is evident, then, that the advent of the licentia docend of the early modern European universities represented a complete break with the particularism of Islamic education whereby one studied with a self-chosen master who conferred his approval on the student.[61] While such a personalized system had its benefits for students, its defect was that no impersonal and objective standards of teaching and evaluation evolved that could serve as a common reference point in the advance of knowledge. It is due to this personalistic and particularistic factor that one finds literally hundreds of schools of law over the centuries, each founded by a faqih who, through the power of his intellect and the magic of his personality, established his own school of law capable of issuing its own rulings (fatwas), unconstrained by a body of precedents and universal legal principles. Thus law, jurisprudence, as the paradigmatic body of knowledge in Islamic civilization, established a model of inquiry antithetical to that required of modern science, that is, a system based on personal authority rather than collective or impersonal collegial standards.

[58] Ben-David, *The Scientist's Role in Society* (Englewood Cliffs, N.J.: Prentice-Hall, 1971), pp. 46–7.
[59] Makdisi, *The Rise of Colleges*, p. 270.
[60] Ibid. p. 271, as well as pp. 129 and 133.
[61] Cf. Madkisi, "Madrasah and University in the Middle Ages," *Studia Islamica* 32 (1970): 255–64.

In contrast, the European universities established an examination system within and between universities. Within each university students were examined orally by a panel of faculty members on a more less standardized set of topics and readings. When doubt arose as to the integrity of a student's training at another university, the student was forced to undergo reexamination. This procedure was most notable in the case of medicine.[62] Moreover, the license establishing the right to teach medicine became an exclusive right to practice medicine, and this development pushed toward a general upgrading of medical practice and medical knowledge. What was critical, however, was the collective appropriation of uniform standards of teaching (and practice) by a professional group located in an institutionally autonomous location – the university, but also in professional guilds – and hence the exclusion of extraprofessional and religious censors as overseers. As noted in Chapter 5, it was to the muhtasib, the religiously oriented market police, not a professional body of physicians, that the duty of regulating medical practice was given in the Islamic world. Similarly, the professional study of astronomy became lodged in the mosque, in the religious office of muwaqqit, not in the madrasas, and, except for short periods, not in the observatory.

Elitism versus communalism

Both Islamic and Judaic cultures contained a strong bias against allowing open access to knowledge by the masses.[63] This derived from a religious injunction to the effect that, if the person were a truly religious person, "he would know that discussion of such things is forbidden."[64] This applied especially to unsettled issues in philosophy, religion, and theology. The penultimate illustration of the effect of this cultural attitude on learning is Maimonides' book, *The Guide of the Perplexed*.[65] In this work Maimonides sought to aid a younger scholar in his search for truth, but knowing that many issues, such as the nature of the universe and its creation, the powers of God, the modes of reasoning, the nature of religious law, and so forth, were forbidden to open discussion (in a book, for example, that anybody might read), he created an elaborate labyrinthine structure. The result has been that many readers of his

[62] See A. B. Cobban, *The Medieval Universities* (London: Methuen, 1975), p. 31; Rashdall, *Universities* 3: 140–145; and Vern Bullough, *The Development of Medicine as a Profession* (New York: Hafner, 1966), pp. 106f.
[63] The classic study of this is Leo Strauss, *Persecution and the Art of Writing*.
[64] Ibn Rushd, *Tahafut al Tahafut* (The Incoherence of the Incoherence), as cited in Barry Kogan, *Averroes and the Metaphysics of Causation*, p. 22.
[65] Maimonides, *The Guide of the Perplexed*, trans. Shlomo Pines (Chicago: University of Chicago Press, 1963), 2 vols.

book have spent their whole lives trying to decipher the meaning and intent of the book. As a purely spiritual exercise, such a discourse might have merit, but in science and natural philosophy, clarity, precision, and brevity of expression are essential values. Accordingly, the method of disguised discussion Maimonides was forced to use is a method of discourse representing enormous inefficiency in the presentation of ideas and information. While it seeks to subvert the norm of secrecy, it takes it as given.

The equally great Arab philosopher Averroes (Ibn Rushd, 1126–98), despite his commitment to the use of reason and the philosophical methods of Aristotle to interpret the Scriptures, enjoined silence on perplexing issues, above all, on those regarding difficult passages of the Scriptures. While the philosopher could in his view get to the true meaning of the Scriptures, he made it clear that this kind of advanced learning and interpretation was meant only for a tiny elite. The masses as well as the dialecticians – that is, the speculative theologians (mutakallimun) – were judged to be incapable of comprehending such matters and therefore were not to be addressed in such writings. "Allegorical interpretations," he argued, "ought not to be expressed to the masses, nor set down in rhetorical or dialectical books, i.e., books containing arguments of these two sorts as was done by Abu Hamid [al-Ghazali]. They should (not) be expressed to this class."[66] Likewise, before Averroes, the legist Ibn Hazm (d. 1064) argued that "knowledge ought to be disseminated, but its dissemination among the untalented and inept people is not only a waste of time, but prejudicial as well, for great harm is done to the sciences by these intruders who pretend that they are scholars, but who actually are ignorant."[67] As a result of this attitude, as we saw in Chapter 2, there were a variety of literary techniques for concealment and disclosure in Arabic-Islamic intellectual life. These included

alluding to certain doctrines only symbolically; scattering or suppressing the premises of an argument; dealing with subjects outside their proper context; speaking enigmatically to call attention to significant points; transposing words and letters; deliberately using equivocal terms; introducing contradictory premises by which to divert the reader; employing extreme brevity to state the truth; refraining from drawing obvious conclusions; i.e., silence; and attributing one's own views to prestigious forebears. (See Chapter 2, this volume, note 125.)

Such techniques, needless to say, run against the grain of the scientific ethos, which seeks brevity and clarity of expression, as well as the norms of universalism and communalism.

---

[66] *Averroes: On the Harmony of Religion and Philosophy*, reprint, ed. and trans. George Hourani (London: Luzac, 1976), p. 66.
[67] Anwar G. Chejne, *Muslim Spain*, p. 168; paraphrasing Ibn Hazm's *Kitabal-akhlaq*.

A practical result of the distrust of the masses, even the literate masses, was the complete rejection of the printing press after it arrived in Europe in the fifteenth century. Here, then, we see a direct assault on the Mertonian norm of communalism. In Merton's original formulation of this imperative, his intent was to point out the communal character of scientific knowledge, and the imperative to make one's findings public.[68] Secrecy was the antithesis of the norm of communalism.

Arabic-Islamic culture was highly ambivalent about the question of disseminating knowledge. On the one hand, the ijaza system maintained the personal link in knowledge transmission, but of course this would not prevent a copyist from making a business of supplying copied manuscripts to scholars and especially political rulers.[69] In addition, from earliest times sectarians and propagandists realized that one of the ways to spread their ideas was to create libraries in which their tracts and legal works could be found. It was a long-standing tradition for Islamic mosques to have a library attached in which books on all subjects were to be found. Of course, this followed the earlier pattern established among Christians in Syria, Persia, Iraq, Palestine, and Egypt.[70] While the intent was originally a religious one – to make knowledge of the Islamic sciences available – these repositories came to house all sorts of items representing the great literary traditions of the past, including Greek philosophy and the foreign sciences. In effect, a literary movement flourished in Islam that was aided and abetted by wealthy patrons who established magnificent libraries with thousands of manuscripts, open and accessible to the literate public.[71] Conversely, during spasmodic periods of sectarian intolerance, these libraries and all their contents were ransacked and burned.

When the techniques and instruments of printing arrived in Islam, they were not put to use for the purpose of printing books for the public. We know that the technique of block printing existed in Egypt and Persia in the thirteenth century, and that a few items were printed, including

[68] Merton, *The Sociology of Science*, p. 273.
[69] On copyists and booksellers, see Johannes Pedersen, *The Arabic Book* (Princeton, N.J.: Princeton University Press, 1984), chap. 4; and A. Demeerseman, "Un étape décisive de la culture et de la psychologie sociale islamiques: Les données de la controverse autour du problème de l'Imprimerie," *Institut des Belles Lettres Arabes* (Tunis) 17 (1954): 113–19.
[70] Ruth S. Mackensen, "Background of the History of Moslem Libraries," *American Journal of Semitic Languages and Literature* 52 (1935/6), p. 104.
[71] See the series of articles by Ruth S. Mackensen: "Four Great Libraries of Medieval Baghdad," *Library Quarterly* 2 (1932): 279–99; "Background of the History of Moslem Libraries," *American Journal of Semitic Languages and Literature* 51 (1934/5): 114–25; 52: 22–33; and 104–10; and "Arabic Books and Libraries in the Umaiyad Period," *American Journal of Semitic Languages and Literature* 52 (1935/6): 245–53; 54: 41–61; and Makdisi and Pedersen, the section on libraries in their article on madrasas, *EI²* 5: 1123.

money.[72] But the use of printing did not take hold, and there was no renaissance of literature nor a bursting forth of printed material.

Likewise, with the invention of the printing press in Europe in the fifteenth century, the invention was rejected in all Islamic lands by Muslims. The banning of the printing press, above all for Islamic religious materials, was done out of fear that such materials might "fall into the wrong hands," a sentiment that persisted in Cairo as late as the early nineteenth century when the English traveler E. W. Lane visited Egypt. "It was argued that God's name, which appears on every page of a Muslim book, could become defiled through this process [of being put into print], and it was feared books would become cheap and fall into the wrong hands."[73]

Within three decades of the appearance of the first printed book and the first German Bible in Europe in 1455, the Muslims banned the printing press.

The Turkish sultan, who was not only the nearest but also the most powerful Muslim ruler, was quick to realize what was happening in Europe, and he feared the consequences this new activity might have among his subjects. A ban on the possession of printed material was proclaimed by Sultan Bayazid II as early as 1485, and was repeated and enforced in 1515 by Selim I, who shortly thereafter became the conqueror of Egypt and Syria, the central lands of Islam.[74]

Consequently, the first Arabic-language books were printed in Europe by Christians in the early sixteenth century, and the ban was not fully removed until the early nineteenth century.[75] In short, there was in Arabic-Islamic civilization a strong distrust of the common man, and efforts were made after the golden age to prevent his gaining access to printed materials. According to Johannes Pedersen, "printing [in Syria] only really became a thriving business in 1834, when American Protestant missionaries moved an Arabic printing press from Malta to Beirut. They inaugurated a new era by printing a lengthy series of books introducing European culture to the Islamic world."[76] Likewise the emergence of a genuine public marked by the presence of daily newspapers had to wait until the mid-nineteenth century. "An official weekly began publication as early as 1832, but it was not until

---

[72] See the article on printing under "matba'a" (printing) in the *EI²* 6: 794–807. I am grateful to Professor Bernard Lewis for bringing this discussion to my attention. Also see T. F. Carter, *The Invention of Printing in China and Its Spread Westward*, ed. L. C. Goodrich (New York: Ronald Press, 1955), chap. 15, "Islam as a Barrier to Printing."

[73] Pedersen, *The Arabic Book*, p. 137; and E. W. Lane, *The Manners and Customs of the Modern Egyptians*, 5th ed. (London: John Murray, 1869), originally composed in the mid-1830s), pp. 281f.

[74] Pedersen, *The Arabic Book*, p. 133.

[75] Ibid., p. 134.

[76] Ibid.

1876 that other papers made their appearance. The first daily newspaper, *al-Muqattam* (named after a hill outside Cairo), was launched in 1889."[77] Despite the availability of the technology to create a free press and to make knowledge public in the Arabic-Islamic world, this development occurred only when Western incursions into the Middle East in the nineteenth century introduced this new form of cultural communication.

Disinterestedness and organized skepticism

These two elements of the scientific ethos seem to represent preeminently modern values and, for that reason, one might suppose that it would be futile as well as anachronistic to look for them in this early period. On the other hand, it would be equally unhistorical to imagine that any form of skeptical and disinterested inquiry had to wait for the age of reason of the seventeenth century to makes its appearance. We should remind ourselves that the philosophical (and scientific) context of the High Middle Ages, in both Islam and the West, was dominated by religiously sanctioned conceptions. And it was in the context of exploring, criticizing, reformulating, and sometimes completely rejecting this traditional cosmology – from the science of the heavens down to the laws of terrestrial motion – that the medievals displayed boldness.

According to Robert Merton, the norm of disinterestedness ought to be seen as "a distinctive pattern of institutional control of a wide range of motives," and "once the institution enjoins disinterested activity, it is to the interest of scientists to conform on pain of sanctions," and even psychological conflict when these norms have been internalized.[78] The most effective means for translating the norm of disinterestedness into practice, Merton argues, is the subjugation of scientists to the accountability of their scientific peers. On the other hand, the norm of organized skepticism, "is both a methodological and an institutional mandate."[79] Methodologically, "the temporary suspension of judgment and the detached scrutiny of beliefs in terms of empirical and logical criteria" is both a mandate and a source of "conflict with other institutions."[80] In other words, the detached scrutiny of all realms of experience, including the epistemological, metaphysical, and social foundations of the natural and social worlds, is bound to be a profoundly disturbing experience for those affected. And those who take up this disinterested course of inquiry are inevitably pushed into conflict with

[77] Ibid., p. 137.
[78] Merton, *The Sociology of Science*, p. 276.
[79] Ibid., p. 277.
[80] Ibid.

other social institutions. At the same time, Merton was suggesting that if the motive force for inquiry was to be effective, it must be an institutional, not just a personal, mandate.

Once alerted to these themes and their dynamics, we can see all of them coalescing in the newly created universities of the twelfth and thirteenth centuries of Europe. The breakthrough to autonomous, self-governing institutional entities in Western Europe was an event of singular importance for the political, social, religious, and intellectual development of Western civilization. The universities of Europe are but one example of this species of self-governing associations that one can find in the medieval period, for they parallel the development of merchant and commercial guilds throughout Europe. Some have even suggested that the sequence of initiation, training, and certification within the universities, parallels the path of certification and advancement within the craft guilds. For example, students were matriculated, initiated as a bachelor upon taking up studies, granted the title of master of arts "upon successful completion of such set intellectual exercises as disputation, determination, or defense of the thesis, and formal inception into the guild of teaching masters." This paralleled "the stages in a craft guild of apprentice, journeyman, and finally master workman following the completion of a perfect piece of work (a shoe, a chest, or the like)."[81]

The differences between teaching masters and master workmen, however, was not just that of intellectual versus manual work. It was the fact that the university masters were granted special privileges, not in the sense of power over others, but in the sense of privileged exemption from civil obligations and duties. They were to become "a sort of intellectual knighthood."[82] Unlike philosophers and intellectuals in Arabic-Islamic civilization, they were especially protected from the ire, warranted or not, of the local townspeople. Beyond that they had economic advantages, such as exemption from local levies, as well as civil exemption from the jurisdiction of the town in which the university was located.[83]

We should also not underestimate the magnitude of the step taken when it was decided (in part, following ancient tradition) to make the study of philosophy and all aspects of the natural world an official and public enterprise. If this seems a mundane achievement, it is due to our Eurocentrism which forgets that the study of the natural sciences and philosophy was banned in the Islamic colleges of the Middle East and that all such inquiries were undertaken in carefully guarded private settings. Likewise, in China,

[81] Pearl Kibre and Nancy Siraisi, "The Institutional Setting: The Universities," in *Science in the Middle Ages*, ed. D. C. Lindberg (Chicago: University of Chicago Press, 1978), pp. 120–1.
[82] Ibid., p. 121.
[83] Ibid.

there were no autonomous institutions of learning independent of the official bureaucracy; the ones that existed were completely at the mercy of the centralized state. Nor were philosophers given the liberty to define for themselves the realms of learning as occurred in the West.[84]

It is therefore a great irony that when the European medievals first discovered the great wealth of intellectual treasures in Arabic, they referred to "our Arab masters," no doubt imagining that they were uninhibited freethinkers, able to pursue intellectual queries wherever they wished.[85] Europeans, having only the evidence of intellectual vivacity at hand, did not have an accurate picture of the institutional arrangements that prevailed in Arabic-Islamic civilization. They knew little of the consistent repression experienced by intellectuals, even such as the great Ibn Rushd, in that civilization.

Accordingly, the new institutional arrangements created in Europe in the twelfth and thirteenth centuries established a situation that was institutionally unthinkable in the Middle East, that is, the unfettered study and public discussion of philosophy and the natural sciences in the legitimate institutions of the realm. The European medievals, fully cognizant or not, created autonomous, self-governing institutions of higher learning and then imported into them a methodologically powerful and metaphysically rich cosmology that directly challenged and contradicted many aspects of the traditional Christian worldview. Instead of holding these foreign sciences at arm's length, they made them an integral part of the official and public discourse of higher learning. By importing, indeed, ingesting the corpus of the new Aristotle and its rigorous methods of argumentation and inquiry, the intellectual elite of the European medievals established an impersonal intellectual agenda whose ultimate purpose was to describe and explain the world in its entirety in terms of causal processes and mechanisms. This disinterested agenda was no longer a private, personal, or idiosyncratic preoccupation, but a shared set of texts, questions, commentaries, and, in some cases, centuries-old expositions of unsolved physical and metaphysical ques-

---

[84] Etienne Balazs, *Chinese Civilization and Bureaucracy* (New Haven, Conn.: Yale University Press, 1964), table 4, p. 147; Needham, *The Grand Titration* (London: Allen and Unwin, 1969), pp. 179; and Sivin, "Why the Scientific Revolution Did Not Take Place in China – or Didn't It?" in *Transformation and Tradition*, ed. E. Mendelsohn (New York: Cambridge University Press, 1984), p. 535. More on this in Chaps. 7 and 8.

[85] Cf. Adelard of Bath's letter to his nephew, regarding the use of reason and the natural world: *Dodi Ve-Nechi (Uncle and Nephew)* with introduction and English translation by Hermann Gollancz (London: Oxford University Press, 1920). Also see Etienne Gilson, *Reason and Revelation in the Middle Ages* (New York: Charles Scribners, 1938) for the ironic sources of reason and rationality found by the Christian medievals in the writings of Averroes and other Arabic writers.

tions that set the highest standards of intellectual inquiry. By incorporating the natural books of Aristotle into the curriculum of the medieval universities, a disinterested agenda of naturalistic inquiry was institutionalized. It was institutionalized as a curriculum, a course of study, in which the study of logic and the exact sciences, especially at Paris and Oxford, took pride of place. Hugh of St. Victor, for example, declared that logic should be the first of the seven liberal arts, since it "provides ways of distinguishing between modes of argument and the trains of reasoning themselves. . . . It teaches the nature of words and concepts, without both of which no treatise of philosophy can be explained rationally."[86]

As a set of intellectual puzzles, this new corpus was a research agenda for the elite of the university. Anyone who reads the naturalistic books of Aristotle, such as the *Physics, Meteorology, On Generation and Corruption,* and so forth, cannot but be impressed with the extraordinary concentration of energy on the naturalistic understanding of the world.[87] This is evident in the *Physics* where Aristotle enunciates the naturalistic framework and makes it evident that the highest form of knowledge is based on principles, causes, or elements, and that it is through acquaintance with these that knowledge and understanding is attained. "For we do not think that we know a thing until we are acquainted with its primary causes or first principles," Aristotle wrote, "and have carried our analysis as far as its elements. Plainly . . . in the science of nature too our first task will be to try to determine what relates to its principles."[88] Nor can one mistake the driving force of the dispassionate search for truth. Consider just the opening passages of Aristotle's *On the Soul:*

Holding as we do that, while knowledge of any kind is a thing to be honored and prized, one kind of it may, either by reason of its greater exactness or of a higher dignity and greater wonderfulness in its objects, be more honorable and precious than another, on both accounts we should naturally be led to place in the front rank the study of the soul. The knowledge of the soul admittedly contributes greatly to the advance of truth in general, and above all, to our understanding of Nature, for the soul is in some sense the principle of animal life. Our aim is to grasp and understand, first its essential nature, and secondly its properties; of these some are

---

[86] As cited in Kibre and Siraisi, "The Institutional Setting," p. 127.

[87] Among others, see Richard McKeon, "Aristotle's Conception of Scientific Method," in *Roots of Scientific Thought*, ed. P. P. Wiener and A. Noland (New York: Basic Books, 1957), pp. 73–89; and Charles Schmitt, "Toward a Reassessment of Renaissance Aristotelianism," *History of Science* 11 (1973): 159–93; as well as William A. Wallace, *Causality and Scientific Explanation* (Ann Arbor: University of Michigan, 1972), especially vol. 1.

[88] *The Complete Works of Aristotle*, rev. Oxford trans., ed. Jonathan Barnes (Princeton, N.J.: Princeton University Press, Bollingen Series, 1984), 1: 315.

thought to be affections proper to the soul itself, while others are considered to attach to the animal owing to the presence of soul.[89]

Furthermore, in the universities there were both standard lectures and unscheduled public discussions (occurring regularly in the afternoon) in which those present recited the arguments that had previously been accepted regarding significant matters of logic and philosophy, and in which new arguments were set forth. In special sessions, the quodlibetal discussions, random questions were invited from the audience, to which the student of a professor would offer the first answer. Then, a day or two later, the professor would offer his fully considered response.[90] The public, open, and communal character of this discourse is unmistakable.

As noted in Chapter 3, the first wave of this new enthusiasm for naturalistic inquiry had grown out of the love affair with Plato's *Timaeus,* producing the various Platonisms of the eleventh and twelfth centuries. It was precisely here, in the area of cosmology (and indeed ontology) that the European medievals displayed intellectual daring and sometimes originality. There were a number of scholars, many associated with the School of Chartres, who followed the Platonism of the time and declared the world to be fully rational and explicable in terms of rational inquiry. What is of utmost importance is the fact that some of these thinkers directly challenged the authority of the Scriptures as a source of knowledge about the physical world. As key figures in the schools and universities of the time, they expressed a daring form of organized skepticism regarding the revealed sources, thereby paving the way for unfettered scientific inquiry as we understand it. Theirry of Chartres (d. ca. 1156) proposed an exegesis of the Bible from a naturalistic point of view. William of Conches took it further by asserting the priority of physical reasoning: "And the divine page says, 'He divided the waters which were under the firmament from the waters which were above the firmament.' Since such a statement as this is contrary to reason let us show how it cannot be thus."[91] Furthermore, it was within the highest intellectual elite of Europe that one found those who argued for the priority of philosophy over theology in matters pertaining to the natural world. The general thrust of such inquiry was clear: "It is not the task of the Bible to teach us the nature of things; this belongs to philosophy."[92]

---

[89] Ibid., p. 641.
[90] Kibre and Siraisi, "The Institutional Setting," p. 131f.
[91] As cited in Tina Stiefel, "Science, Reason, and Faith in the Twelfth Century: The Cosmologists' Attack on Tradition," *Journal of European Studies* 6 (1976): 1–16 at p. 7.
[92] Chenu, *Nature, Man, and Society* (Chicago: University of Chicago Press, 1968), p. 12. Attentive readers will again recognize here the early medieval sources of Galileo's famous seventeenth-century statement, "The Intention of the Holy Spirit is to teach one how to go to heaven, not how the heavens go." See Chap. 3, note 50.

From this point of view, the organized skepticism that we associate with present-day, modern views of things has a long history in the West, and it begins not later than the twelfth and thirteenth centuries when the superiority of rational thought over biblical literalism was asserted by the moderni of the schools and universities. As Tina Stiefel put it, "The most daring of all intellectual enterprises" of this time was the work of a few scholars, including William of Conches, Thierry of Chartres, and Adelard of Bath,

all of whom wrote in the first half of the twelfth century, and all of whom were concerned with the strict application of critical, analytical thinking to all aspects of natural phenomena, whether astronomy or physiology. Relying on their faith in natural causation and in the atomist structure . . . of the cosmos, they both postulated and attempted to formulate a rational methodology for the investigation of rerum natura: they invented for themselves a new discipline – natural science.[93]

Of course, the new Aristotle presented a formidable challenge to Christian theology, yet it too was incorporated into the new curriculum.[94] As we saw above, the new Aristotle presented a new and imposing edifice of scientific and secular learning, and it was accompanied by Arabic commentaries. Together these two rational currents – the Platonism of the twelfth century and the new Aristotle – aided the foundation of an unending disinterested agenda of inquiry that became institutionalized in the curriculum of the European universities. Aristotle's natural books were not, as in Arabic-Islamic civilization, sequestered in private homes and carefully controlled intimate discussion groups. They were given center stage. As Professor Grant puts it, "For the first time in the history of Latin Christendom, a comprehensive body of secular learning, rich in metaphysics, methodology, and reasoned argumentation, posed a threat to theology and its traditional interpretation."[95] And it should be noted that these philosophical justifications of the natural study of the world (whether Platonist or Aristotelian) were far more sophisticated and powerful than their counterparts to be found in China in the thirteenth century. The neo-Confucian spokesmen of that period spoke rather vaguely of "the investigation of the nature of things" (more on which in Chapter 8).[96]

---

[93] Tina Stiefel, "'Impious Men': Twelfth-Century Attempts to Apply Dialectic to the World of Nature," in *Science and Technology in Medieval Society*, ed. Pamela Long (New York: New York Academy of Sciences, 1985), pp. 187–8.

[94] For more discussion on university curricula, see Chap. 5, section entitled "Universities and the West," and Chap. 9, section entitled "Science, Learning, and the Medieval Revolution."

[95] Edward Grant, "Science and Theology in the Middle Ages," in *God and Nature: Historical Essays on the Encounter Between Christianity and Science*, ed. David C. Lindberg and Ronald L. Numbers (Berkeley and Los Angeles: University of California Press, 1986), p. 52.

[96] See Needham, *SCC* 2: 455ff, and Wing-Tsit Chan, *Chu Hsi: Life and Thought* (New York: St. Martin's Press, 1987), pp. 136–7, 44.

In the West, however, this new metaphysics created an intellectual space within which men could entertain all sorts of questions about the constitution of the world. And this the medievals did. They asked whether the world had a beginning or whether it had always been in existence; whether there were other worlds and, if so, would the same physical laws obtain in such worlds. In that domain concerning speculation about time, space, and motion, they asked questions about the existence of a vacuum and its properties. Could God cause the earth to be instantaneously accelerated in a straight line, and if he could, would this produce a vacuum? "Would the surrounding celestial spheres [in a vacuum] collapse inward instantaneously as nature sought to prevent formation of the abhorred vacuum? Indeed, could an utterly empty interval, or nothingness, be a vacuum or space? Would a stone placed in such a void be capable of rectilinear motion? Would people placed in such vacua see and hear each other?"[97] These and dozens more such questions were asked. It is difficult to imagine a more crowded agenda of disinterested inquiry and organized skepticism among natural thinkers in any other period of time, or any other civilization.

Of course this outbreak of disinterested inquiry and freethinking within the academy did not go unremarked by Christian traditionalists. The allure of disinterested inquiry quickly projected scholars into conflict with the vested interests of the traditional religionists. Efforts were made to condemn certain arguments and assumptions that appeared to place limits on God's powers. Most notably, this took the form of the condemnation of 219 suspect ideas by the Bishop of Paris in 1277. This reaction, however, was too little and too late, for the teaching of Aristotle's works, including the sequence and timing of that instruction, had already been established by statute in Paris in 1255. In fact, the view is now firmly established that the condemnation of 1277 actually encouraged the philosophers and natural scientists of the fourteenth century to redouble their efforts in order to delineate the autonomy of their inquiries. The condemnation spurred philosophers to conduct a great variety of thought experiments, to imagine the impossible in the service of reconciling Aristotelian thought with Christian theology. This meant imagining non-Aristotelian possibilities, with the result that such speculations paved the way for the overthrow of the Aristotelian worldview in the sixteenth and seventeenth centuries.[98]

---

[97] Grant, "Science and Theology," p. 57.
[98] See Grant, "Science and Theology," pp. 55–9; and Grant, "The Condemnation of 1277, God's Absolute Power, and the Physical Thought in the Late Middle Ages," *Viator* 10 (1979): 211–44, among others. While this thesis is consistent with Duhem's insistence on the importance of the medieval precursors of Galileo and modern science, it does not agree with

At the same time, the philosophers of the universities asserted their right to continue their inquiries on a number of grounds, not least of all that of pursuing truth. As Mary McLaughlin pointed out, "Few phrases are repeated more frequently by the masters" of the thirteenth and fourteenth centuries, "than the words, 'friend of truth,' by which the philosopher's duty is described."[99] The legitimation for this was to be found not only in Aristotle and his commentators but in the Bible itself: "And ye shall know the truth, and the truth shall make you free."[100] Altogether, there were multiple sources, philosophical and religious, in Europe that served to create a new foundation for the study of the natural world and for challenging the Scriptures as the sole authority regarding that world. In sum, this probing philosophical and scientific curriculum based on the work of Aristotle was the main thrust of European universities for over four hundred years, from 1200 to 1650. During that period all those trained for the master of arts studied this curriculum, and it served thereby to shape and instill a major commitment to the norms of disinterestedness and organized skepticism that are at the heart of modern science.

Turning our attention to Arabic-Islamic civilization, such values as disinterested inquiry and organized skepticism, to the degree that they can be found in small circles of physicians and faylasufs (philosophers) outside the centers of learning in Arabic-Islamic civilization, received no validation by the mediatorial elite of Islam. The madrasas were closed to the teaching of science and philosophy, and official Islamic law, fiqh, denied that all men had reason in the Greek and Platonic sense. Nor did Islamic jurisprudence have any place for the idea of conscience, that inner moral agency that could guide the actor in moral dilemmas. Moreover, there was no room for organized skepticism within Islamic thought. The true believer was enjoined to show that whatever true opinions exist either are to be found in the Quran or are fully consistent with it.[101] This was at the root of the opposition to the

---

Duhem's assigning of 1277 as the birthday of modern science. According to Grant, "Despite the exaggerated and indefensible character of Pierre Duhem's claim that the Condemnation of 1277 was 'the birth of modern science,' he was right to emphasize the special significance of two articles, 34, which made it mandatory to concede that God could make more than one world, and 49, which compelled assent to the claim that God could move the heavens, or world, with a rectilinear motion even though such motion might leave behind a vacuum," p. 217.

99 Mary McLaughlin, *Intellectual Freedom and Its Limits in the Twelfth and Thirteenth Centuries*, reprint (New York: Arno Press, 1977), p. 308; and Grant, "Science and Theology," pp. 55–9.

100 John 8:32.

101 Of course, there were powerful voices of skepticism during the tenth and eleventh centuries in Islam, such as al-Farabi and Ibn Sina, who were not above criticizing the authority of the

teaching of philosophy in the madrasas. The view was widespread that studying philosophy made one impious, and those who took up its study were often attacked, as Goldziher has pointed out.[102] In Spain, "falsafa and tanjim were only cultivated in secret, as those who studied them were branded as *zindik* [a heretic], even stoned or burned."[103] For these reasons there was no opening for the public study of the natural sciences, and support for these inquiries under the law of religious charities, under the law of waqf, was not permissible. Exceptions to the violation of this rule, we have seen, were short-lived.

Furthermore, the idea of innovation in general implied impiety if not outright heresy. A variety of attitudinal structures served to inhibit the pursuit of certain lines of inquiry, and when these lines were transgressed by bold and original thinkers, there were no mechanisms by which those promising paths could be singled out and incorporated into ongoing and publicly legitimated forms of inquiry. Conversely, those who pursued innovative lines of inquiry were likely to provoke the ire of the religious traditionalists, and this was a further reason to avoid all publicity. One tradition of the Prophet in wide use claims that the "worst things are those that are novelties, every novelty is an innovation, every innovation is an error and every error leads to Hell-fire":

In its extreme form this principle has meant the rejection of every idea and amenity not known in Western Arabia in the time of Muhammad and his companions, and it has been used by successive generations of ultra-conservatives to oppose tables, sieves, coffee and tobacco, printing-presses and artillery, telephones, wireless, and votes for women.[104]

The role of the scientist, above all as the innovator, was neither institutionally permissible nor culturally tolerated in Arabic-Islamic civilization during this period.

Quran and the Prophet, but this had to be done with the utmost subtlety. On the other hand, it would seem that as the Arabic-Islamic empire became fully Islamized, this inclination to skepticism waned, as did the freedom of intellectuals to speak out. On the course of conversion to Islam, see Richard W. Bulliet, *Conversion to Early Islam: An Essay in Quantitative History* (Cambridge, Mass.: Harvard University Press, 1979). Also see Glick, *Islamic and Christian Spain*, pp. 33ff, and passim.

102 Ignaz Goldziher, "The Attitude of Orthodox Islam Toward the Ancient Sciences," in *Studies in Islam*, ed. Merlin Swartz (New York: Oxford University Press, 1981 [1916]), pp. 185–215.

103 J. Pedersen, "Madrasa," in *Shorter Encyclopedia of Islam*, ed. H. A. R. Gibb and J. H. Kramer (Ithaca, N.Y.: Cornell University Press, 1960), p. 306, following Makkari.

104 Bernard Lewis, "Some Observations on the Significance of Heresy in the History of Islam," *Studia Islamica* 1 (1953): 52.

Conclusion

In the foregoing discussion I have outlined some of the institutional and attitudinal impediments that prevented the rise of modern science in Arabic-Islamic civilization. From this it is evident that Islamic law was a major factor setting limits on the development of spheres of autonomy, perpetuating a particularistic and personalistic educational system. The cast of Islamic legal thought inhibited the development of universalistic laws and impersonal norms of evaluation and generally circumscribed the areas within which innovation could be practiced without fear of being labeled impious or heretical. Most importantly, Islamic law had no place for autonomous organizations, professional associations, or civic institutions. The one legally protected instrument of association was that of waqf – religious endowment – but this was directly tied to religious prescriptions and prohibitions. Beyond that, Islamic law possessed none of the legal mechanisms of change and transformation essential to the law of corporations.

As recently as the mid-1980s, concerned Muslim intellectuals lamented the fact that modern science had not taken root in the Islamic countries of the world, with the consequence that there is no Islamic Hong Kong, Singapore, or Japan.[105] According to Abdus Salam, president of the Third World Academy of Science, "of all civilizations of this planet, science is weakest in the lands of Islam."[106] The reason given for this is that "'secular', 'Western', and 'Eastern' science and technology have no basis in [the] Islamic ethos and Muslim culture. Their adoption makes Muslims less Islamic."[107] More simply put, modern science is perceived as un-Islamic, and those who embrace it are thought to have taken a first and fatal step toward impiety. It was for this reason that like-minded Muslims founded the Muslim Association for the Advancement of Science in 1985. According to these Muslims, all scientific ideas must be shown to be consistent with, if not derived from, the Shariʿa. The contemporary Islamization project in Pakistan (launched under the Zia ul-Haq government) has turned into a natural site for state-supported conferences on Islamic science and the presentation of a variety of pseudoscientific writings designed to show that the findings of modern sci-

---

[105] S. W. A. Hussaini, "Toward the Rebirth and Development of Shariyyah Science and Technology," *MAAS Journal of Islamic Science* 1, no. 2 (1985): 81–94, at p. 83.
[106] As cited in Pervez Hoodbhoy, *Islam and Science: Religious Orthodoxy and the Battle for Rationality*, with a foreword by Mohammed Abdus Salam (London: Zed Books Ltd., 1990), p. 28.
[107] Hussaini, "Toward the Rebirth and Development of Shariyyah Science, Technology," p. 83.

ence are, after all, discoveries found in the Quran, the sacred book of Islam.[108] It remains to be seen whether a new science can be produced through
an effort that must always find harmony between the Islamic sacred law and
the findings of dispassionate science.

[108] Pervez Hoodbhoy, as a Western-trained scientist and engineer, has heroically attempted to
point out in his book, *Islam and Science,* the extraordinary naïveté and wrongheadedness of
much of this new Islamic science.

7

Science and civilization in China

The problem of Chinese science

Due to the publication of Joseph Needham's profound and monumental study, *Science and Civilisation in China*, the question of why modern science arose in the West but not in the East has focused on a comparison of Europe and China. The implicit suggestion has been that Chinese science came closest to paralleling Western scientific achievement, and therefore China probably came closer than any other civilization to giving birth to modern science. As we saw in Chapters 2 and 5, however, the path leading to the scientific revolution in Europe was paved most significantly by Arabic-Islamic scientists. Not only had the Arabs developed, discussed, and deployed several aspects of the experimental method, but they had also developed the mathematical tools necessary to reach the highest levels of mathematical astronomy. Furthermore, the work undertaken at the Marâgha observatory in the thirteenth and fourteenth centuries, culminating in the work of Ibn al-Shatir (d. 1375), resulted in the development of new planetary models of the universe that have often been described as the first non-Ptolemaic models along the path to modern science. It was these planetary innovations, moreover, that were to be adopted by Copernicus.[1] The missing ingredient was the heliocentric anchoring, not mathematical or other scientific ingredients. It was the failure to make this metaphysical leap from a geocentric to a heliocentric universe that prevented the Arabs from making the move "from the closed world to the infinite universe."

In the case of China, however, the disparity between the state of Chinese

---

[1] See Chap. 2, section entitled "The Achievements of Arabic Astronomy," for the details of the equivalence between the models of Copernicus and the Marâgha school of astronomy.

science and that of the West – but also the disparity with Arabic science – was far greater in regard to the theoretical foundations upon which the scientific revolution was ultimately launched in Europe. The superiority of China to the West, to which Needham refers, was primarily a technological advantage, conveyed in Needham's claim that from the first century B.C. until the fifteenth century, "Chinese civilization was much *more* efficient than occidental [civilization] in applying human natural knowledge to practical human needs."[2] This superiority was wholly of a practical and technological nature, not one of theoretical understanding. If we focus our attention on natural science (as distinguished from technology), then the puzzle regarding "the great inertia" of Chinese science becomes all the more perplexing.[3]

For if one takes the point of view that science is above all a system of error detection, not a set of skills for building machines, mechanical or electronic, then attention must be directed toward those abstract systems of thought and explanation that give higher order to our thinking about the natural realm. Science at its heart is systematic and theoretical knowledge about how the world is and how it works. It is *episteme* as opposed to *techne*. It is speculative in that it is always conjecturing the existence of new entities, processes, and mechanisms, not to mention possible new worlds. Its task is to determine which of these ideas and entities have a real existence in the world. Popper's description of this process as "conjectures and refutations" aptly captures this dynamic.[4] From such a point of view, science is about how to describe, explain, and think about the world and is not concerned with how to make labor easier or how to control nature. The science of medicine, on the other hand, is the paradigmatic anomaly in which the desire to control nature – to improve health and extend life – is bound up with scientific advance. But this is only to suggest that medicine cannot generally be taken as a useful case in debates about the nature of science.[5]

---

[2] Needham, "Science and Society in East and West," in *The Grand Titration* (London: Allen and Unwin, 1969), pp. 190 and 214; hereafter cited as *GT*.

[3] Wen-yan Qian, *The Great Inertia: Scientific Stagnation in Traditional China* (London: Croom Helm, 1985). This very personal discourse contains many insights regarding Chinese science and civilization, but as a contribution to a sociological understanding of the problem, it contains only the enigmatic phrase "software decides."

[4] Karl Popper, *Conjectures and Refutations* (New York: Harper, 1968).

[5] For a more philosophically probing discussion of the logic of discovery, the problems of the theory-laden character of observation, and other such philosophical issues, see the introduction to Toby E. Huff, *Max Weber and the Methodology of the Social Sciences* (New Brunswick, N.J.: Transaction Books, 1984), especially "The Rise and Fall of Logical Positivism," pp. 2–8. As a guide to this vast literature in the philosophy of science, I only mention the following: N. R. Hanson, *Patterns of Discovery* (Cambridge: Cambridge University Press, 1958); F. Suppe, ed., *The Structure of Scientific Theories*, 2d ed. (Chicago: University

Having said that, it may be noticed that technological inventions are almost always (and certainly before the twentieth century) devoid of those philosophical and metaphysical implications which are an inherent part of the scientific quest. It might therefore be suggested that the inventiveness of the Chinese (despite their failure to fully utilize the many technological devices that they did invent)[6] is the outcome of a lack of intellectual freedom to pursue science (the game of disputing the nature of the world) and the displacement of this energy and intellectual curiosity into intellectually safe domains where metaphysical questions would not be raised.

If we consider the main fields of scientific inquiry that have traditionally formed the core of modern science – namely, astronomy, physics, optics, and mathematics – it is evident that the Chinese lagged behind not only the West but also the Arabs from about the eleventh century. By the end of the fourteenth century in the areas of mathematics, astronomy, and optics, there was a considerable debit on the Chinese side, despite the fact that there had been many chances for the Chinese to benefit from Arab astronomers and to borrow or assimilate the Greek philosophical heritage through constant interchanges between the Arabs and the Chinese.[7] During Yüan times (ca. 1264–1368) Needham tells us, the Arabs or, more probably, the Persians played a significant role in bringing new mathematical ideas to Chinese science and this role paralleled that played by Indians in T'ang times.[8] Although the Chinese did make many advances in mathematics (especially algebra) and astronomy, those advances were not on the path to modern astronomy as it actually unfolded in Islam and the West. Even those who claim virtues for Chinese astronomy do so almost wholly on the basis of

---

of Chicago Press, 1977); Imre Lakatos, "The Methodology of Scientific Research Programs," in *Criticism and the Growth of Knowledge,* ed. A. Musgrave and I. Lakatos (New York: Cambridge University Press, 1970), pp. 91–196; and Larry Laudan, *Progress and Its Problems* (Berkeley and Los Angeles: University of California Press, 1977).

[6] Four examples of this are given by Derk Bodde in *Chinese Thought, Society, and Science: The Intellectual and Social Background of Science and Technology in Pre-Modern China* (Honolulu: University of Hawaii Press, 1991), p. 362: the invention of the seismograph; Chu Tai-yü's discovery of equal temperament in music; magnetism; and the invention of an astronomical clock by Su Sung (ca. 1090). To these one might add the invention of movable type (ca. 1041–8). For attempts to explain the failure to employ the last of these, see *SCC* 5/1: 220ff.

For more on this, see the recent discussion by Kenneth R. Stunkel, "Technology and Values in Traditional China and the West," *Comparative Civilizations Review* no. 23 (199): 75–91; and no. 24 (1991): 58–75. Some of the restrictions on Chinese thought are forcefully presented by Harry White in "The Fate of Independent Thought in Traditional China," *Journal of Chinese Philosophy* 18 (1991): 53–72.

[7] Needham has sketched out this Arab influence on the Chinese, but there is still much to be done. See especially *SCC* 3: 372–82.

[8] *SCC* 2: 49.

empirical observations, namely, ancient but quite precise observations of astronomical phenomena not elsewhere recorded.[9] Geometry as a systematic deductive system of proofs and demonstrations was virtually nonexistent in China, as was trigonometry.[10] These of course were the special branches of mathematics needed to advance astronomical model building. Moreover, as a practical matter, the chief astronomers in the official Chinese Bureau of Astronomy at the opening of the Ming dynasty (1368–1644) were not aware of the significance that changes in geographic location make in astronomical calculations. Thus Ho Peng-yoke reports:

> It was not until the year 1447 that the Director of the Bureau [of Astronomy] . . . reported to the Emperor that the north polar angular distance and the times of sunrise and sunset in Peking differed from those in Nanking, and the same should also apply to the length of the day and night in winter and in summer, and pointed out that the time-indicating rods of the clepsydra [waterclock] in Peking were all based on those used in Nanking. The Emperor had to order that these rods be re-made and re-calibrated.[11]

Conversely, the Arab astronomers had prepared many zij tables recording planetary coordinates for many different locations throughout the Middle East precisely because of the differences in time produced by geographic location.[12]

---

9 Ho Peng-yoke, *Modern Scholarship on the History of Chinese Astronomy* (Camberra: Faculty of Asian Studies, the Australian National University, 1977).

10 According to Needham, China "never developed a theoretical geometry independent of quantitative magnitude and relying solely for its proofs purely on axioms and postulates as the basis of discussion." SCC 3: 91. Likewise, Ulrich J. Libbrecht, *Chinese Mathematics in the Thirteenth Century* (Cambridge, Mass.: MIT Press, 1973), says, "it must be admitted that the proficiency of Chinese mathematicians in this field [geometry] was not of a high standard," p. 36; and "Chinese geometry cannot be compared with Greek geometry because the Chinese did not have the slightest conception of deductive systems. All we find in their mathematical handbooks are some practical geometrical problems concerning plane area and solid figures," pp. 96ff. In his chapter on Chinese trigonometry, he says that "the title of this chapter must be qualified, because trigonometry was unknown in China," p. 122.

11 Ho Peng-yoke, "The Astronomical Bureau of Ming China," *Journal of Asian History* 3–4 (1969): 139–53 at p. 146. What is significant about the phrase "the emperor had to order" is the fact that nothing could be done in the realm of astronomy (or astrology) without the permission of the emperor, as astronomical knowledge was treated as a state secret due to Chinese conceptions of the linkages between the natural and social orders, i.e., the linkage between "the mandate of heaven" and all terrestrial and super-terrestrial events.

12 See E. S. Kennedy, "A Survey of Islamic Astronomical Tables," *Transactions of the American Philosophical Society*, n.s., 46, pt. 2 (1956): 165ff; and David A. King, "The Astronomy of the Mamluks," *Isis* 74, no. 274 (1983): 532. The tenth-century astronomer Ibn Yunus "prepared a substantial number of tables for timekeeping by the sun and for regulating the astronomically determined times of prayer, all computed for the latitude of Cairo." By 1250 the Cairiene perfection of these tables produced universal timekeeping tables containing thousands of entries, including one by Najm al-Din al-Misri "which gives the time since the rising of the sun or a star as a function of its altitude for all declinations and terrestrial

Within mathematical astronomy per se, it has been pointed out by G. E. Lloyd that from the time of Eudoxus (ca. 400–ca. 350 B.C.) it was supposed in the Greek (and later in the Arab) world "that some geometrical model would provide the solution to the problem of celestial motion";[13] but, as Needham, Nathan Sivin, Christopher Cullen, and others have noted, this assumption could not be made for Chinese science.[14] Rather than being based on geometrical models, Chinese astronomy was an algebra-based point estimation system that relied upon numerical calculations rather than geometrical analysis.[15] Their deficiencies in this area, moreover, led the Chinese to employ Muslim astronomers in the Chinese Bureau of Astronomy continuously from the thirteenth century onward. Indeed, in 1368 a special Muslim Bureau of Astronomy was established in China that was still functioning at the time of the arrival of the Jesuits in the sixteenth century.[16] Upon the arrival of the Jesuits, there were four competing astronomical systems: the traditional Chinese system; that of the Muslims (based on the lunar calendar); the new European; and that of the so-called new Eastern Bureau.[17]

For these reasons Needham notes that "there can be no doubt but that there was every opportunity for Arabic and Persian mathematical influences (as from the observations of Marâgha and Samarqand) to enter Chinese traditions."[18] Even more tantalizing are the reports that a Mongol ruler in China, Mangu (d. 1257; the brother of Hulagu who ordered the construction of the Marâgha observatory) is said to "have mastered difficult passages of Euclid by himself."[19] In what language was this version of Euclid, and why is it that Mangu's successor – Khubilai Khan – did not suggest the learning of Euclid to the court officials surrounding him?[20] These facts make

latitudes [and which] contains over a quarter of a million entries." King, "On the Astronomical Tables of the Islamic Middle Ages," *Colloquia Copernicana* 3 (1975): 37–56 at pp. 44–5. For a brief introduction to Islamic timekeeping, see King, "Ibn Yunus' 'Very Useful Tables' for Reckoning Time by the Sun," *Archive for the History of the Exact Sciences* 10 (1973): 345–7.

13 G. E. Lloyd, "Greek Cosmologies," in *Ancient Cosmologies*, ed. C. Blacker and M. Loewe (London: Allen and Unwin, 1975), as cited in Christopher Cullen, "Joseph Needham on Chinese Astronomy," *Past and Present* no. 87 (1980): 39–53, at p. 40.

14 Cullen, "Joseph Needham on Chinese Astronomy," p. 40, and Needham, *SCC* 3, sec. 20, pp. 229ff.

15 Cullen, "Joseph Needham on Chinese Astronomy," p. 40, who also cites Sivin, "Cosmos and Computation in Early Chinese Mathematical Astronomy" (Leiden: E. J. Brill, 1969).

16 Needham, *SCC* 3: 49–50.

17 Ho Peng-yoke, "The Astronomical Bureau in Ming China," p. 151.

18 Needham, *SCC* 3: 50.

19 Aydin Sayili, *The Observatory in Islam* (Ankara: Turkish Historical Society Series 7, no. 38, 1960), p. 189.

20 No doubt a major part of the answer to this question is to be found in the rise of the

it all the more puzzling why it was that the Jesuits are credited with having introduced Western astronomy to the Chinese (albeit incompletely because of the Galilean controversy just then unfolding) as well as geometry, when the Marâgha models clearly assumed all the fundamentals of Western astronomy at that time except the heliocentric orientation.[21] In other words, given the direct contact in the capital city between some of the best Muslim astronomers of the time and the Chinese astronomers in the official Bureau of Astronomy, the Chinese ought to have had nearly two centuries to translate Euclid's *Elements* and to assimilate the Ptolemaic models (as perfected by al-Tusi, al-ʿUrdi, al-Shirazi, and Ibn al-Shatir) before they were transformed into the Copernican models by Europeans in the sixteenth and seventeenth centuries.

We may also note that the science of optics was vital for scientific theory in the West, above all in connection with the development of the telescope and the microscope – instruments that played major roles in the development of astronomy and medicine.[22] But it was the Arabs, especially in the work of Ibn al-Haytham (d. ca. 1040), who laid the foundation for modern optics. Although Needham suggests that the Chinese of the early medieval period "kept more or less abreast" of the Arabs in optics, he concedes that they were "greatly hampered by the lack of the Greek deductive geometry," which the Arabs had inherited,[23] and consequently "never equalled the highest level attained by the Islamic students of light such as Ibn al-Haytham."[24] The most important school of optics among the Chinese was the ancient one of the Mohists (ca. third and fourth centuries B.C.). Moreover, it ought to be noted that the Arab experimental tradition in optics, especially in connection with the rainbow as an optical phenomenon, only truly began with al-Haytham and that it ran from him to Qutb al-Din al-Shirazi (d. 1311), to his

neo-Confucianists who remade Chinese education (and the examination system) so that its sole focus was again on the Confucian classics, exclusive of all science and natural philosophy. More on this below. See William de Bary, *Neo-Confucian Orthodoxy and the Learning of the Mind-and-Heart* (New York: Columbia University Press, 1981), chap. 1, as well as the discussion of Khubilai Khan's powerful neo-Confucian adviser, Hsü Heng (1209–81), pp. 131ff.

21 Among others on this subject, see Joseph Needham, *SCC* 3: 437–61; Nathan Sivin, "Copernicus in China"; and John B. Henderson, *The Development and Decline of Chinese Cosmology* (New York: Columbia University Press, 1984), pp. 144, 150 and passim.

22 Needham subsumes this area of inquiry under the study of light, since it was not a field of inquiry with its own identity in Chinese science. The specifically Chinese sciences are listed by Sivin in "Science and Medicine in Imperial China – The State of the Field," *The Journal of Asia Studies* 47, no. 1 (1983): 43.

23 *SCC* 4: xxiii.

24 Ibid., p. 78.

student Kamal al-Din al-Farisi (d. ca. 1320), and from thence to the Europeans, that is, Roger Bacon (d. 1292), Pecham (d. 1292), Witelo (d. after 1275), and Theodoric of Freiburg (d. ca. 1310).[25] Likewise, it has been argued that Kepler's theory of retinal image was directly influenced by Ibn al-Haytham's optics.[26] And, not least of all, it has been said that Newton performed the same experiments as his predecessors regarding refracted light in vials of water.[27]

Although we think of physics as the fundamental natural science, Needham concluded that "the Chinese had very little systematic thought in this domain."[28] While one can find "Chinese physical thought," "one can hardly speak of a developed science of physics."[29] Chinese physical thought was wave- rather than particle-oriented according to Needham,[30] and this view seems consistent with Manfred Porkert's translation of *wu hsing* (five elements) as "the Five Evolutive Phases."[31] This interpretation of wu hsing, according to Nathan Sivin, mercifully lays "to rest the idea that they are material elements."[32] In short, in terms of powerful systematic thinkers in physics, there "is no one to correspond to the so-called 'precursors of Galileo,' men such as Philoponus and Buridan, Bradwardine and Nicholas d'Oreme, and hence no dynamics and no cinematics."[33] Although I have said little of Arab achievements in physics and the dynamics of motion, it should be remembered that at least in the eleventh and twelfth centuries in Islamic Spain, Arab physical thought was highly developed. In fact, Ernest Moody long ago showed that there is a fairly direct connection between Ibn Bajja's (d. 1138/9) commentaries on Aristotle and Galileo's theory of free fall.[34] Indeed, Moody ascribed to Ibn Bajja (Avempace) an essential role that

[25] David C. Lindberg, "Lines of Influence on Thirteenth-Century Optics: Bacon, Witelo, and Pecham," *Speculum* 46 (1971): 66–83.

[26] David C. Lindberg, *Theories of Vision from al-Kindi to Kepler* (Chicago: University of Chicago Press, 1976), pp. 86 and 190ff.

[27] A. I. Sabra, *Theories of Light from Descartes to Newton* (London: Oldbourne, 1967); Sabra, "Ibn al-Haytham," *DSB* 5: 189–210.

[28] Needham, *SCC* 4/1: 1.

[29] Ibid.

[30] Ibid.

[31] Manfred Porkert, *The Theoretical Foundations of Chinese Medicine: Systems of Correspondence* (Cambridge, Mass.: MIT Press, 1974), pp. 9ff.

[32] Sivin, foreword to Porkert, *Theoretical Foundations of Chinese Medicine*, p. xiii.

[33] *SCC* 4/1: 1.

[34] See Ernest Moody, "Galileo and Avempace: Dynamics of the Leaning Tower Experiments," in *Roots of Scientific Thought: A Cultural Perspective*, ed. Philip P. Wiener and A. Noland (New York: Basic Books, 1957), pp. 176–206; and Moody, "Galileo and His Precursors," in *Galileo Reappraised*, ed. Carlo Golino (Berkeley and Los Angeles: University of California Press, 1966), pp. 23–43.

"enabled Galileo to generalize Buridan's impetus theory and transform it into a general inertial dynamics."[35]

Finally, in addition to these comparative benchmarks, we should mention the fact that the Arabs contributed significantly to discussions of scientific method. This influence is seen most clearly in the effect Ibn Sina's *Canon* had on discussions of scientific method in the European Middle Ages,[36] and the *Canon* had an equal influence on medical theory and practice in Europe from the fourteenth to the sixteenth centuries.[37]

Although it seems doubtful that early Chinese methodological discussions were equivalent to those of Aristotle and Plato, it must be said that in the work of Mo-tzu (fourth century B.C.) there are keen methodological insights that, in Needham's words, "could have become the fundamental basic conceptions of natural science in Asia."[38] Perhaps one could even agree with Needham that the Mohists "sketched out what amounts to a complete theory of scientific method."[39] The problem is that the Mohists and their thought faded into Chinese history and apparently had little influence on Chinese natural thinkers and none at all on Western thought. Despite the promising beginnings one sees in Mohist philosophical thought, it never gained much influence in the Chinese thought world. The result is, Nathan Sivin reminds us, that there was no overall, coherent natural philosophy such as one finds among the Greeks, Arabs, or medieval Europeans. This follows from Sivin's reminder that the sciences in China were a heterogeneous mixture of inquiries far wider in scope than those of the Western tradition. China "had sciences but no science, no single conception or word for the overarching sum of them all."[40] What is more, "Philosophers were in no position to

---

35 "Galileo and Avempace," p. 40.
36 A. C. Crombie, "Avicenna's Influence on the Medieval Scientific Tradition," in *Avicenna: Scientist and Philosopher*, ed. G. Wickens (London: Luzac, 1952), pp. 84–107; and Crombie, "The Significance of Medieval Discussions of Scientific Method for the Scientific Revolution," in *Critical Problems in the History of Science*, ed. Marshall Clagett (Madison, Wis.: University of Wisconsin Press, 1959), pp. 79–101.
37 Nancy G. Siraisi, *Avicenna in Renaissance Italy: The Canon and Medical Teaching in Italian Universities after 1500* (Princeton, N.J.: Princeton University Press, 1987); Siraisi, *Medieval and Early Renaissance Medicine* (Chicago: University of Chicago Press, 1990), pp. 48ff.; Charles H. Talbot, "Medicine," in *Science in the Middle Ages*, ed. David C. Lindberg (Chicago: University of Chicago Press, 1978), pp. 391–428.
38 *SCC* 2: 182. Also see A. C. Graham, *Later Mohist Logic, Ethics, and Science* (Hong Kong: Chinese University Press, 1978); and *Disputers of the Tao* (LaSalle, Ill.: Open Court Press, 1989), as well as Benjamin Schwartz, *The World of Thought in Ancient China* (Cambridge, Mass.: Harvard University Press, 1985), pp. 164–8.
39 *SCC*, 2: 182.
40 Sivin, "Why the Scientific Revolution Did Not Take Place in China," in *Transformation and Tradition in the Sciences*, ed. E. Mendelsohn (New York: Cambridge University Press, 1984), p. 533.

define a common discipline among them, as Aristotle and his successors had done in Europe, and so philosophers had practically no influence on the development of these pursuits."[41]

For all the foregoing reasons we might find more validity in the claim of some historians of science (as well as Chinese scholars) that we should not have expected Chinese scientific thought to culminate in a "modern scientific revolution."[42] On the other hand, one might ask why the Chinese did not keep abreast of scientific thought as manifested in the Arab-Islamic world prior to the sixteenth century. Such a query suggests that the protest against asking why modern science did not emerge in China is ill-considered.

This protest is additionally unpersuasive if we grant the supposition that at least some men in all societies have universally in all ages sought the truth about man and nature, and that those conjectures about such things which have withstood the demands of rational criticism and the strictures of empirical comparisons represent a converging order of universal truths, available to all peoples. If we conjoin this assumption with the caveat that this pursuit is always *unended,* then we can without prejudice seek to identify those factors – social, religious, philosophical, legal, economic, and political – that have either facilitated or inhibited intellectual progress in scientific thought in the various societies and civilizations of the world. Whether or not this sort of study focuses on the question of why there was no scientific revolution in China (or Islam) is largely a question of emphasis. In light of the previous China–Islam comparison, this need not be a strictly East and West comparison. We in the West – with hindsight – have discovered notable disjunctures of intellectual outlook at various points in the history of the West and have become persuaded that these disjunctions were indeed revolutionary. At the same time, the absence of such intellectual ruptures leading to progressive advance in the civilizations of Islam and China (and the absence of the consequent intellectual innovations in those civilizations) has clearly created a perception that there have been dramatic differences of intellectual outlook regarding the study and explanation of the natural world between the civilizations of the world. These differences in cultural outlook, societal organization, and economic performance, being of more than passing interest, are surely legitimate phenomena for social scientific investiga-

---

[41] Ibid., p. 535.
[42] Among others, this view has been expressed by Nathan Sivin in his many papers noted above; also see A. C. Graham, "China, Europe, and the Origins of Modern Science: Needham's *The Grand Titration,*" in *Chinese Science: Explorations of an Ancient Tradition,* ed. Nathan Sivin and S. Nakayama (Cambridge, Mass.: MIT Press, 1973), pp. 45–69; Wing-Tsit Chan, "Neo-Confucianism and Chinese Scientific Thought," *Philosophy East and West* 6 (1957): 309–32; as well as Chung-Ying Cheng, "On Chinese Science: A Review Essay," *Journal of Chinese Philosophy* 4 (1977): 395–507.

tion and explanation. They are in principle no different from internal domestic questions equivalent to "why do Hispanic Americans, or Portuguese Americans, and so forth, have much lower levels of educational attainment than other southeastern Europeans in the United States?" Conversely, many other social scientists have asked why it is that individuals of Asian descent (especially Chinese, Vietnamese, and Korean) have such high levels of educational attainment (and consequently high levels of economic success) in the United States. On a different level, others have asked why it is that such Chinese societies as Taiwan, Hong Kong, and Singapore have done so much better economically and technologically than mainland China. These are hardly questions analogous to that of asking why the house next door did not catch on fire, or "why your name did not appear on page 3 of today's newspaper," as Nathan Sivin suggests regarding the question why modern science failed to emerge in China.[43] To disallow such inquiries as to why one social group or another – one society or civilization or another – did not follow a particular line of cultural and economic development, above all one leading to higher levels of scientific achievement and economic performance, is little more than moral censure.

There are numerous indications that over the course of the past four hundred years or so Chinese scientists have attempted to embrace those universal components of modern science that have emerged and to reevaluate their own traditional intellectual resources in terms of the outlook of the world's (still evolving) modern science. According to Needham, the first of the sciences in China to achieve this fusion into ecumenical science were mathematics and astronomy. In his view, by 1644, "the end of the Ming dynasty, there was no longer any perceptible difference between the mathematics, astronomy, and physics of China and Europe; they had completely fused, they had coalesced."[44] While there may be some disagreement about the precise timing of such fusions and how fully any of the sciences of China became fused with those of the West, events of the past several decades suggest that contemporary Chinese leaders have in fact decided that the advancement of science and technology is an indispensable ingredient in China's efforts to modernize. They have apparently been persuaded that agricultural and labor reform and the stimulation of capital investment is not

---

[43] Sivin, "Why the Scientific Revolution Did Not Take Place in China," p. 536. And also A. C. Graham, "China, Europe, and the Origins of Modern Science."

[44] Needham, "The Evolution of Oecumenical Science. The Roles of Europe and China," *Interdisciplinary Science Reviews* 1, no. 3 (1976): 203. For additional details of this "quiet revolution" in cosmology, see John Henderson, *The Development and Decline of Chinese Cosmology;* as well as Sivin, "Why the Scientific Revolution Did Not Take Place in China"; and Sivin's "Science and Medicine in Chinese History," in *Heritage of China*, ed. Paul S. Ropp (Berkeley and Los Angeles: University of California Press, 1990), pp. 164–96.

enough to transform China into a modern society. To accomplish that goal they must directly encourage and promote modern science and technology, with all the political consequences that such a decision necessarily entails. [45]

An illustration of the effect of this policy can be seen in the case of pharmaceutics, one of the oldest indigenous sciences in China. Officials in the People's Republic of China have now taken the position that there is only one universal science of pharmaceuticals and China's long and rich past of studies in this area are being rethought in the light of the rationale of modern science. According to Paul Unschuld, "In addition to the experience of people, modern science is now recognized as the one and only basis of knowledge. As a result, traditional Chinese *materia medica* has been reevaluated over the past decades in accordance with contemporary scientific assumptions as to the active ingredients of . . . herbal substances and their most effective mode of application in therapy."[46]

Quite recently the Chinese leadership declared that "the Chinese people should take an active part in the coming science and technology revolution." The real competition is in science and technology, and the harnessing of these would yield large gains in productivity, it was claimed. They further recognize that "science and technology belong to the whole of mankind, but for historical and social reasons, many developing countries were lagging far behind in the field compared with the developed countries."[47] Nor should we overlook the contemporary fact that the two largest groups of foreign students studying the sciences and technology in the United States are from Taiwan and mainland China.[48]

In a word, though Needham's timetable for the fusion points of Chinese science into universal science may be overly optimistic by some accounts,[49]

---

[45] These issues have been explored in some depth in *Science and Technology in Post-Mao China*, ed. Denis Fred Simon and Merle Goldman (Cambridge, Mass.: Harvard University: The Council on East Asian Studies, 1989). It goes without saying that the first conflict spawned by that decision was the pro-democracy movement of the late 1980s, which culminted in the 1989 Tiananmen shootings. Still, the Chinese remain commited to the Four Modernizations.

[46] Paul Unschuld, *Medicine in China: A History of Pharmaceutics* (Berkeley and Los Angeles: University of California Press, 1986), p. 285.

[47] From the *People's Daily* as reported in *China Daily*, May 4, 1991, p. 2.

[48] Between the early 1980s and 1991, the number of students from China studying in the United States increased dramatically from less than 10,000 students to 39,600 – the largest increase for any foreign place of origin, Marianthi Zikopoulos, ed., *Open Doors, 1990–91* (New York: Institute of International Education, 1991), table 2.4, p. 21. It should also be noted that 8 percent of Chinese students in this country are graduate students, ibid., p. 74. Insofar as the study of the sciences is concerned, 33 percent of Chinese graduate students (as compared to 29 percent of undergraduates) are in the life, physical, and so-called science-technology fields, a far larger proportion than any other group, Marianthi Zikopoulos, ed., *Profiles* (New York: Institute of International Education, 1991), table 2.10.

[49] Jonathan Spence, for example, in a review of Needham's volumes argues that "China cannot

contemporary Chinese leaders have recognized that modern science and technology contain elements of indispensable truth well worth having. As Needham put it some time ago, without this knowledge, "plagues are not checked, and aircraft will not fly. The physically unified world of our own time has indeed been brought into being by something that happened historically in Europe, but no man can be restrained from following the path of Galileo and Vesalius."[50] Put differently, "Man has always lived in an environment essentially constant in its properties, and his knowledge of it, if true, must therefore tend towards a constant structure."[51] Although this may be a statement of faith, it appears that such a faith has a universal appeal.

If this be the case, then men universally may be supposed to be working on a set of common natural problems, and their many conceptual modes of undertaking that quest are as so many alternative and tentative guesses (conjectures), which over time are bound to be refined as mistakes and "garden paths" are eliminated and better conjectures fashioned in their stead. In this light the sociological problem is to understand the social and institutional impediments that have gotten in the way of the free, open, and unended quest for the best conceivable scientific descriptions of the structural properties and natural processes that govern the world we inhabit.

Given the enormous scope of Chinese civilization and its different metaphysical grounding, the present discussion cannot presume to achieve the mastery of the subject matter required of a sinologist. Nevertheless, given the preceding account of the successes and failures of scientific development in Islam and the West, it may be useful to extend our analysis to the case of Chinese science. It goes without saying that an inquiry such as the present one would be unthinkable without the monumental achievement of *Science and Civilisation in China* and those of many other sinologists.

China and the comparative context

As I suggested in Chapters 4 and 5, Europe in the twelfth and thirteenth centuries experienced a profound social and intellectual revolution that placed social life on an entirely new footing. At the center of this revolution was a legal transformation that redefined the nature of social organization in all its realms – political, social, economic, religious. For present purposes

---

be seen as having entered the world of universally valid science during that same long period [the time following the arrival of the Jesuits] in any meaningful way." "Review Symposia: Science in China," *Isis* 75, no. 1 (1984): 180–9, at p. 185.

50 Needham, *SCC* 3: 448f.
51 Needham, *SCC* 5/2: xxi.

the development of the law of corporations was foremost among these changes. At the time when the Christian church declared itself to be a corporation, a "whole body" for all legal purposes, it granted such status to a variety of other collectivities, such as residential communities, cities, towns, universities, economic interest groups, and professional guilds. As a result, each of these collectivities of individuals was granted legal autonomy to make its own internal laws and regulations, to own property, to sue and be sued, and to have legal representation before the king's court. In effect, this social revolution transformed the church into a *Rechtsstaat* – a social body ruled by law – and it created the model of relationships between and within all other social realms. It created a model of legally grounded social organization which was to prevail in the city, the political state, in academic sittings (that is, the university), and in the economic sphere. Nor should we forget that this revolution sharply demarcated the religious domain – the moral and the ethical – from the secular state. This was a revolution that was not carried out in Islamic law and theology and is related to the reasons why the Muslim countries of the Middle East had to borrow Western legal codes at the end of the nineteenth century in order to enter the international community of constitutional regimes that recognize legally protected human and political rights. Likewise, to anticipate later discussions, this was a revolution not carried out in Chinese law. For the emperor and his officials (in addition to being above the law) were supposed to be paragons of the good, exemplars of moral correctness.

European legal scholars (the Romanists and canonists), furthermore, worked mightily to establish the idea of a universal law, law that would apply equally to all individuals whatever their national origins. The guiding idea here was that of natural law. It was a conception of law as divinely given, embedded in man and nature, and effected by reason operating in and through all rightly guided individuals. In the final analysis, the natural law stood as a point of judgment, for even the divine law, that is, the law of Scripture and church council (decretals), had to be reconciled to it. A concerted effort was therefore made to reconcile all the realms of law so that they would conform to the demands of natural law.

In addition, I have suggested that it was this social and intellectual revolution which opened the doors to intellectual freedom, above all by creating autonomous universities with their own internally established intellectual agendas. And, not least of all, there was in the universities a breakthrough in what some would call the "logics of decision."[52] Here I refer to the new dialectics, the new method of analysis and synthesis, which aimed at the

[52] This is Benjamin Nelson's term; see *On the Roads to Modernity*, pp. 71–2.

resolution of contradictions, pioneered by Peter Abelard and the legists.[53] In the course of their disputations, they evolved a method of questioning that aimed at stating a proposition, raising contradictory points of view, and then finding a new synthetic formulation (synthesis) that resolved the contradictory points of view, thereby advancing knowledge and understanding. This was not a presentation of "the new in the form of the old," but a genuine effort to advance knowledge and understanding by overcoming the contradictions of the past. This method of dialectical discourse had built into it a means (a method) of inducing "progressive problem shifts," to use current terminology in the philosophy of science.[54] It freed intellectual discourse from perpetually resorting to bland commentaries upon and restatements of the wisdom of the ancients.

While we are concerned with the nature of the scientific role, we again remind ourselves that in the end, the question regarding the rise of modern science is not whether the Chinese made intellectual advances over the Greeks or Arabs in one or another field such as optics, physics, or mathematics, but whether or not breakthroughs occurred in the directive structures of discourse and inquiry that opened the door of the institutional structures to freedom of inquiry. To understand the social and cultural context of science in China, we must completely alter our angle of vision, for as different as Islam and the West are religiously and metaphysically, they share an underlying starting point that I shall call atomism. That is, both Christians and Muslims assume that whatever processual flow there may be in the universe, life and the physical cosmos are composed of indivisible atoms. This should not be surprising since both the Christian West and the Islamic Middle East either built their metaphysical and philosophical structures on Greek-inspired atomism, or, as in the case of Islamic theology, borrowed the conceptual language of atomistic thinking while constructing a theocentric philosophy that reserved for God all powers of shaping and regulating the visible and invisible worlds. The philosophy of Islamic occasionalism which emerged from that effort, above all, in the work of al-Ash'ari (d. 935/6), assumed the atomistic essence of nature, but strenuously denied any pattern of causality in the human and natural worlds. It reserved for God all powers of moving and controlling nature. The apparent regularity of human and natural worlds were but expressions of the habit of God in nature, and at any moment the pattern could change, just as a willful person could change his mind. Indeed, the world was held together at every moment by the active

[53] See above, Chaps. 4 and 5, and Berman, *Law and Revolution* (Cambridge, Mass.: Harvard University Press, 1983), pp. 131–42.

[54] Imre Lakatos, "The Methodology of Scientific Research Programs," and Larry Laudan, *Progress and Its Problems*.

agency of God. This was the faith of al-Ghazali (d. 1111) and his powerful voice ensconced this worldview in Islam as never before his time.[55]

But when we enter Chinese civilization, this metaphysical outlook is almost wholly absent. In place of the Western atomism governed by laws of nature, or the Islamic occasionalism governed by God's will, we find an organic world of primary forces (yang and yin) and the five phases (metal, wood, water, fire, and earth) constantly shifting in recurrent cycles.[56] Within this cosmos there is no prime mover, no high God, no lawgiver. Of course it is assumed that there is a pattern to existence and all things and that there is a unique way (tao) for all things, but the explanation of the patterns of existence is not to be sought in a set of laws or mechanical processes, but in the structure of the organic unity of the whole. Moreover, Chinese cosmological thought came to stress the harmonious unity of natural and human patterns. That is, the patterns of the natural world were studied in order to find correlative correspondences between the patterns of heaven and those of human society below. This search for correspondence was manifested on all levels, on the social, the political, and even the individual levels, but the pivotal concern focused on the correspondence between the conduct of the emperor and the patterns of the heavens.[57] The explanation for the disruptions of the social order was sought in the shifting patterns of the heavenly cosmos, and the imperative was to bring the social order into conformity with the natural order. If this was done, the social order would be set aright and the ruler's mandate from heaven would return, empowering him to achieve his political and social objectives. Conversely, the organic harmony of the natural and social worlds could be upset by the misconduct of the emperor. Too much rain, destruction of crops, meteor showers, and other astronomical occurrences were said to be caused by the misconduct of the emperor, and it was his duty to reform his conduct so that the mandate of heaven would return.

At a more fundamental level, it has been said that the correlative mode of thought is a primitive but natural instinct of mankind to think of the world

---

[55] On atomism among the Muslims, see W. M. Watt, *The Formative Period of Islamic Thought*, p. 301 and passim, as well as the references in note 94, Chap. 4, above.

[56] It should be noted that many sinologists prefer the term *five phases* whereas Joseph Needham translates it as five *elements*. According to Derk Bodde, the Chinese term *wu hsing* really means "the five active entities." See Bodde, *Chinese Thought*, pp. 100–1, for this and alternative translations. In the context of scientific thought, see Needham, *SCC* 2: 232ff.

[57] For expressions of this metaphysics in science (astronomy and medicine), see John Henderson, *The Development and Decline of Chinese Cosmology*, chap. 1; Manfred Porkert, *The Theoretical Foundations of Chinese Medicine*, pp. 9–54; A. C. Graham, *Yin-Yang and the Nature of Correlative Thinking* (Singapore: Institute of East Asian Philosophies, National University of Singapore, 1986); and Derk Bodde, *Chinese Thought*, pp. 97–103. Needham's pioneering discussion of this topic is in *SCC* 1: 386–7.

in pairs, especially binary pairs. This gives rise to such elementary classifications as light versus darkness, heat versus cold, heaven versus earth, and so on. While these antimonies imply opposites, they do not imply antagonisms, but natural and inevitable complementarities, each arising and following its course as a natural procession. Such processions of natural patterns may be called inscrutably patterned, for they do not follow a set of laws of nature. In more complicated and advanced symbol systems found in China, these correlative patterns fall into threes, fours, fives, and even nines. But in the view of many sinologists, China never outgrew this correlative way of thinking and thus did not embark on the path of causal thinking as did the West.[58]

With these differences of background, as well as the related social and philosophical outlooks in mind, we may turn to the emergence of imperial China on the eve of the European twelfth-century renaissance.

The emergence of imperial China

To make the comparison between China and the West as temporally relevant as possible, we must look at the social institutions that emerged in China during the European High Middle Ages, that is, the period of time when the legal revolution in the West was being launched and the universities as well as cities and towns were emerging as autonomous entities. In China this corresponds to the rise of the Sung dynasty (960–1279) and the emergence of the imperial state during the Ming dynasty (1368–1644). Between the Sung and the Ming dynasties there was a brief Mongolian rulership called the Yüan (1264–1368).

The claim has been made that Sung life in China experienced an unprecedented era of growth – economic, cultural, and political – the net result of which was a new and reinvigorated China with many scientific and technological achievements to its credit.[59] The case for asserting that China experienced a renaissance paralleling that of the West (of the fourteenth and fif-

---

[58] See Bodde, *Chinese Thought*, p. 99. For more on correlative thinking, see Chap. 8 below.

[59] The social, economic, and demographic changes of this era are richly detailed in Robert M. Hartwell, "A Revolution in Chinese Iron and Coal Industries in the Northern Sung, 960–1127 A.D.," *Journal of Asian Studies* 21, no. 2 (1962): 153–62; and Hartwell, "Demographic, Political, and Social Transformation in China, 750–1550," *Harvard Journal of Asiatic Studies* 42 (1982): 365–445. Another overview can be found in Mark Elvin, *The Pattern of China's Past* (Stanford, Calif.: Stanford University Press, 1973), as well as Etienne Balazs, "The Birth of Capitalism in China," in *Chinese Civilization and Bureaucracy* (New Haven, Conn.: Yale University Press, 1964), chap. 4. Needham's work, *SCC* 2: 493ff., however, establishes the foundation for these judgments regarding the flowering of science and technology.

teenth centuries) has been made by Jacques Gernet. According to him, "the educated Chinese of the eleventh and twelfth century was as different from the T'ang [618–907] predecessors as Renaissance man from Medieval man."[60] What is strikingly evident, he says,

is the advent of a practical rationalism based on experiment, the putting of inventions, ideas, and theories to the test. We also find curiosity at work in every realm of knowledge – arts, technology, natural sciences, mathematics, society, institutions, politics. There was a desire to take stock of all previous acquisitions and to construct a synthesis of all human knowledge. A naturalist philosophy which was to dominate Chinese thinking in the following ages developed in the eleventh century and attained its definitive expression in the twelfth.[61]

While there is some doubt that a full-fledged renaissance took place in science as well as in Chinese culture in general,[62] it is agreed that the eleventh-century intellectual awakening, including the rediscovery of the classics, "produced an array of brilliant intellectuals unequalled in number in any other period of Chinese history."[63] In mathematics, the Sung dynasty appears to has been the most brilliant ever. The intellectual foundations for this leap forward have generally been ascribed to the work of the neo-Confucians, to Chu Hsi (1130–1200) and his predecessors, although neo-Confucianism as a state ideology was only established much later. It was Chu Hsi who championed the idea that one should investigate the nature of things and inspired others to follow their own inclinations in the scientific realm, though the roots of this philosophy go further back.[64] There is some doubt, however, whether Chu Hsi meant to encourage the investigation of the natural order instead of the social and moral orders. Charles Hucker suggests that he did not: "'The investigation of things' that it advocates unquestionably has similarities with the modern scientific spirit. However, the 'things' that Chu Hsi and his followers principally emphasized were not natural laws, but ethical virtues traditionally espoused by Confucianism – filial piety, loyalty, and humankindness."[65] In addition, the fact that the neo-Confucian orthodoxy, which remodeled the civil service examination, ex-

---

[60] Jacques Gernet, *A History of Chinese Civilization* (Cambridge: Cambridge University Press, 1982), p. 330.

[61] Ibid.

[62] Bodde, *Chinese Thought*, p. 185 and n32, is skeptical of this claim.

[63] Charles O. Hucker, *China to 1850: A Short History* (Stanford, Calif.: Stanford University Press, 1978), p. 107.

[64] For the importance of Chu Hsi for scientific thought, see Needham, *SCC* 2: 455ff. For a similar appraisal of Chu Hsi's influence in Sung medicine, see Paul Unschuld, *Medicine in China: A History of Ideas* (Berkeley and Los Angeles: University of California, 1985), pp. 166, 195.

[65] Hucker, *China to 1850*, p. 118.

cised all elements of science from it suggests that the neo-Confucian slogan was not interpreted as encouraging the pursuit of science.

For the moment, however, we should concentrate on the rise of the new imperial governmental structures that uniquely took shape during the Sung dynasty and later greatly expanded and solidified the governing apparatus under the Ming dynasty (1368–1644). The rise of imperial China, or as some prefer to call it, "Gentry China," began with the reign of Emperor T'ai-tsu (r. 960–76). It was during this period that the foundations of the centralized, bureaucratic, and autocratic rule of modern China were laid and the reins of state power placed securely in the hands of the imperial person. As a result, imperial rulership over the kingdom – though its dominions were less than the traditional bounds of everything under heaven (*t'ien-hsia*) – was ingeniously tied to a centralized bureaucratic structure that was solely dependent upon the will and largess of the emperor. Through an extraordinary act of diplomacy, Emperor T'ai-tsu persuaded his military commanders to retire and surrender their commands throughout the empire. He then filled their posts with appointed literary scholars as officials who were directly dependent upon the continuing pleasure of the throne. By this means, the emperor transformed the nature of rulership from a system based on hereditary power and patronage to a meritocratic system based upon selection from among those who had passed a standardized examination administered by the central government.[66] Of course, elements of an examination system existed earlier, and the selection of all personnel through examination was never complete, but the Sung era saw a radical extension of it, as well as a major restructuring of the examination itself.

To make this system of literary rulership possible, the civil service examination system was put into effect more thoroughly than ever before.[67] By this means the major avenue of access to office was that of the official examination system. The net effect was to thoroughly displace the rich and powerful (hereditary and military) families as political adversaries and to destroy their power to affect official appointments. Henceforth their major

---

[66] Charles O. Hucker, *Dictionary of Official Titles in Imperial China* (Stanford, Calif.: Stanford University Press, 1985), pp. 49ff; and Jack Dull, "The Evolution of Government in China," in *Heritage of China*, pp. 55–85 at pp. 72ff; Winston W. Lo, *An Introduction to the Civil Service of Sung China* (Honolulu: University of Hawaii Press, 1987), pp. 59ff.

[67] On that subject see Ho Ping-ti, *The Ladder of Success: Aspects of Mobility in China 1368–1911*, rev. ed. (New York: Columbia University Press, 1967); John Chaffee, *The Thorny Gates of Learning in Sung China* (Cambridge: Cambridge University Press, 1985); Edward Kracke, *Civil Service in Early Sung, 960–1067* (Cambridge, Mass.: Harvard University Press, 1953); and Thomas H. C. Lee, *Government Education and Examinations in Sung China* (Hong Kong: Chinese University Press, 1985).

avenue of ascent to power was through the examination system – regular or irregular.[68]

The social organization of China for a long time had been based on hierarchical principles that extended authority from the emperor and his officials down through the bureaus to the provinces, the prefects, and thence to the districts (counties). The lowest level of administrative responsibility was the district (*hsien*), which was generally composed of a central city, surrounded by towns and smaller villages. This administrative area was managed by the district magistrate.[69] Such an official had a broad administrative mandate that made him "chief legal officer, financial official, and guardian of public security."[70] As such, district managers were but "junior members of a complex chain of command that reached above them to the prefects, past the prefects to the provincial governor, and through them to the ministries in Peking and the emperor himself."[71]

Above the district magistrate was the prefecture. This was the largest territorial administrative unit in the imperial command during the Sung and was composed of several adjacent districts. While the prefectual officials had considerable discretion to manage their territories, they were in turn fashioned into a "circuit" administered by other central government officials who linked the provinces to the imperial capital. In earlier times, the provinces exercised military authority, but in the Sung they were stripped of this power.[72]

To ensure the official management and control of all the territories under the sway of imperial edict, the new Sung emperor established a series of overlapping appointments at the district and prefectural level. "In order to suppress regional separatist inclinations and to establish firm control over local government units," the early Sung emperors made irregular appointments, drawn from a broad pool of candidates whose job it was to administer various units and whose title was "manager of affairs" of such-and-such prefecture or district, instead of the more regular titles of "prefect" or "dis-

---

68 As we shall see, there were always exceptions to the examination system, from *yin privileges* extended to the sons and other relatives of the top echelon of government officials, to the holding of special examinations for the sons of officials and noble families. From time to time offices and official titles were sold to raise revenue or to appease powerful families within the bureaucracy. For the Sung manifestations of this deformity, including cheating, see Chaffee, *Thorny Gates,* chap. 5.

69 T'ung-Tsu Ch'ü, *Local Government in China under the Ch'ing* (Cambridge, Mass.: Harvard University Press, 1962); John R. Watt, *The District Magistrate in Late Imperial China* (New York: Columbia University Press, 1972); and Lo, *Introduction to the Civil Service,* chap. 2.

70 Jonathan Spence, *The Death of Woman Wang* (New York: Viking, 1978), p. xiii.

71 Ibid.

72 Lo, *Introduction to the Civil Service,* chap. 1.

trict magistrate."[73] Later titles became more settled and regularized. Initially they were used precisely to prevent any identification of the acting official with the territory he administered and which could become a political power base.

This was only one level of supervision of the country put into place. The emperor also sent out other officials "as virtual spies on the prefectural Manager of Affairs."[74] These officials were empowered to memorialize, that is, to officially address the emperor in writing regarding any and all activities of the official of the administrative area in question, and without the knowledge or consent of the nominal official in charge. This covert but higher official was given the title of "controller general."[75]

Still another level of control was instituted between the prefectures and the central government – the so-called circuits. Here again fear of ceding any administrative space to potential warlords induced the emperor to nominate officials whose office was to coordinate communication between the district and prefectural level and the office of the emperor. In this manner, the Sung dynasty created the most complex and bewildering system of bureaucratic management ever devised – though this was only the initial stage of a development that was to extend into the twentieth century.

On the highest level there were several bureaus and departments that managed state affairs at the top. These included the Bureau of Military Affairs, the Grand Councils (whose members met regularly with the emperor), the State Finance Commission, and the Censorate – the overall surveillance bureau (on which more below). The grand councillors were in a direct line from the emperor to the Department of State Affairs, below which there was the Six Ministries (Personnel, Revenue, Rites, War, Justice, [Public] Works), through which authority extended downward to the circuits, the prefectures, and the districts (Figure 7).

Finally, we should note the censorial system.[76] This was an office of surveillance that was given the broadest mandate to investigate the activities – both public and private – of all officials in the administrative structure. In the Sung dynasty the Censorate was confined to activities within the capital, but in the succeeding dynasties, it was active in surveilling all levels of official administration. Although it was far more highly focused on official governmental activities, the office of censor had its parallel in Arabic-Islamic civilization in the office of the *muhtasib* – the market police – though these

---

[73] Hucker, *Dictionary*, p. 44.
[74] Ibid.
[75] Ibid., p. 45.
[76] Charles O. Hucker, *The Censorial System of Ming China* (Stanford, Calif.: Stanford University Press, 1966).

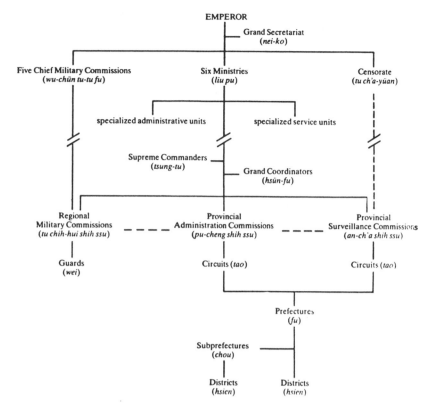

Figure 7. Beginning with the Sung dynasty (960–1279), the Chinese rulers developed a new and powerful centralized administration of authority throughout China. In the Ming dynasty (1368–1644), this administrative reform was powerfully carried forward so that centralized authority was invincibly lodged in the hands of the emperor. As is apparent in this diagram, the power of the emperor extended down through an efficient chain of command to the village level of everyday life. Each of these offices received instructions from central headquarters in Nanking, or, after 1421, in Peking. (Reprinted from *A Dictionary of Official Titles in Imperial China* by Charles O. Hucker with the permission of Stanford University Press, 1985.)

latter were strictly religious officials whose surveillance was confined to activities outside the official government. However, insofar as public officials could be observed engaged in un-Islamic activities, their behavior could fall under the censure of the market police. This cadre of spiritual enforcers clearly persists in contemporary Saudi Arabia in the office of mutawwiʿ, the obediencer (collectively, the *mutawwin*).

The Chinese censorial office differed in the sense that it was specifically established to serve as watchdog over bureaucratic officials, but its failure to distinguish between official and unofficial (public versus private) activities created an unlimited power of moral and ethical censorship that transcended all boundaries. During certain periods, the carping of the censors incapacitated officials altogether, leading emperors to complain that the only officials at work were the military and the censors.[77] Furthermore, in the Yüan dynasty (thirteenth and fourteenth centuries), the Censorate was authorized to directly apprehend and punish officials accused of wrongdoing, as well as to suggest correct public policy.[78]

We see then that at the very time Europe was decentralizing its administrative powers – first by separating religious and moral authority from the secular state via the investiture controversy and then by encouraging the establishment of legally autonomous cities, towns, professional guilds, and universities – China was embarked upon an unprecedented centralizing program. All power and authority was increasingly centralized in the hands of the emperor – who was himself above any form of legal regulation – with a vast network of overlapping and countervailing officials who served to guarantee the centralized and ultimately autocratic rulership of all Chinese territories. Consequently, in the succeeding dynasty (Ming, 1368–1644), the structure and style of government that the founding emperor (Ming T'ai-tsu) fashioned upon Sung foundations, "rooted power securely and unchallengeably in the throne. It required that the emperor be actively in charge and did not permit the emergence of any power center independent of the emperor. Moreover, it inclined the emperors toward capricious and ruthless exercise of their authority over the officialdom."[79] In a word, while the Europeans desacralized the office of kings and monarchs, thereby making them secular rulers, the Chinese intensified the sacralizing of the office of the emperor by reaffirming its heavenly mandate and establishing one jurisdiction: that of the throne. The official ideology continued to maintain that peace and harmony in the empire could only be maintained if the emperor rightly aligned himself with the forces of nature and correctly ordered his moral relationships, and if all his subordinate subjects followed the prescribed path of filial piety and submission to the will of the throne.

We should also note the existence of a religiously sanctioned social order at the lowest level of authority and administration, namely, at the district level. For there too the sanction of strong religious forces served to reinforce the quasi-divine status of the district magistrate himself. In all the cities of

[77] Ibid., p. 299.
[78] Hucker, *Dictionary*, p. 61.
[79] Hucker, *The Ming Dynasty: Its Origins and Evolving Institutions* (Ann Arbor: University of Michigan Center for Chinese Studies, 1978), p. 96f.

China there were city gods with temples to which the district magistrate made his way upon arrival and to whose spiritual authorities he paid his respects.

City gods, or *ch'eng-huang shen*, occupied a place of considerable importance in the official religion of Ming and Ch'ing China. Every county and prefectural capital had its *ch'eng huang miao* [city-god temple], which was maintained in part by state expenses. There the newly-appointed magistrate or prefect paid his respects upon arrival, and there he carried out a regular program of sacrifices during his term in office.[80]

So intimate was the connection between city gods, magistrate, and bureaucracy that city gods in the T'ang and Sung were perceived as "divine bureaucrats, and if a magistrate died in office, or was particularly effective in his duties as magistrate or prefectural administrator, he would often turn into a city god of the city in which he was based upon his death."[81] The web of government and spiritual forces was so intimate that the traffic between the earthly and spiritual worlds was constantly trod by officials speaking as gods and spirits speaking as officials:

Thus a city god was conceived of not only as the relatively civilized spirit of a human being (rather than of a white tiger or a defeated general . . .), but also as a person with a particular political and social role: a local official. This had important ideological implications. It made the city gods subordinate to higher divine authority, and subject to the rules and regulations of the heavenly bureaucracy. . . . But it also had a profound effect on the conception of local officials. After all, a man who governs in partnership with a god must have some divine qualities himself. Thus prefects and magistrates began to take on some of the sacred authority that characterized monks, priests, and shamans.[82]

In such a manner, the divine sanctioning of the authority of the emperor had its counterparts on the local level, and these served to reinforce the autocratic structure of government and local administration.

Because of the encompassing administrative structure in imperial China, neither cities nor towns were the locus of autonomous administrative units. All power was lodged in the hands of officials (with overlapping jurisdictions) tied to the central administration in Peking. Throughout Chinese history and into the twentieth century the officials lowest on the administrative ladder, the district magistrates (*hsien chih*), ruled over territories composed of several villages, areas similar to our counties. In the Ch'ing dynasty

[80] David Johnson, "The City-God Cults of T'ang and Sung China," *Harvard Journal of Asiatic Studies* 45 (1985): 363–457 at pp. 363ff.
[81] Ibid., p. 438.
[82] Ibid., p. 443.

(1644–1912), districts were composed of a central walled city "surrounded by a few towns and several score villages or hundreds of villages, the size of which varied."[83] Each area (town and village) had its head man who was appointed by the district magistrate. Households were organized into security units, headed by another appointee of the magistrate. In the early seventeenth century, there was a constable (*ti-pao*) who was likewise appointed by the magistrate, to keep an eye out for crime and disorder and to report everything to the magistrate. It was also his responsibility to supply the government with anything it needed by way of provisions or arrangements (as an inquest).[84] It was his job to enforce the collection of taxes, to investigate murders, and to arrest thieves.[85] Failure to fulfill his obligations resulted in a beating. But in no sense were any of these agents of the central government representatives of the community: they were simply appointed and dismissed by the magistrates or his superiors. "There was no autonomy in the *chou* [prefect], the *hsien* [district] or in the towns and villages that constituted them. In fact no formal government of any sort existed below the chou and hsien levels."[86]

All power and authority radiated down from the emperor and his officials to the local level. Although the magistrate was in charge of the daily operations of his district, he was rarely empowered to make momentous decisions. Indeed, all major decisions had to be approved by higher authorities. The magistrate "was not empowered to make major decisions. Excepting in certain routine matters like the handling of minor civil cases, which were under his jurisdiction, the magistrate had to report to his superior and secure approval on most details of administration. This situation led Ku Yen-wu to conclude that 'the magistrate possessed the least power among all officials.'"[87] On virtually all issues of governance, the emperor and his officials had the last word. In regard to legal jurisdiction, the magistrate had virtually none, and none that could not be overridden by higher officials. The same could be said of prefectural and provincial officials who could be countermanded by officials in the capital. In regard to precedents it was extremely rare for them to be set by an action taken by a magistrate on the local level. Even in a situation where a previous legal case might be referred to regarding the length of a term of banishment, for example, that case could not be cited as a precedent until the decision had been circulated by the highest authori-

---

[83] Ch'ü, *Local Government*, p. 4.
[84] Ibid.
[85] Ibid.; and Watt, *The District Magistrate*, p. 190.
[86] Ch'ü, *Local Government*, p. 1.
[87] Ibid., p. 193. Also Winston Lo, *Introduction to the Civil Service System*, p. 39.

ties, even though they had earlier written and approved of the sentence.[88] In general, no decision was originated by the local magistrate that could in any sense become a precedent, since all such decisions had ramifications beyond the local scene and would be reviewed with the utmost attention and would then be decided by the highest officials.[89]

Similarly, as a matter of law, all capital offenses had to be reviewed by the emperor before the sentences could be carried out.[90] These were frequently delayed for a year or so, and the hope was that the emperor would declare a jubilee and pronounce a general amnesty whereby all prisoners would be released.[91] In a word, there was no room for autonomy at any level. The singularity of the all-encompassing power of the emperor restricted jurisdiction to one domain, and hence the legal forms of local autonomy (incorporation, for example) that preserved local jurisdiction could not develop. There was no legal space, accordingly, for the rise of cities, towns, and universities with the autonomy that such institutions enjoyed in the West.[92] The mandate of heaven, which guaranteed social harmony, was something that was thought to be lodged solely in the singular authority of the emperor.

Plainly, the revolution in law and self-government that occurred in Europe

---

[88] E. Alabaster, *Notes and Commentaries on Chinese Criminal Law and Cognate Topics, with Special Relation to Ruling Cases, Together with a Brief Excursus on the Law of Property* (London: Luzac, 1899), p. 14.

[89] This of course was different from the situation in English common law, where decisions by any lower court are taken (and this remains true today in the United States) as precedents in future cases. See Mirjan R. Damaska, *The Faces of Justice and State Authority* (New Haven, Conn.: Yale University Press, 1986). Moreover, during early Roman times and in northern Europe until the French Revolution, decisions (1) were made independently of the emperor/prince, and (2) were legitimate sources of legal authority, i.e., case law, for future litigation. See John Dawson, *The Oracles of the Law*, reprint (Westport, Conn: Greenwood Press, 1978).

[90] George T. Staunton, ed. and trans., *Ta Tsing Lü Li: Being the Fundamental Law of the Penal Code of China* (London: Cadell and Davies, 1810); Derk Bodde and Clarence Morris, *Law in Imperial China*, reprint (Philadelphia: University of Pennsylvania Press, 1973), pp. 41–2, 113, and passim.

[91] Escarra, *Chinese Law*, trans. Gertrude W. Browne (W. P. A. University of Washington; photo-mechanical reproduction by Harvard University, Harvard University Asian Research Center, 1961), pp. 350–2. This procedure was connected to the spiritual calendar according to which autumn was the season for dying.

[92] Although merchant guilds existed in China, they were not endowed with the legal powers of corporations that existed in the West. T'ung-tsu Ch'ü, *Local Government*, pp. 168ff, asserts that they had too little power as well as respect to be important in local government; while H. B. Morse says, "In China gilds have never been within the law. They grew up outside it." *The Gilds of China* (New York: Russell and Russell, 1967), p. 29. Likewise, Peter Golas, "Early Ch'ing Guilds," in *The City in Late Imperial China*, ed. G. William Skinner (Stanford, Calif.: Stanford University Press), pp. 555–80, reaches a similar conclusion. On p. 559 (note) Golas stresses their dissimilarity from the guild of the Western type.

in the twelfth and thirteenth centuries, whereby cities and towns were granted the right to make their own laws, to create courts of adjudication, to raise taxes, to own property, to sue and be sued, to establish their own standards for weights and measures, and to print money,[93] did not take place in China – neither in the twelfth and thirteenth centuries nor in the seventeenth or early twentieth centuries. "No Chinese communities ever established themselves as municipalities possessing defined powers of independent jurisdiction."[94]

To better understand this failure to develop autonomous and self-governing social institutions, we should consider the nature of Chinese law more directly.

Chinese law

The Chinese conception of law contains many nuances that make it different from both Western and Islamic law. It is very unlike Islamic law in that it is not associated with the idea of commandments from God, although it does contain the idea of rites and traditions that are sanctified by virtue of their ancient origins in the practice of sage-kings of the past. Once the distinction between positive law (*fa*) and propriety (or sacred rites, *li*) is made, then it can be seen that the li of the ancient way (tao) is a sacred source of law, that is, a source or model of human behavior rooted in the nature of things. As Benjamin Schwartz put it, one can "agree with Herbert Fingarette that the entire body of *li* itself, even when it involves strictly human transactions, somehow involves a sacred dimension and that it may be entirely appropriate to use a term such as 'holy rite' or 'sacred ceremony' in referring to it."[95] Hence one starting point of Chinese law is

the idea that heaven and earth were governed by one principle – called tao, the way, the creative principle of natural order. Any act contrary to this order in human society resulted in a disruption of the harmony between heaven and earth, and might lead to such calamities as flood, drought, internal disorder. So that the order might be preserved, heaven chose men of outstanding virtue, te, and gave them the mandate, ming, to rule their fellow creatures.[96]

---

[93] Berman, *Law and Revolution.* chap. 12.
[94] Sybille van der Sprenkel, "Urban Control in Late Imperial China," in *The City in Late Imperial China,* pp. 609–32 at p. 609. And see Etienne Balazs, "Chinese Towns," in *Chinese Civilization and Bureaucracy,* pp. 66–78.
[95] Benjamin Schwartz, *The World of Thought,* p. 67.
[96] M. Meijer, *The Introduction of Modern Criminal Law in China* (Batavia: de Unie, 1950), p. 2.

While the Chinese acknowledge a type of positive law enacted by men, their greater commitment is to *li*, to the sacred rites of the past, and this commitment is rooted in powerful interlocking assumptions.

On the one hand, li is connected to the idea of correct patterns of conduct in the sense of the tao, the underlying patterns of nature.[97] As such, these patterns could be said to be eternal and never-changing. At the same time, Confucius himself apparently believed that the ancient sage-kings had in fact achieved perfection of conduct, that "the tao had in its essential features been realized in the Chinese past and that concrete knowledge of the Tao in its last embodiment during the early Chou period . . . was available."[98] In other words, the wise Chinese ancestors had achieved the ideal of human propriety and we have empirical knowledge of it in the five classics and other documents of the past.[99]

On the other hand, whatever this conduct of the ancient sages was, it was rooted in the natural order of things – it was a reflection of the true ordering of nature. Such a conception of the ideal (and hence sacred), realized in the past, rooted in nature, and about which we have evidential knowledge, is a formidable idea that must have had a powerful holding force in Chinese thought. As such it seems likely that it served as an inhibiting force which prevented the rise of philosophy as an autonomous discipline such as was the case among the Greeks and medieval Europeans.

We should point out, however, that this so-called conception of natural law is different from that of the European medievals. First of all, the Chinese conception is exceedingly concrete and tied to a single group (the ruling elite of China) during a single period of time (the past). It is in this regard ethnocentric despite the fact that China has always known other ethnic groups (usually called barbarians). For example, there were Mongolians, Khatians, Koreans, Indians, indeed, Buddhists, Vietnamese, Malaysians, Javanese (who were conquered by Khubilai Khan), Muslims, and many others whose conceptions of law were notably different from the purely Chinese.. But because of the concrete Chinese view of these matters, the Chinese did not seek to formulate a concept of natural law that was truly

[97] Schwartz, "On Attitudes Toward the Law in China," in *The Criminal Process in the People's Republic of China: An Introduction*, ed. Jerome A. Cohen (Cambridge, Mass.: Harvard University Press, 1968), pp. 62–70 at p. 62.

[98] Schwartz, *The World of Thought*, p. 85.

[99] The five Confucian classics consist of *The Books of Documents, Book of Poetry, Book of Rites, Book of Changes*, and *The Spring and Autumn Annuals*. See James Legge, trans., *The Chinese Classics*, reprint (Hong Kong: Hong Kong University Press, 1960), 5 vols. These are a composite of idealized history, ethical exhortations, and prescriptions for proper conduct, as well as metaphysical speculation.

universal, which transcended both the greatness of the ancient sages and that of other ethnic groups. In other words, the Chinese were never impelled to work out a science of law. It is therefore a sleight of hand to dismiss China's parochialism on this score by saying that it has always been isolated from others. It can only be said that China was isolated from European influences, not from contact with the approximately fifty-five different ethnic groups that currently are officially registered within China.[100] Chinese legal thought did not move to that higher level of abstraction which considered local (regional and national) variation in customs (li) and which simultaneously postulates a higher level of sacred order, eternal and even divine, which is associated with natural law in the West. In other words, the natural law of the West refers above all to the higher level that transcends the differences between the Italians, the French, the Germans, and so forth. It is precisely because it is above these local variations that it is thought to be natural and imbedded in nature.[101]

An illustration of the lack of an integrated system of law and institutions within the borders of China even in the nineteenth century can be seen in the differential treatment of Chinese nationals of different ethnic persuasion for example, Chinese Muslims and Mongolians (Tartars). Because of the lack of a natural law concept in the Western sense and thus the lack of a uniform legal system, two legal codes existed – one for the Chinese and another for Mongolians (and other Muslims):

> Nor is the difficulty an imaginary one. If the Tartars claimed that Tartar law should hold in Tartary, while Chinese law was permitted in China, then things might . . . be accommodated; but mixed in both countries as the Chinese and Tartar now are, it is inconvenient, to say the least, to have (as often happens) to apply two codes to the same case – with the result, that of two criminals equally guilty of the [same] offense charged, the one escapes with a whipping, and the other, besides being bambooed, is transported for life. The Chinese insist on *even* justice; and with unequal laws, it is impossible to obtain it.[102]

Likewise in the case of Muslims, the Judicial Board (in Peking)

> ruled that the special laws relating to Muslims applied to Muslims only, and not to Chinese concerned with them. So if two Muslims and a Chinese commit a robbery,

---

100 Kevin Sinclair, *The Forgotten Tribes of China* (Missisauga, Ont.: Cupress, 1987), p. 9.
101 As Max Weber put it, in China, "a divine, unchangeable law of nature existed only in the form of sacred ceremonies, the magical efficacy of which had been tested since time immemorial, and in the form of sacred duties toward the ancestral deities. A development of natural law of modern occidental stamp, among other things, could have pre-supposed a rationalization of the existing law which the Occident had in the form of Roman law." *The Religion of China*, trans. Hans Gerth (New York: Free Press, 1951), p. 158.
102 Alabaster, *Notes and Commentaries on Chinese Criminal Law*, p. li.

the Muslims under the Muslim law "where three or more . . . etc.," incur the penalty of military servitude, and the Chinese, under the Chinese law, "where several persons are concerned . . . etc.," incur transportation only.[103]

This failure to work out an integrated system of law within the territories ruled by Chinese authorities can also be seen as an expression of the norm of recognizing the extreme particularities of Chinese law. Thus, before punishing any offender, the legal officials were to mediate on the "eight deliberations,"[104] that is, the peculiar conditions of social status which exempt categories of individuals from punishment. These include the status of a privileged person by virtue of official rank, birth, age, and so on.[105] This way of proceeding gives the appearance of attempting to find an excuse for not applying the law because of the peculiarity of one's social status.

The traditional Chinese conception of li (propriety) – in opposition to the idea of uniform law – has remained so strong that Chinese officials have only with great difficulty been able to transcend the traditional concepts of propriety through legislation in areas vital to economic commerce. A famous case in point is one involving merchants in Shanghai confronting a Supreme Court decision contrary to their wishes. We have the dialogue between a Mixed Court judicial officer from Peking who was called upon to question a local member of a merchant guild in Shanghai (in a mixed court there) regarding the authority of the Chinese Supreme Court at the turn of the century. The insights revealed by this case make reproduction of the extended dialogue well worth presenting here:

Assessor: Do you respect your Supreme Court?
Witness: Naturally, as a Chinese organization we ought to respect it, but if a decision of the Supreme Court is not good, then we merchants have no way of changing it.
Assessor: Would you say that you are better able to decide whether a judgment of the Supreme Court, rendered by some of the most competent men in China, is good or bad?
Witness: No, we merchants usually do what the other merchants do.
Assessor: In short, do you obey the decisions of your Supreme Court or not?
Witness: If the judgment is reasonable, I obey. If it is not, I do not obey it.
Assessor: Then you constitute yourself as a higher judge?
Witness: Not I alone.

---

[103] Ibid., p. lii. I have changed "Mohamedan" to "Muslim."
[104] Escarra, *Chinese Law,* p. 348.
[105] Derk Bodde discusses reflections of this notion in the imperial codes under the headings of "Let the Punishment Fit the Crime," "Privileged Social Groups," and "Differentiation within Family," in "Basic Concepts of Chinese Law," in *Essays on Chinese Civilization,* 2d ed. (Princeton, N.J.: Princeton University Press, 1981), pp. 184f. This essay is also reprinted as chap. 1 of *Law in Imperial China.*

Assessor: Let us interpret the order of merchants: "We are right in what we say and the Supreme Court is wrong, even in the business of law."
Witness: That is quite so.[106]

The word *reasonable* in the text is apparently a translation of the word *li*, meaning conforming with the sacred rites (or traditions).[107] It seems fair to say that the Chinese retain a conception of law that contains an aura of sacred and eternal, so that "positive law is accepted only as it represents a custom itself judged in accord with the law of nature."[108]

Given this broader context, one may notice that attention shifts to the five relationships, that is, father–son, emperor–subject, husband–wife, older brother–younger brother, friend–friend. If these relationships were put in order, Confucius argued, then all other relations would be in accord, social order would prevail, and the emperor could continue to enjoy the mandate of heaven. We may also note that like Islamic law, Chinese law made no distinction between moral/ethical relationships and those of positive law more broadly. Law and moral conduct were in effect synonymous.[109]

Historically Chinese legal thought has been of two schools – the Confucian and legalistic. The Confucians (following the master) believed in the rulership of wise men (sage-kings) and thought that the best means of ruling was that achieved by benevolence and exemplary conduct. While the authority of the emperor was clearly enforced by a complex hierarchy of officials, it was always assumed that the personnel of Chinese officialdom, like the emperor himself, were paragons of moral virtue, who by this very quality were most qualified to lead. And since Confucian ideology explicitly sanctioned a hierarchical social order, it was thought to be perfectly correct that rulers should rule and the ruled should unquestioningly obey. This was but another expression of Confucian filial piety.[110] One the of the clearest expressions of this view is in the words of Confucius's most important follower, Mencius (ca. 371–ca. 289 B.C.). He argued: "Some labor with their minds and some labor with their physical strength. Those who labor with physical strength are ruled by others. Those who are ruled are sustained by others, and those who rule are sustained by others. This is a principle universally recognized."[111] It was only by the display of exemplary conduct

---

[106] The case was originally reported in the *North China Herald* (July 31, 1926) and was published in a preface by A. Padoux to a translation of a Chinese treatise on law, in French, by Jean Escarra and R. Germain, *La conception de la loi*. This text is from Escarra's *Chinese Law*, pp. 113–14.

[107] Joseph Needham makes these points also in *SCC* 2: 259.

[108] Escarra, *Chinese Law*, p. 113.

[109] Ibid., p. 99.

[110] Staunton, *Ta Tsing Lü Li*, p. xviii.

[111] As cited in Ho Ping-ti, *The Ladder of Success*, p. 17.

on the part of the ruler and his officials (who served to guide him toward the correct path) that the people could be guided toward appropriate conduct in their affairs. If the emperor tried to rule by issuing laws, then the people would soon discover the nature of the laws and find ways around them. According to this view, the issuing of law would only lead people to become more contentious and litigious. Instead of government by laws there should be government by li, propriety, of rightly ordered ritual relationships. Laws are only as good as the people who draw them up, it is argued. One should therefore study li, the ancient customs and traditions of the wise rulers, from which one could ascertain the wisdom to rule by virtue.[112]

In contrast, the legalist school asserted that only the enunciation of uniform and severely enforced laws would bring about the harmonious state that is sought. This view was vigorously defended by Han fei-tzu (d. 233 B.C.).[113] He asserted that, though wise men may be found occasionally, such men are rare because most men act in a self-interested fashion. It is for this reason that punishments are necessary: to coerce people into doing what is right rather than what is solely in their self-interest. If the government is to be strong, it must eliminate factionalism and privilege, and the way to do this is to create a uniform system of laws, publicly announced and earnestly enforced. Times change, moreover, and therefore the laws of society must also change. The customs and traditions (li) of the past were adequate to their time, but they are so no longer.

The legal spirit that ultimately grew out of this clash of views was one which is at once severely punitive and highly particularistic. Instead of applying uniformly to all citizens, it is riddled with exceptions that acknowledge privileges for all sorts of groups: privileges of economic class, of family origin, of intellectual achievement, of age (and youth), and multifarious other conditions. For example, in the last dynastic code of the empire (published in 1647, reprinted in 1725 and later), which goes back to the Ming dynasty (1368–1644), astronomers who were members of the Imperial Board of Astronomers in Peking are singled out for special treatment. Likewise, throughout imperial times, candidates who had successfully passed the state examinations were exempt from certain taxes, from yearly forced labor service, and from corporeal punishment (which was the most common punishment for crime). Likewise, those successful candidates who became offi-

---

[112] See Schwartz, "Legalism," in *The World of Thought*, chap. 8; Escarra, *Chinese Law*, pp. 66–80; as well as Derk Bodde, "Basic Concepts," pp. 178ff.

[113] In this discussion I have followed Escarra, *Chinese Law*, and Derk Bodde, "Basic Concepts," pp. 171–94 in *Essays on Chinese Civilization*. Additional discussion and text of Han fei-tzu are in Wing-tsit Chan, *A Source Book of Chinese Philosophy* (Princeton, N.J.: Princeton University Press, 1963), pp. 251–61, as well as Schwartz, *The World of Thought*.

cials in the imperial service could not be arrested or prosecuted until the emperor had first been informed and given his permission for the prosecution to proceed. In such a manner the Confucian ethic, which recognized differences of natural ability and moral quality, became ensconced in the law.

The penal character of Chinese law and the absence of a distinction between public and private law (and between criminal and civil offenses) are seen in the prescription of beatings with a bamboo stick for virtually all offenses, even those of an essentially civil nature. For exceeding the official usury rate (36 percent per annum in the Ch'ing), the punishment was forty blows with the bamboo.[114] No consideration was given to the restitution of the money illegally obtained, nor was any attention directed to the rights of the parties involved. The Chinese legal mind, one might say, was not interested in these private matters of right and wrong, nor in the abstract question of justice per se, but in the consequences the disruption implied for the social order. As Needham stressed in his study of Chinese law, the issue is not who is right and who wrong but "what has happened." In Chinese law the focus is on what has happened to the natural order: it was the injury – the failure to perform the rites, to exhibit the right attitudes, possibly the failure of the emperor in his duties – which has caused an official to commit a grievous error, and so forth.[115] In such a system of collective responsibility there can be no law of negligence that affixes blame for wrongs committed by one person against another in terms of an omission to act prudently, thereby causing harm to another.[116] The consequences of these views for

---

[114] Staunton, *Ta Tsing Lü Li*, p. 150.

[115] Escarra, *Chinese Law*, pp. 108–9.

[116] Van der Sprenkel notes that this effort to assign collective responsibility for the disorder that results from a legal violation "might include punishment for a person who would according to English law be judged innocent," *Legal Institutions in Manchu China: A Sociological Analysis* (London: Athlone Press, 1966), p. 71. Herbert Fingarette's study of the Confucian *Analects* contains what seem to me penetrating insights regarding the absence "of any sense of moral responsibility as ground for guilt and hence punishment as moral retribution," *Confucius: The Sacred as Secular* (New York: Harper and Row, 1972), p. 27. He further claims that we should not equate the Confucian sense of shame with Western guilt (p. 30). This seems to me wholly consistent with the view that there is an absence of the Western concept of *conscience* or inner moral agency, in Chinese philosophy, which I discussed in Chap. 4. Consequently, this conceptual vacuum could not contribute to the elaboration of a fully developed moral and legal concept of individual responsibility on the one hand and negligence on the other. Likewise, the absence of a conception of an inner moral agency capable of arriving at moral truths (unaided by revelation) seems related to the absence of an independent source of reason and rationality such as was available in Western thought, above all in Christian theology (see my discussion in Chap. 4). For a discussion of the concept of conscience in Chinese philosophy, see Chung-Ying Cheng, "Conscience, Mind and Individual in Chinese Philosophy," *Journal of Chinese Philosophy* 2 (1974): 3–40; and Edmund Leites, "Conscience and Moral Ignorance: Comments on Chung-Ying Cheng's 'Conscience, Mind, and Individual in Chinese Philosophy,'" *Journal of Chinese Philosophy*

legal theory are not at all minor, as Derk Bodde has noted: "The law was only secondarily interested in defending the rights – especially the economic rights – of one individual or group, and not at all in defending such rights against the state. What really concerned it . . . were acts of moral or ritual impropriety or of criminal violence which seemed to Chinese eyes to be violations or disruptions of the social order."[117] Thus the paramount concern for the cosmic order and its interconnections with the social order kept attention focused on the hierarchy of order in the realm.

For these reasons, the official law always operated in a vertical direction from the state upon the individual, rather than on a horizontal plane directly between two individuals. If a dispute involved two individuals, individual A did not bring suit directly against B. Rather he lodged his complaint with the authorities, who then decided whether or not to prosecute individual B.[118]

What is more, there was a great paucity of legal experts, both public and private. There were no private lawyers available to advise the plaintiff, and even if there had been, it would have been a great (and dangerous) affront to challenge the court (the local district magistrate) and quite impossible to claim any right to do so. As Jean Escarra put it, "To admit that one's application and interpretation [in court] could be the subject of discussion, that the judge could be contradicted, would attest intolerable disorder. . . . [T]here is no place for an advocate in the traditional Chinese judiciary organization. That would have been singularly dangerous!"[119] Indeed, those who attempted to defend others by preparing legal briefs for relatives, friends, or clients were accused of being "litigation tricksters" and were generally given three years' penal servitude for their trouble.[120] This outcome may be seen as a result of the Confucian ethic that stressed the need to maintain outward obedience and respect for all authorities, especially public authorities. To Chinese eyes such public displays (as challenging the word of authority figures) are unforgivable signs of disrespect and dissension and are the ultimate betrayal of filial piety, of family and clan, and, above all, the betrayal of the principle of *jang*, yieldingness.[121] The Confucian ethic of

2 (1974): 67–78. Benjamin Schwartz's commentary on Fingarette, in *The World of Thought*, especially pp. 71–5, does not differentiate modern Western psychology from this higher level of moral and ethical agency that comes out of the fusion of Greek philosophy and Christian theology.

[117] Derk Bodde, "Basic Concepts," p. 171.

[118] Ibid., p. 172.

[119] Escarra, *Chinese Law*, p. 348.

[120] Bodde and Morris, *Law in Imperial China*, pp. 413, 180, and 190 n26.

[121] Needham, *SCC* 2: 61ff; Sprenkel, *Legal Institutions of Manchu China*, p. 81; Tu Wei-ming, "The Confucian Tradition," in Paul S. Ropp, ed., *Heritage of China*, pp. 112–37 at pp. 116ff.

jang-jen (benevolence and human kindness) encourages one to yield to others at all times in order to avoid discord and dissension. Jang is expressed through a form of piety and self-effacement that elevates others and effaces the self. Joseph Needham bears witness to these traits of yielding to others in contemporary China when he refers to "the difficulty of passing through any doorway with a group of people," and the self-effacement of scholars who are observed "positively struggling for the least honorable places at a dinner-party."[122] Public decorum and deference to others are thus the epitome of the Confucian worldview.

Thus, filial piety and, more concretely, the ethic of jang enjoined the individual to obey all those above him in authority, and this differential worldview was ensconced in the law. In any case of assault in the *Ta Ch'ing Lü-Li* (*The Imperial Penal Code and Statutes* of the Ch'ing), there must always be a determination of senior and junior, so that greater punishment can be automatically directed toward the junior person.[123] Still another expression of the value of filial piety is the enforced period of mourning – up to twenty-seven months – enforced on state officials and stated in many dynastic codes, including the T'ang, Sung, Ming, and Ch'ing.[124] In short, the Confucian stress on obedience stifled the development of all forms of contentiousness in public forums.

On another level, the lack of expert legal wisdom is seen in the district magistrate himself, who was appointed because of his successful passing of official literary examinations and thus had no legal expertise. It is true that during some periods of the Sung, separate legal exams were held under the influence of Wang An-shih (1021–86), and that there was even a school of law located at the capital. However, the law school did not survive the move south (in 1127) when the Sung dynasty split into its northern and southern kingdoms. During the T'ang dynasty and the early Sung there was a designation for those who had passed a special law examination (*ming-fa*),[125] but it soon lost favor and disappeared. At the same time it must be remembered that no school at any level was ever authorized to grant a degree: the degrees of Chinese education were but certifications of having passed various state-sponsored civil service examinations, not diplomas signifying completion of a comprehensive course of study. During the short life of the official law

122 Needham, *SCC* 2: 62.
123 Staunton, *Ta Tsing Lü Li*, p. 343.
124 Bodde and Morris, *Law in Imperial China*, p. 39.
125 Hucker, *Dictionary*, "ming-fa," no. 4009. Also see Lee, *Government Education*, chap. 6. With 8,291 separate entries, Hucker's *Dictionary* is a gold mine of information that serves rather like the *Encyclopedia of Islam*. I have given the number of the entry for each set of Chinese characters for a particular title, institution, or concept.

school, it admitted mainly those who had already received the official degrees through the examination system, and a study of the dynastic codes (which were reissued with major or minor modifications when each new ruling dynasty came to power) does not constitute a serious introduction to jurisprudence.[126] In short, there were no legal institutes or manuals to study and analyze for none of these existed.[127] Thus if legal expertise existed, it was located at the very top of the imperial hierarchy – in the Bureau of Justice – and had been gotten through on-the-job training. As Escarra and others have pointed out, the triumph of the uniform literary examination (toward the end of the Sung) "killed technical [legal] studies in China."[128] As a result, Chinese legal history shows very little change from the T'ang and Sung codes through the reprinting of the Ming codes in the nineteenth and twentieth centuries.

In sum Chinese legal thought was polarized between those who believed that society ought to be ruled by an elite vanguard of exemplary individuals, who would set the right example for people, and those who believed that "the people's nature is such that they delight in disorder" (Han fei-tzu)[129] and that therefore severely enforced uniform laws must be put in place. At the same time, Chinese thought stressed the importance of preserving exemplary traditions that reflected the harmonious realization of the tao through collective responsibility. While all people are called upon to lead exemplary lives, the emperor and his officials have the primary duty to rightly order their conduct (and state affairs) to facilitate the correct ordering of the social world in harmony with nature.

The inherent stress on hierarchy and collective responsibility inhibited the development of any autonomous spheres of social action or sovereignty. The emperor's paramount responsibility for maintaining the mandate of heaven

---

[126] The continuity between the dynastic legal codes of China from the T'ang to the Ch'ing is discussed in Bodde and Morris, *Law in Imperial China*, pp. 5–63.

[127] Escarra, as cited in Needham *SCC* 2: 524–25; cf. Bodde and Morris, *Law in Imperial China*, pp. 68–75, and Escarra, *Chinese Law:* Chinese law "is above all . . . a non-systematic discrimination of concrete cases. There are no attempts at reduction to unity, or almost none," p. 106.

[128] Escarra, *Chinese Law*, p. 466, who cites Pelliot. This vacuum in legal expertise, above all on the local level, gave rise to the private legal secretary (*mu-yu,*) in the early Ming (mid-fourteenth century), who was hired by the district magistrate. The legal secretary was often someone who had not yet succeeded in passing all the state examinations or, possibly, a failed official, who privately specialized in legal learning, and thus came to advise the local magistrates, who personally hired them. The legal clerks (another group of subalterns hired by the magistrate) were less informed and could receive information but not conduct the trial itself. Likewise, the private secretary, whose knowledge of legal matters might be better than that of the magistrate, could not appear in court, although he might sit behind a screen to whisper advice to the magistrate. See Ch'ü, *Local Government*, chaps. 3 and 6.

[129] As cited in Schwartz, "Legalism," p. 323.

meant that he should always maintain order from the top down by asserting his sovereignty. To allow groups of individuals (villagers, townsmen, or other social groups) the freedom to pursue independent courses of action of their own choosing would be to show signs of having lost the heavenly mandate. Conversely, for any individual (or group) to proceed on a course that neglected his (or their) duties to filial piety – to show disrespect toward ancestors, elders, and other authorities – was likewise to display impiety. Accordingly, there is an absence of legal thought spelling out any theory of autonomous spheres of limited jurisdictions, that is, a theory that treats collective actors as a whole body, or a corporation, with corresponding rights to internal self-regulation and external representation. There was also an absence of a theory of public and private spheres to articulate the difference between ownership and jurisdiction, that is, rightful legal authority to judge matters of internal regulation versus ownership of the collective assets of the entity in question. Likewise, there was an absence of a theory of representation that grants to lawfully constituted groups or individuals powers of representation – both in the strict legal sense and in the political sense (which derives from the former). It may also be noted that the law carefully circumscribed legitimate interests (that is, rights) to be involved with others so that these were tied to degrees of familial relatedness. Anyone not related to another person and who attempted to intervene in a legal matter was subject to special punishment. This is seen most poignantly in the case of the volunteer legal aide who draws up legal papers for another person and is consequently treated as a "litigation trickster" and, as we noted earlier, sentenced to penal servitude. By this means Chinese authorities nipped in the bud all efforts to establish any form of public representation for groups or individuals. There is an absence of the European legal maxim "What touches all should be considered and approved by all (*quod omnes tangit . . .*)."[130] Indeed, we seem to find the opposite: what affects all must be decided by the emperor (and his officials). As Derk Bodde sums it up, it must be remembered that

the judicial system of imperial China, like the governmental system as a whole, was a centralized monolith with no division of powers; that there was no private legal profession; that on the lowest level of the *hsien* (district) or *chou* (department), where all cases originated (save those in the capital or in frontier regions), the magistrate rarely possessed any specialized legal training and handled cases that came before him simply as one of many administrative duties; that, however, he often personally employed a non-civil service private secretary who did possess specialized knowledge of the law; and that all but minor cases automatically went upward from

---

[130] Gaines Post, *Studies in Medieval Legal Theory*, pp. 62f, 90, 163ff, and 175.

the *hsien* or *chou* to higher levels for final ratification, some as high as the emperor himself.[131]

There was no intellectual revolution in Chinese legal thought in Sung, Ming, or later dynasties.[132] The existing legal patterns and structures continued into the twentieth century with very minor changes in basic principles, above all, with the ancient stress on penal law and the corresponding absence of civil liberties and civil rights. What was most lacking in Chinese thought was the impulse to synthesize long-standing Chinese practice into abstract principles and the creation of a real science of law independent of the dictates of the emperor and the positive law in existence. What was absent in China, Jean Escarra concluded, was

that tradition of jurisconsults succeeding one another through the centuries, whose opinions, independent of the positive law, and whatever its practical applications might be, built up, on account of their methodical, doctrinal, and scientific character, the "theory" or speculative part of law. China had no "Institutes," manuals, or treatises. A jurisconsult such as Tung Chung-shu, litergilogists like the elder and younger Tai, codifiers like Chhangsu wu-chi . . . did not accomplish works parallel to those of a Gaius, a Cujas, a Pothier or a Gierke.[133]

Nor, I should add, did they produce a great integrative work equivalent to Gratian's *Concordance of Discordant Canons*, and whose task was to produce a legal structure and doctrine universal in conception.[134]

The absence of a legal theory of autonomous jurisdiction – that is, some form of corporate entity – within a hierarchy of nested jurisdictions with powers of legislation (that is, internal governance), adjudication, and representation must be seen as one of the most serious weaknesses of Chinese civilization. For without some spheres of autonomy, no groups can emerge as professionals, that is, legitimate specialists who represent the highest levels of thought and action in a particular sphere of human endeavor. Historically speaking, the first nonecclesiastical professionals to emerge are probably judges and lawyers. We saw in the case of Islamic legal theory that jurisprudence (that is, religious law, fiqh) actually had priority over the

---

[131] Bodde and Morris, *Law in Imperial China*, p. 113.
[132] Escarra, pp. 110–11.
[133] Escarra, as cited in Needham, *SCC* 2: 524–25.
[134] The study by David Buxbaum, "Some Aspects of Civil Procedure at the Trial Level in Tanshui and Hsinchu from 1789 to 1895," *Journal of Asian Studies* 30 (1971): 255–80, does not alter the assessment of the nature of Chinese law set out by Derk Bodde and Clarence Morris, above all, as it pertains to the issues we have been discussing. Even if there were greater procedural regularity of Chinese law in the Ch'ing at the trial level, it does not alter the fact that Chinese law lacked the kind of systemization that Western law achieved in the twelfth and thirteenth centuries, nor does it alter any of the other comparative assessments I have offered regarding structures of autonomy, the laws of corporations and trusts, etc.

religious specialists whom we would call theologians (mutakallimun). But it too failed to undergo the kind of higher level elaboration seen in Western law. On the other hand, judges and lawyers who were strictly nonecclesiastical figures did not emerge in Islam until the penetration of the Middle East by the Western powers in the nineteenth century. Thus lawyers (advocates) as a professional group were also missing in traditional Islamic society.

In the West, on the other hand, there was a highly respected lay tradition of judges who adjudicated disputes and established law. Indeed, under the Romans, they created not just a system of case law but a vast nonecclesiastical codification of the law – the *corpus juris civilis*. In the ecclesiastical domain, we may also remind ourselves, theology – that is, the systematic and scientific exposition of the canons of faith – was a highly regarded discipline, which, in its university location, created a body of professionals free from the dictates of the state. Once the universities arose, theology as an intellectual discipline took its own path, but even philosophy was able to assert its rights in the academy in defiance of the religious hierarchy, that is, the bishops. Likewise, in the twelfth and thirteenth centuries, law (jurisprudence) was probably the most developed science in Europe, and of course some universities (such as Bologna) were known for their law faculties. These professionals too were free from the dictates of the state. The tradition of the independent legal professional who defines, expands, comments upon, and codifies the law continues down to the present, both in Europe and the United States.

I should also mention medicine, which similarly produced its own professionals who were able to establish legal standards, not just for teaching medicine but also for its practice – something China and Arabic-Islamic civilization were unable to do. In the case of Islam, this again was because of the absence of an adequate legal theory of limited jurisdiction manifested in a theory of corporations.

In the case of China, we see more sharply that its legal structures and legal theory were inadequate to the task of fashioning limited zones of legal autonomy for any professionals – philosophical, scientific, juridical, or medical. The lack of professional development in Chinese medicine seems to have continued well into the modern era, perhaps even into the nineteenth century. Even in the Ming dynasty it appears that there was no official definition of or regulation of physicians. One source refers to the fact that the lack of government regulation of medical practice resulted in "one grand free-for-all profession with no registration code or ethics whatsoever."[135]

---

[135] K. Chimin Wong and Wu Lien Teh, *A History of Chinese Medicine*, 2d ed. (Shanghai: National Quarantine Service, 1936), p. 141, citing Morse, *The Three Crosses in the Purple Mist*.

Similarly, Paul Unschuld points out that in China, "while a detailed mal-practice legislation was elaborated from T'ang times on, no comprehensive governmental regulation of the vocational practice of health care, and above all, almost no supervision of the qualifications of the physicians and pharma-cists as a whole were introduced during the Imperial Age."[136] In the realm of pharmacopoeia, there was no legally binding canon such as existed in the West from the sixteenth century.[137] In part this was the result "of Confucian policy not to allow experts with a specialized expert knowledge to rise socially as a group[, as] this might have led to social tensions, crises, and even restructuring."[138] Thus we get the paradoxical situation in which all positions of elevated learning in Chinese civilization were officially con-trolled by the Chinese state through the examination system, yet the control of vocational activities was largely unregulated. This was due, on the one hand, to the fear that acknowledging and thus encouraging experts might result in their gaining power and, on the other hand, to the lack of a concep-tion of public law. In general, however, it must be said that Chinese officials were extremely fearful of the emergence of autonomous social groups, espe-cially professionals such as private lawyers (but also teachers), as well as economic experts such as merchants.[139]

I will now consider the civil service examination system, the central bat-tleground on which the issue – whether there could be autonomous spheres of learning independent of the bureaucratic state officials – was decided.

Education and the examination system

The nature and function of the Chinese examination system with its three degrees ("cultivated talent," "recommended man," and "presented scholar")

---

[136] Paul Unschuld, *Medicine in China: A History of Pharmaceutics,* p. 3.
[137] Ibid., p. 5. Unschuld points out that in the West the term *pharmacopoeia* is used to specify a type of pharmaceutical literature that "describes a selected series of medications for which the way of production, the quality, and the analytic methods of testing, have been deter-mined, while the book as a whole has the force of law thus subjecting the drug trades to its rules," p. 5. According to the same author, in China the various pharmaceutical works "were not binding for any vocational group."
[138] Ibid., p. 4. Joseph Needham's fascinating thesis that China and Islam were the source (or even a source) for "qualifying examinations" in the West from this point of view seems highly conjectural. In Chap. 5 I outlined the contrasting attitudes and structures of exam-inations in Islam and the West. The official examination of Islamic physicians under Caliph al-Muqtadir, to which Needham refers, appears to have been an isolated case that gave no impetus to the establishment of uniform examinations for practicing physicians in Arabic-Islamic civilization – though it did eventually give rise to manuals designed for testing physicians by the public. For other comments on this, see Chap. 5 above and note 116.
[139] The extraordinary restrictions placed on merchants in China are discussed in Balazs, *Chi-nese Civilization and Bureaucracy,* chap. 4 and passim.

has been known for some time and was discussed by Max Weber in *The Religion of China*.[140] What has not received adequate attention, however, is the fact that the Chinese educational system was both rigidly controlled and focused on literary and moral learning, while the European universities were both autonomous and self-controlled as well as centered on a core curriculum that was essentially scientific.[141] The implications of these institutional contrasts for the development of science, as in the case of Arabic science, can hardly be overstated. For if science is to flourish over the long run, it must have official approval as well as public support – something that was rarely available in China. The fields of astronomy and mathematics did benefit from state support, but just as often there were state-sponsored interdictions of the study of astronomy and mathematics.

The traditional examination system was a unique system and must be regarded as one of the greatest successes and greatest failures of any such system in the world history of education. According to Ho Peng-ti, it "entailed a wastage of human effort and talent on a scale vaster than can be found in most societies."[142] This inefficiency is revealed in the fact that it was "not uncommon for a scholar to have failed a dozen or more times in high-level examinations which were usually held at a three-year interval. The whole life of such luckless scholars was thus wasted in their studies and examination halls."[143]

It was a great success to the degree that it recruited into the literate world of officialdom the best and brightest men of China from all ranks of society for more than a thousand years. But even this claim has been challenged by more recent scholarship. It has been asserted that if one follows the rise of families into the bureaucracy, especially extended families, the evidence of open recruitment by virtue of the examination itself is not so striking. Robert Hartwell has put this thesis most forcefully:

---

[140] Max Weber, *The Religion of China*, pp. 115–16.

[141] For the latter, see Chap. 5 above and Edward Grant, "Science and the Medieval University," in *Rebirth, Renewal and Resilience: Universities in Transition, 1300–1700*, ed. James Kittelson and Pamela Transue (Columbus, Ohio: Ohio State University Press, 1984), pp. 68–102.

[142] Ho Peng-ti, *The Ladder of Success*, p. 259.

[143] Ibid. It might also be said that the extreme centralization of the administration of the examinations (in district, provincial, and imperial capitals) represented a colossal inefficiency both in terms of the great distances of travel required of the candidates and the unbelievably large mass of examination papers (all essays) generated by as many as thirteen thousand test-takers (in one place). These were even copied in duplicate (!) to assure anonymity. For a nineteenth-century account of the whole exam-taking system see Edward Harper Parker, "The Educational Curriculum of the Chinese," *The China Review* 9 (1879/80): 1–13. For an example of the "Eight-legged essay," see F. S. A. Bourne, "Essay of a Provincial Graduate with Translation," *The China Review* 8 (1879/80): 352–6.

There is not a single documented example, in either Su-Chou [province] or in the collective biographical material on policymaking and financial officials, of a family demonstrating upward mobility solely because of success in the civil service examinations. Indeed, in every documented case of upward mobility, *passage* of the examination *followed* intermarriage with one of the already established elite gentry lineages.[144]

From such a point of view the examination system was not a device in and of itself that lifted commoners from all walks of life into positions of power. It can be said, nevertheless, that it did measurably succeed in creating a meritocracy of government constantly infused with new blood and thereby prevented the permanent bureaucracy of China from becoming a completely hereditary rulership.[145] This was so because generally an individual could study for and take the state-sponsored civil service examinations, and if he passed the exams, he would automatically receive the examination certificate along with legal and social privileges. These included an official position in the Chinese bureaucracy.

On the other hand, the rigidity of the educational content of the examination system – virtually unchanged from early Ming (ca. 1368) to the twentieth century[146] – and its absolute uniformity, bordering on political indoctrination, make it a colossal failure insofar as science, innovation, and creativity are concerned. As in the case of Arabic-Islamic civilization, the modest Chinese successes and advances made in the sciences (not technology) were achieved in spite of, not because of, the official forms of education and examination. The recruitment system, created by the universally applied examination among a population of some 100 to 115 million people during the High Middle Ages (ca. 1200–1500, roughly double the population of Europe at the same time),[147] did not produce great systematic thinkers equivalent to Averroes, Peter Abelard, Gratian, Aquinas (and others in law), Buridan, Ockham, Copernicus, Galileo, Kepler (in the natural sciences), and so forth. This is not to say that China produced no great thinkers; it clearly

---

[144] Hartwell, "Demographic, Political, and Social Transformation in China, 750–1150," pp. 365–445 at p. 419. John Chaffee in *The Thorny Gates*, pp. 12ff, qualifies but does not challenge this basic thesis.

[145] It should also be noted that the yin privileges and other devices, e.g., special examinations as well as facilitated examinations, allowed powerful families to subvert the system; see Chaffee, *Thorny Gates*, chap. 5: "When competition became acute in the Southern Sung [960–1127], relatives of officials used their right to take special examinations to subvert the essential fairness of the system," p. 17.

[146] Thomas Lee, *Government Education*; W. Franke, *The Reform and Abolition of the Traditional Chinese Examination System* (Cambridge, Mass.: Harvard University Press, 1963), p. 8.

[147] Albert Feuerwerker, "Chinese Economic History in Comparative Perspective," in *Heritage of China*, pp. 224–41 at p. 227.

did. One thinks here especially of the neo-Confucians Ch'eng I and Ch'eng Hao, Chu Hsi (1130–1200), and even naturalist thinkers such as Shen Kua (1031–95). But it did not encourage or tolerate thinkers who were essentially disputatious and critical of the intellectual status quo (as Abelard was, for example), and who attempted to develop and systematize the intellectual tools necessary to push the life of the mind forward. There was no Chinese equivalent to the scholastic method of disputation,[148] no canons of logic à la Aristotle, and no mathematical methods of proof such as one finds in Euclid's geometry. Derk Bodde points out, "Throughout its history Confucianism has deprecated the use of debate as a means of advancing knowledge."[149] This is further signified by "the virtual absence in ancient Chinese philosophy of anything resembling the Socratic dialogue (meaning a reasoned discourse between two individuals pursued in order to approach closer to clarity and truth)."[150] One might say, therefore, as Jean Escarra and Pelliot did in regard to law, jurisprudence, and technical legal studies,[151] that the civil examination system also killed off scientific theory (natural philosophy) as a coherent account of the world. It did that simply by standardizing the civil service examination around Confucian literary studies focused on moral and ethical issues of governing and by disallowing any state-sponsored scientific education (other than astronomy and mathematics, both of which were carefully controlled) to be part of the examination system.[152] The imperial state standardized the subject matter for the awarding of educational certificates (by focusing on literary and moral studies based on the Confucian classics) and created encyclopedias and primers expressly designed for examination takers throughout the whole empire.[153] Only for short periods of time did the civil service examination (during the Wang An-shih reforms of the Sung) include questions on law. On the other hand, there were occasional special examinations held to recruit mathematicians and astronomers to official posts, but since these were not part of a

---

[148] Hajime Nakamura, *Ways of Thinking of Eastern Peoples*, rev. trans. Philip P. Wiener (Honolulu: East West Center, 1964), p. 188; Fung Yu-lan, *A History of Chinese Philosophy*, trans. Derk Bodde (Princeton, N.J.: Princeton University Press, 1968) 1: 193–4; 257–8.

[149] *Chinese Thought*, p. 178.

[150] Ibid., p. 179.

[151] Jean Escarra, "Chinese Law," *Encyclopedia of the Social Sciences* 5 (1933): 251; Pelliot, "Notes de Bibliographie chinois: droit," *Bulletin de l'Ecole française de l'extrême Orient*, p. 27, as cited in Escarra, *Chinese Law*, p. 466.

[152] As Needham put it, the examination system "was entirely based on literary and cultural subjects and did not include subjects which could in any sense be called scientific." *GT*, p. 179.

[153] See Balazs, *Chinese Civilization and Bureaucracy*, pp. 143–7.

general course of study, nor regularly scheduled, the study of science was not encouraged. As a result, specialized knowledge such as mathematics, astronomy, and medicine tended to be confined to the families of the elite officials. This conclusion seems to be reinforced by Hartwell's data showing that passage of the state examinations tended to follow rather than precede marriage into a gentry family.

To understand the impact of the Chinese educational system, one must remember that in the Sung dynasty the emperor and his advisers sought to establish a universal educational system. To that effect, a program of establishing local schools in all provinces, prefects, and districts was undertaken early in the eleventh century.[154] With the advent of block printing, the government supplied every school with a set of the officially approved versions of the Confucian classics (and commentaries) for the students to study and memorize.[155] The officials attached to these schools, however, since they were government functionaries, did little actual teaching, in part because of their other state duties, the need to perform the Confucian ceremonials, and their duty to administer the state-sponsored exams.[156] Over time these schools became largely places where the official examinations were held and instruction was provided either in private academies (shu-yüan) or by personally engaged teachers. The academies were wholly oriented to the examination system, and they were staffed mostly by present or former government officials.[157]

The subject matter of the examinations was of course the Confucian classics, poetry, and official histories. From an early age young boys were taught to memorize the Confucian classics (initially without even knowing the meaning of what they were memorizing), to write Chinese calligraphy, and to write classic poetry.[158] The examinations largely asked students to recite passages from the classics, to comment upon selections from them, or at more advanced levels to write about the appropriate course of conduct for the wise and virtuous ruler. The so-called eight-legged essay, which emerged in early Ming times, was a composition based on a quotation from the

---

[154] Chaffee, *Thorny Gates,* p. 75; Thomas Lee, "Sung Schools and Education before Chu Hsi," in *Neo-Confucian Education: The Formative Period,* ed. John Chaffee and William de Bary (Berkeley and Los Angeles: University of California Press, 1989), 105–36.

[155] Lee, *Government Education,* p. 23.

[156] Chaffee, *Thorny Gates,* pp. 73ff; Lee, *Government Education;* Franke, *Reform.*

[157] T. Grimm, "Academies and Urban Systems in Kwangtung," *Chinese Government in Ming Times,* ed. Charles O. Hucker (New York: Columbia University Press, 1968); John Meskill, "Academies and Politics in the Ming Dynasty," in ibid., pp. 149ff; and Chaffee, *Thorny Gates,* pp. 89–94.

[158] Lee, *Government Education,* chap. 1.

classics that was to be presented in rigid stylized form. It has been described as an exercise similar to composing a fugue based on a few introductory notes:[159]

To cite an example of this style – the Ming Examination of 1487 had set as a topic a six-character quotation from the *Mencius: Lo t'ien che, pao t'ien-hsia*. The standard translation . . . reads: "He who delights in Heaven, will affect with his love and protection the whole empire." (A literal translation would be "Love Heaven person, protect Heaven-below.") In his eight-legged essay the candidate would be expected to proceed as follows – make a preliminary statement (three sentences), treat the first half ("love Heaven person") in four "legs" or sections, make a transition (four sentences), treat the second half ("protect Heaven-below") in four "legs," make a recapitulation (four sentences), and reach a grand conclusion. Within each four-legged section, his expressions should be in antithetic pairs, such as con and pro, false and true, shallow and profound, each half of each antithesis balancing the other in length, diction, imagery, and rhythm.[160]

The examinations were first held at the district level and then successful candidates went to the prefectural examinations. Those who passed became *sheng-yüan* (cultivated talents). Following that, successful students every three years could take the provincial exams and if they passed they were awarded the *chü-jen* title, "recommended man." During some periods of time this qualified the candidate for a position in the government such as district magistrate, or possibly a lower-level position in the capital under the supervision of a higher official.

The chü-jen certificate is sometimes equated with the Western bachelor's degree, and the successful candidates for a provincial chü-jen examination in 1669 performed the following feats:

[They] had pondered three passages chosen that year by the Shantung examiners; they had placed them in their correct context and explicated them. From the Confucian *Analects* there was the phrase "They who know the truth" from Book VI, chapters 17 and 18: "The Master said, 'Man is born for uprightness. If a man lose his uprightness and yet live, his escape from death is the effect of mere good fortune.' The Master said, '*They who know the truth* are not equal to those who love it, and they who love it are not equal to those who delight in it.'" From the *Doctrine of the Mean* came the phrase "Call him Heaven, how vast he is!" From the closing sentences of Book XXXII, on the man of true sincerity: "Shall this individual have any being or anything beyond himself on which he depends? Call him man in his ideal, how earnest he is! Call him an abyss, how deep he is! *Call him Heaven, how vast is he!*" And from the *Book of Mencius* there was "By viewing ceremonial ordinances"

[159] Franke, *Reform*, pp. 19–20.
[160] J. K. Fairbank, Edwin O. Reischauer, and A. M. Craig, eds., *East Asia: The Modern Transformation* (Boston: Houghton-Mifflin, 1965), p. 122.

from Book II, Part I, where Mencius quotes Confucius's disciple Tzu-Kung in his absolute praise of his teacher (and of the historian's power): "Tzu-Kung said, '*By viewing the ceremonial ordinances* of a prince, we know the character of his government. By hearing his music, we know the character of his virtue. After the lapse of a hundred ages I can arrange, according to their merits, the kings of a hundred ages – not one of them can escape me. From the birth of mankind till now, there has never been another like our master.' "[161]

Next the candidate could take the metropolitan examination in the capital. Although there were quotas allotted for each province, those who passed this examination had reached the top, except for the final palace examination administered by the emperor himself. Although the palace examination was pro forma, emperors could and did fail candidates as well as assign them final rankings among each other.[162] Successful candidates were called presented scholars (*chin-shih* or, literally, advanced scholars),[163] and this was the most prestigious award available, sometimes equated to the doctorate.

It is obvious that such a system of universal examinations, based on examination questions created by a board of senior bureaucrats, established an extraordinary uniformity of attitude and opinion. Because it standardized the very texts that were to be studied in addition to the exams themselves, this educational system, particularly from Ming times (1368–1644) on, created a virtual state dogma, "an unparalleled uniformity of thought [that] was enforced not only among the officials but throughout the whole leading class. . . . There remained almost no opportunity for the development of original ideas, for any deviation from the orthodox interpretation led certainly to failure."[164] But this is not to say that the system was incapable of producing and selecting experts with technical knowledge. Robert Hartwell, for example, has shown that the Northern Sung dynasty (970–1127) experienced unprecedented economic growth and development and that the services of financial experts were necessary to accomplish this result. Accordingly, during the eleventh century "nearly ninety percent of the chief financial officials were brought into the administration through examination,"[165] a portion of which involved problems of policy analysis. It seems

---

[161] Spence, *The Death of Woman Wang* (New York: Viking Press, 1978), p. 16.

[162] Franke, *Reform*, p. 6; Kracke, *Civil Service*, pp. 60–7; as well as Robert M. Hartwell, "Financial Experience, Examinations, and the Formulation of Economic Policy in North Sung," *Journal of Asian Studies* 30 (1971): 381–314 at 300–2. Chaffee, *Thorny Gates*, pp. 23, 49.

[163] See Hucker, *Dictionary*, "chin-shih," no. 1148; Adam Ynen-Chung Liu, *The Hanlin Academy, 1644–1850* (Hamden, Conn.: Archon Books, 1981), chap. 1; as well as Ho Ping-ti, *The Ladder of Success*, pp. 12–14.

[164] Franke, *Reform*, p. 13; and Derk Bodde, *Chinese Thought*, pp. 185, 193.

[165] Hartwell, "Financial Experience," p. 300.

that financial experts played a role not only in the setting of economic policy but in formulating questions for the civil service examination at both provincial and imperial (metropolitan) levels. To make the system yield the experts it needed, a set of complex administration problems were added to the examination at the palace. One might also note that in other areas, such as history, law, ritual, and the classics, no original compositions were required, since the exam depended entirely on memory recall and the elucidation of passages of text. In a word, even though one might say that practical problems in the domain of statecraft were part of the examinations during certain periods of time, it cannot be said that the system encouraged scientific interests in general – it did not.

This whole system was administered by the Directorate of Education (*kuo-tzu chien*).[166] Although it was the nominal agency in charge of educational matters, it was not powerful and generally lost out in political battles. Its most important duty was running the official printing office which turned out the officially sanctioned Confucian classics and commentaries.[167] It was also in charge of providing education for the sons of higher-ranking state officials and therefore ran a directorate school. In addition to that, it established (or reestablished in the Sung) an imperial university (or, perhaps better, an imperial academy, since it had few of the characteristics of the independent college known to Islam or the Western university). As William de Bary put it, "Though the *kuo-tzu chien* is commonly referred to as the National University, it had a relatively small staff of instructors and student body, and it is misleading to suggest that it was either 'national' or a 'university' in the modern sense."[168] Likewise, it is misleading in the medieval European sense, since it was not an autonomous corporate entity. Its faculty members (*po-shih*) were "merely salaried functionaries with special interest in scholarship and teaching" on assignment for approximately three years.[169] They had no ability to control the curriculum and the academy had no authorization to grant any degree, much less a license to teach (*licentia docendi*) or any other universally recognized designation of scholarly

---

[166] See Hucker, *Dictionary*, s. v. *kuo-tzu chien*, no. 3541; as well as Lee, *Government Education*, chap. 4 and passim. There are several terms which are used to refer to both the supreme school (*t'ai-hsüeh*) and the Directorate of Education, *kuo-tzu chien*, and the *kuo-tzu hsüeh* (school for the sons of state) which later was merged with the imperial academy (t'ai-hsüeh). But these designations often shifted, with the result that there has not been a uniformity of usage of these terms in English. Lee, however, makes the point, as does Hucker, that there was a distinction between the Directorate of Education which ran various schools for officials, and the national university itself.

[167] Lee, *Government Education*, chap. 4.

[168] De Bary, *Neo-Confucian Orthodoxy*, p. 225 n167.

[169] Lee, *Government Education*, p. 103.

achievement. In that regard, even the Islamic madrasas, as lawful endowments given in perpetuity to religious charity, had more autonomy. Furthermore, the ijaza (the authorization to transmit or to teach law) could only be awarded by the scholar personally; no degree titles or licenses to teach law, for example, were awarded to individuals by the Islamic state. This was universally recognized to be the exclusive purview of scholars who were masters in their own field.

In the Chinese case, an unending succession of examinations placed one higher on the list for preferred positions in officialdom but did not grant one an honorific degree that would be universally recognized as the culmination of a specific course of study. During the Sung reforms, the Directorate of Education and the schools under its administration were closed, reopened, expanded, reshuffled, or merely abolished.[170] Although separate schools of law, medicine, and mathematics were opened between 1073 and 1104, none (including the imperial academy) survived in continuous existence through the next century. Instead of granting each of the disciplines represented by these schools autonomous existence, each was either abolished or subordinated to some higher-level government office which strictly controlled what was taught and learned.

In sum, within the government hierarchy, no tradition of independent learning emerged and no agency was given autonomous control of a curriculum of education. Everything centered around passing the civil service examinations, and consequently, students were only interested in mastering the material required for the state examinations. One learned official wrote in 1042, "When the examination year comes, the Directorate School is flooded with more than a thousand students . . . and then when the examination is over, they all disappear, and teachers find nothing to do except sit in their chairs."[171] The official civil service examination system created a structure of rewards and incentives that over time diverted almost all attention away from disinterested learning into the narrow mastery of the Confucian classics. Astute scholars recognized this but were powerless to change it. In the thirteenth century another official wrote that "schools are considered to be the business of officials and examinations are considered to be the vocation of scholars, alas!"[172] Since the examination material was well known in the form of the official versions of the classics, including standard commentaries – and model examples of the essays were available in book-

[170] These Byzantine struggles are spelled out in some detail in Lee, *Government Education*, especially chap. 4. Likewise, vignettes of the histories of each of these government organizations are presented in Hucker's *Dictionary*.

[171] Lee, *Government Education*, p. 76.

[172] This by Yeh Shih, as cited in Chaffee, *Thorny Gates*, p. 88.

Figure 8. A bookseller's stall in Sung China. (Reprinted from *Science and Civilisation in China,* vol. 5, pt. 1, by Joseph Needham and Tsien Tsuen-hsuin, Cambridge University Press, 1985, with permission.)

shops (Figure 8) – there was no incentive to go beyond the standard material. Furthermore, private academies (shu-yüan) grew in number and superseded the state schools as places of instruction. Likewise, the imperial academy itself was of little significance because the way to the top of the ladder of success was through the examination, and attending the imperial academy was an aid only insofar as it increased one's contacts with officials who could help one get a better position after passing the examinations.[173] Of course, there was a third alternative, and that was to buy a title.[174] But aside from the illicit (or even periodically official) selling of titles, the institution known as *yin privilege* was long established in Chinese culture. The yin privilege was the custom of granting to the relatives of officials, primarily the sons of officials, a position in government "without undergoing other qualification tests, or with exemption from other qualification tests. . . . This was considered one of the 'proper paths' for attaining official rank throughout Chinese history." And, Hucker continues, it probably "yielded half or more of the total civil service personnel."[175]

Reprise

The preceding overview of the legal and institutional arrangements of China during the Sung and early imperial empire (the Ming) – the period corresponding to the European High Middle Ages – suggests that there were no official efforts to encourage autonomy of thought or action. Furthermore, the educational system, as a contingent set of rewards, was set against the pursuit of scientific inquiry per se. The ideal education for the imperial state was founded on the pursuit of moral and humanistic education and the study of exemplary historical figures. In many respects it represented the cultivation of the enlightened amateur, that is, a person well versed in moral uprightness, not a person skilled in the mundane rudiments of either office holding, bureaucratic management, or scientific inquiry. The ideal orientation of the literati during the Sung, Yüan, and Ming dynasties was "the cultivation of moral perfection" in the service of sagehood.[176] Derk Bodde has summed up this aspect of the moralistic thrust of classical Chinese culture and education. In his words, this orientation resulted in "the evaluation of literature, art, and music according to whether or not they convey a high moral message or have been created by persons of high moral stature; the evaluation of history according to whether its events have been dominated by good or bad persons rather than by analyzing the impersonal facts of geography, economics, and institutions." The unrelieved dominance of neo-Confucian ideology served to thwart "the creation of a suitable meth-

---

[173] See Lui, *The Hanlin Academy*, chap. 1; Watt, *The District Magistrate;* as well as Lee, *Government Education,* and Ch'ü, *Local Government,* pp. 18–22.
[174] See the sources above.
[175] Hucker, *Dictionary,* s.v. *yin* (protection privilege), no. 7971.
[176] Benjamin Elman, *From Philosophy to Philology: Intellectual and Social Aspects of Change in Late Imperial China* (Cambridge, Mass.: Harvard University Press, 1984), p. 3.

odology for studying the phenomena of nature."[177] The pursuit of scientific subjects was thereby relegated to the margins of Chinese society.

Before we can fully assess the channeling currents and the inhibitory effects of these cultural and institutional imperatives, we must consider some additional aspects of Chinese modes of thought and their stylistic expression.

[177] Bodde, *Chinese Thought*, p. 307.

8

Science and social organization
in China

We saw in the preceding chapter and earlier that from about the eighth to the fourteenth century, the Arabs had the most advanced science in the world. Consequently, in the fields of astronomy, mathematics, optics, and physical experimentation – which led directly to modern science – Chinese science was second to that of the Arab-Islamic world. Present scholarship regarding Chinese science suggests, moreover, that China developed along lines quite independent of the West and the Arabic Middle East. The Chinese knew nothing of Aristotle, Euclid, Ptolemy, or Galen. Nevertheless, there were areas in which the Chinese did accomplish great things, though in almost no case was there continuous and progressive development.

In the famous Chinese work *The Nine Chapters on the Mathematical Procedures* (from about the first century A.D.), discussion can be found of arithmetic fractions, the statement of formulas for the computation of areas and volumes, the solution to systems of simultaneous equations, and procedures for square and cube extraction.[1] During the Sung dynasty (ca. 960–1279) Chinese mathematics underwent a period of vigorous growth in algebraic computation.[2] However, the Chinese system of representation and positional notation, as well as its techniques of computation (with counting rods), were cumbersome and not nearly as generalizable and easy to use as the Arabic-Hindu numeral system. This system, which was located in a decimal place value system, had been available in al-Khwarizmi's work since about 825.[3] In contrast, the course of development of mathematics in China

---

[1] "Mathematics in China and Japan," *Encyclopedia Britannica* 23 (1991): 633b–633e.

[2] Ibid., and see Lǐ Yan and Dù Shíràn, *Chinese Mathematics: A Concise History* (Oxford: The Clarendon Press, 1987), pp. 109ff, as well as Needham, *SCC* 3: 38ff.

[3] E. S. Kennedy, "The Exact Sciences [The Period of the Arab Invasion to the Suljugs]," *The Cambridge History of Iran* 4 (1975): 380. And see Michael Mahoney, "Mathematics," in

required a move from computation with counting rods to the use of the abacus (generally in about the sixteenth century)[4] and the incorporation and use of the zero (in the thirteenth and fourteenth centuries). Only in the seventeenth century was the method of paper and pen calculation (and hence recorded arithmetic operations) introduced into mathematics with the arrival of the Jesuits.[5]

The main defects of Chinese mathematical and scientific thought were both substantive and logical. In regard to logic, Chinese thought lacked the logic of proof as well as the concept of mathematical proof as constructed in Euclid's *Elements*. It likewise lacked Hindu-Arabic numerals and the zero until about the thirteenth century.[6] Perhaps most important, the Chinese had no trigonometry, an essential part of mathematical astronomy. As noted earlier, to compensate for this, the Chinese employed Arab astronomers in the Chinese Astronomical Bureau in Peking from the thirteenth century onward.[7]

Substantively, Chinese astronomy lacked Ptolemy's planetary models (as contained in the *Almagest* and his *Planetary Hypotheses*), and it is difficult to imagine the leap to the Copernican worldview and modern astronomy without the intermediate stage of the relatively simple geometrical and circular world that was understood in the Middle East and the Occident even as early as Eudoxus (ca. 400–350 B.C.). The Chinese did not make this transition until the seventeenth century under the influence of the Jesuits. Some writers criticize the Jesuits for not providing full information regarding the Copernican hypothesis at that time, as well as Galileo's latest work, due to the official ban on dissemination of these ideas.[8] But the fact is that geometric and Ptolemaic astronomy (on which Copernicus had based his innovations) had been widely available in Arabic-Islamic civilization for centuries, and in Joseph Needham's view, the Chinese astronomers had every chance to learn

---

Science in the Middle Ages, ed. David C. Lindberg (Chicago: University of Chicago Press, 1978), pp. 151f.

[4] I am not claiming that the Chinese had to go through an intermediate phase of using the abacus but only pointing out that since they did move from counting rods to the use of the abacus, there was a need for another technical innovation that would allow pen and paper calculation within a system of notation at least as potent and generalizable as the Hindu-Arabic system. As it happened, the use of the abacus continued for a long time, and its use throughout China is still prevalent. See Lî Yan and Dù Shíràn, *Chinese Mathematics,* p. 176.

[5] Ibid., p. 191.

[6] Needham, *SCC* 3: 10, 43.

[7] Nathan Sivin, "Wang Hsi-Shan," *DSB* 14: 159–68, at p. 159. Also see Sivin, "Why the Scientific Revolution Did Not Take Place in China – or Didn't It?" in *Transformation and Tradition in the Sciences,* ed. E. Mendelsohn (New York: Cambridge University Press, 1984), pp. 531–54.

[8] Nathan Sivin, "Copernicus in China," *Studia Copernicana* 6 (1973): 63–122.

about it,[9] for the Chinese had been in direct communication with the astronomers working at the Marâgha observatory during the thirteenth and fourteenth centuries.

In optics, which in early science probably played something like the role of physics in modern science, the Chinese, in Needham's words, "never equalled the highest level attained by the Islamic students of light such as Ibn al-Haytham."[10] Among other reasons, this was a reflection of the fact that the Chinese were "greatly hampered by the lack of the Greek deductive geometry" which the Arabs had inherited.[11]

It may also be added that though we think of physics as the fundamental natural science, Needham concluded that there was little systematic physical thought among the Chinese.[12] While one can find Chinese physical thought, "one can hardly speak of a developed science of physics." Powerful systematic thinkers were lacking, thinkers who would correspond to the so-called precursors of Galileo, represented in the West by such names as Philoponus, Buridan, Bradwardine, and Nichole d'Oresme.[13]

These facts are not presented here as the main reasons Chinese science failed to gestate modern science, but rather as symptoms of an outcome that is itself the product of an antecedent cultural setting and its institutional arrangements. One can consider them major *internal* factors that inhibited the development of modern science.

In this chapter I want to explore the *external* factors that were rooted in the cultural and institutional foundations of China and were powerfully inhibitive with regard to the development of original thought and the pursuit of scientific inquiry. We must approach the problem from two angles: that of the institutional role arrangements of Chinese intellectual life and that which constituted the cultural imperatives and the symbolic technology of Chinese civilization. The latter refers to the stylistic patterns of language use and the modes of thought typical of Chinese culture and civilization. When these two domains are viewed together – the cultural and the institutional – we see powerful inhibitions impeding the rise of modern science.

The thesis that Chinese thought styles were deeply implicated in the failure of China to give birth to modern science has been given its most persuasive formulation in Derk Bodde's recent book, *Chinese Thought, Society, and Science*.[14] This work is probably the most penetrating study of

---

[9] Needham, *SCC* 3: 50.
[10] Ibid., p. 78.
[11] Needham, *SCC* 4/1: xxiii.
[12] Ibid., p. 1.
[13] Ibid.
[14] Derk Bodde, *Chinese Thought, Society, and Science: The Intellectual and Social Background*

the subject undertaken since Needham's own investigations appeared, and it is surely one of the most insightful studies to have appeared since Marcel Granet's *La pensée chinoise*.[15]

In his analysis, Professor Bodde examines the whole range of forms of symbolic communication, including grammar and punctuation, styles of thought and conceptual organization (of time, space, and things), the nature and effects of correlative thinking, and the influence of patterns of authority, social classes, and religion on scientific thought. He also considers the influences of morals and gender and includes those images of nature – at least seven – found in Chinese thought. In the concluding chapter of his study, which reviews the many Chinese views of mankind and nature, Bodde reexamines the question of the presence (or absence) of the idea of laws of nature, which is a previously published dialogue between Derk Bodde and Joseph Needham.[16] In all of these domains he finds far more impediments to the rise of modern science than he finds causes for thinking that Chinese cultural patterns and habits of thought were supportive of scientific inquiry.

Professor Bodde's study is particularly useful for the light it sheds on Chinese conceptual thought and the ways in which its modes shaped scholarly discourse and disinterested inquiry. By such means it adds further depth to our understanding of the sources and conceptions of reason and rationality in China. In his analysis of the Chinese language and its singular effects on thought and communication, we see still another level of the inhibitory effects of Chinese symbolic technology on the pursuit of scientific inquiry.

Some problems of written Chinese

It must be said at the outset that in the domain of linguistic analysis and the effects of language on thinking and modes of thought, it is exceedingly difficult to draw conclusions regarding the relative advantage or disadvantage of one language or another. The peoples of the world are especially possessive of their native tongues and this not infrequently leads to chauvinism and ethnocentrism. At the same time it is nearly impossible to demonstrate that the use of one language in particular prevents the thinking of

---

of *Science and Technology in Pre-Modern China* (Honolulu: University of Hawaii Press, 1991).

[15] Granet, *La pensée chinoise* (Paris: Albin Michel, 1934). Derk Bodde himself does not employ the phrase "modes of thought" but this way of thinking about these problems has a considerable history. See *Modes of Thought: Essays on Thinking in Non-Western Societies*, ed. Robin Horton and Ruth Finnegan (London: Faber and Faber, 1973).

[16] Derk Bodde, "Chinese 'Laws of Nature': A Reconsideration," *Harvard Journal of Asiatic Studies* 39 (1979): 139–55.

certain thoughts. During this last century such issues have been vigorously debated and it is clear that they hinge on exceedingly fine-grained philosophical analyses of words, concepts, and semantic nuance. Moreover, the very rise and existence of modern science and its now global study and use by peoples of widely diverse native cultures and languages suggest that mankind is not a prisoner of its available languages for more than brief periods. Nevertheless, there is considerable value in attempting to assess the relative advantages and disadvantages of linguistic forms on scientific thought and inquiry. From our point of view the really central issue is the degree to which any language possesses something called a universalizing potential that manifests itself in a universal appeal to the diverse language speakers of the world.

In his analysis of the Chinese system of written communication, Derk Bodde points to the many weaknesses of the Chinese language as an instrument of clear and unambiguous communication. These include its ancient lack of punctuation, the habit of ignoring paragraph indentations, capitalization of proper names (or the use of other signifiers), the lack of continuous pagination, as well as the absence of a system of alphabetization.[17] The importance of the last of these as an aid to the organization of knowledge can hardly be overestimated. This state of affairs is itself related to the absence of Chinese grammarians until the twentieth century.[18]

Professor Bodde also notes that Chinese characters tend to be monosyllabic and although they have undergone relatively little morphological change, they are capable of taking on very different meanings. Indeed, alternative translations (which are grammatically correct) may produce diametrically opposite meanings (on which more later). On another level, Bodde accentuates the tendency of writers of literary Chinese to use a great variety of archaic metaphors, allusions, clichés, and notoriously unmarked direct transcriptions from ancient authors. This state of affairs is obviously one filled with many pitfalls for the unwary reader or unfortunate translator.[19]

The ambiguity of Chinese words and their use is illustrated by the following example. A simple phrase from Confucius is composed of eight terms: *Kung hu yi tuan ssu hai yeh yi.* This phrase, Professor Bodde tells us, could be given two literal translations which are apposite: "Attack on strange shoots this harmful is indeed" or "Study of strange shoots these harmful are

---

[17] Bodde, *Chinese Thought*, chap. 2, but especially pp. 90–2.
[18] Ibid., p. 94f.
[19] A good example of the many levels of ambiguity and allusion, compounded by the omission of punctuation, appears in Bodde, *Chinese Thought*, on pp. 55–8.

indeed."[20] Given a fluid English translation, this phrase has four equally correct translations according to Bodde:

1. "To attack heterodox doctrines: this is harmful indeed!"
2. "Attack heterodox doctrines [because] these are harmful indeed!"
3. "To study heterodox doctrines: this is harmful indeed!"
4. "Study heterodox doctrines [because] these are harmful indeed!"[21]

These declarations are strikingly diverse in their meaning and implications and seem to me far more ambiguous than comparable translations of French into English or German into English or Arabic into English (though I am an inadequate judge of linguistic matters). The choice of translation is determined by the ideological context of the writer. In this case, the first translation is clearly unlike the manner of Confucius, as is the fourth option. The third rendition, "To study heterodox doctrines: this is harmful indeed!" Bodde tells us, is the one chosen by most translators and commentators, yet Bodde finds himself drawn equally to the second rendition, "Attack heterodox doctrines [because] these are harmful indeed." Whether or not one draws the conclusion that the Chinese language is more ambiguous than the Indo-European languages, it can surely be said that it is a language far from ideal for scientific communication. Its deep ties to the original Chinese characters and their idiomatic use makes Chinese far more tied to ancient, and what in English and other European languages would be regarded as archaic, usages.

While Bodde's assessment of the defects of written Chinese as a vehicle of communication must be accepted, it is difficult to place as much weight on those defects as he does. For example, in his comparative analysis of the development and use of alphabetization – something that began in Indo-European languages as early as 200 B.C.[22] – it turns out that this absence of alphabetization (and the lack of punctuation, indentation, capitalization, and so forth) was also typical of classic (literary) Arabic until the present century.[23] Moreover, it is worth pointing out that virtually all of these elements of alphabetization, punctuation, indentation, capitalization, catchwords, and continuous pagination were in place in the written languages of Europe, especially Latin, by the thirteenth century. This fact reinforces our earlier characterization of the revolutionary transformation of European thought and society in the twelfth and thirteenth centuries.

We have seen that Arabic science was far more advanced than Chinese

---

[20] Ibid., p. 40.
[21] Ibid., p. 41.
[22] Ibid., p. 62.
[23] Ibid., pp. 61 and 65.

science. It is probably true that the Arabic language is slightly less ambiguous than Chinese, since it is an inflected language in which word morphology is important and it is largely based on a triliteral root system. On the other hand, Arabic like Hebrew, especially in classical texts, omits the diacritical marks, that is, the vowels. Similarly, as in Chinese, there is no way of indicating proper names through capitalization or other markers. The tenses of verbs can also be ambiguous, as between "I am going," "I will go," or "I went." Furthermore, there was a widespread practice in Arabic of printing an additional book (or books) on a completely different subject in the margins of a manuscript.

Still, one may also note that early on the Arabic language found many grammarians who went to great lengths to shape Arabic into a uniform and systematic language. Indeed, from early times great debates raged as to whether it was better to know grammar or logic, and the view that grammar was better and more intellectually powerful generally prevailed.[24]

It is true that the Arabs inherited a great deal of information from the Greeks, and borrowed considerable amounts from the Indians. It must also be said, however, that at the beginning of Islamic civilization a very deliberate policy of translating almost all the great works of the Greeks and many others was undertaken, whereas such a policy seems never to have held sway in China.[25] Of course, there were important social psychological forces operative in the Arabic case that were absent in the Chinese. That is, when Arabic-Islamic civilization emerged in the seventh century A.D., it was almost wholly an oral (as opposed to a scribal) culture, centered in nomadic tribes and devoid of classical learning. In such a setting where the remaining Hellenic, Hebrew, and Christian cultures presented long histories and written traditions, Arabic and Islamic leaders no doubt perceived the disparities of learning and set about to rectify the situation. This led to great efforts to assimilate the treasures of the surrounding cultures and civilizations.

In contrast, the Chinese in the seventh century A.D. had a thousand-year-old scribal tradition that considered non-Chinese barbarians. For this reason, the Chinese were exceedingly cautious and selective in their borrowings

---

[24] See D. S. Margoliouth, "The Discussion Between Abû Bishar Mattâ and Abû Saʿid al-Sîrafî on the Merits of Logic and Grammar," *Journal of the Royal Asiatic Society* (1905): 79–129; and M. Mahdi, "Language and Logic in Classical Islam," in *Logic in Classical Islamic Culture*, ed. G. E. von Grunebaum (Wiesbaden: Otto Harrassowitz, 1970), pp. 51–83.

[25] For the translation movement into Arabic, see Max Meyerhof, "Von Alexandria nach Baghdad," *Sitzungsberichte der Prüssischen Akademie der Wissenschaften* no. 23 (1930): 389–429; Meyerhof, "On the Transmission of Greek and Indian Science to the Arabs," *Islamic Culture* 11 (1937): 17–29; as well as F. E. Peters, *Aristotle and the Arabs* (New York: New York University Press, 1968); and Richard Walzer, *Greek into Arabic* (Columbia, S.C.: University of South Carolina Press, 1962).

from other cultures. This was quite noticeable in the domain of science. As we have seen, even when major scientific improvements and innovations were passed to the Chinese – whether via the Indians or the Arabs – these innovations were either put aside altogether or only adopted after the lapse of many centuries. This was the case with the appearance of the Indian zero in the eighth century and with regard to the Ptolemaic system as used by the Marâgha observatory in the thirteenth century.[26]

The case for the view that the Chinese language, in and of itself, has been a great impediment over the centuries, by being a primary distorting medium of foreign ideas, has been put most forcefully by Arthur Wright. Reflecting on the experiences of the many foreign individuals who came to China over the centuries and who attempted to bring their ideas forth by translating them into Chinese, Wright sums up as follows:

Thus the monks from medieval India, the Jesuits from Renaissance Europe, emissaries of modern scientific thought such as Bertrand Russell, and representatives of the Comintern all spoke inflected polysyllabic languages. . . . Structurally Chinese was a most unsuitable medium for the expression of their ideas, for it was deficient in the notations of number, tense, gender, and relationships, which notations were often necessary for the communication of a foreign idea.

Moreover, Chinese characters as individual symbols had a wide range of allusive meanings derived from their use in a richly developed literary tradition. . . . Further, the Chinese was relatively poor in resources for expressing abstractions and general classes or qualities. Such a notion as "Truth" tended to develop into "something that is true." "Man" tended to be understood as "the people" – general but not abstract. . . . These characteristics of the Chinese language reduced many proponents of foreign ideas to despair. . . . Kumarajiva (344–413), devoted Buddhist and stouthearted missionary, . . . was moved to sigh: "But when one translates the Indian [Buddhist texts] into Chinese, they lose their literary elegance. Though one may understand the general idea, he entirely misses the style. It is as if one chewed rice and gave it to another; not only would it be tasteless, but it might also make him spit it out."[27]

As a general characterization of the Chinese language this seems rather harsh, yet it reflects an informed assessment based on representative experi-

---

[26] Regarding the contacts between the Persians and the Chinese, see Needham, *SCC* 3: 372ff. Regarding the Indian transmissions, Needham claims that the Indian zero was introduced into China in the eighth century, along with trigonometry, as well as a more accurate division of the circle, but these innovations, above all the use of the zero, were to lie dormant for four more centuries; *SCC* 3: 202–3.

[27] Arthur Wright, "The Chinese Language and Foreign Ideas," in *Studies in Chinese Thought* (Chicago: University of Chicago Press, 1953), p. 287, as abridged in Bodde, *Chinese Thought*, p. 30.

ences of many different contexts in which the language itself seemed to stand in the way of communication and the reception of new or foreign ideas.

It is probably true, moreover, that Chinese is more ambiguous than Arabic and that it is less rich in abstract and generalized categories – classifications that Arabic writers were especially good at formulating.[28] As we saw in earlier chapters, when a set of new scientific terms, namely, *experiment, experimenter,* and *experimentation,* was introduced by the Arabic writer Ibn-al-Haytham in the eleventh century, the Latin translators of the terms did indeed translate them properly.[29]

In the case of Chinese there are surely many examples that might be cited which illustrate the conceptual problems that derive from alternative cosmologies and philosophical stances. Joseph Needham, on the other hand, believes that the limitations of the ideographic nature of Chinese "is generally grossly overrated," and that in his work "it has proved possible . . . to draw up large glossaries of definable technical terms used in ancient and medieval time for all kinds of things and ideas in science and its applications."[30] Still, this leaves unanswered the question regarding the degree to which Chinese as opposed to, say, Arabic was relatively more difficult to work with, and the degree to which the Chinese themselves did not exert the energy required to produce the glossaries (not to mention the standard grammatical guides) to which Needham refers.

At the same time one should note that when the Chinese literati did begin to introduce higher standards of evidential scholarship, for example, in the *k'ao-cheng* movement of the seventeenth and eighteenth centuries, the whole thrust of the movement was directed toward a return to the classics. It was a movement of revivalism and fundamentalism that sought to purify Chinese thought by returning to the unsullied Confucian past of the five classics. It thereby rejected the neo-Confucian four books. To accomplish this task it deployed an extreme method of textual analysis and adopted the ancient rallying cry of the "rectification of names." According to this, all the proper names of things must be understood and adhered to, and if this were done,

---

[28] The penchant of Arab writers to classify the forms of knowledge, for example, is well illustrated in Franz Rosenthal's books: *Knowledge Triumphant* (Leiden: E. J. Brill, 1970) and *The Classical Heritage in Islam* (Berkeley and Los Angeles: University of California Press, 1975). This leaves open the question of how much all this was influenced by Greek conceptions and habits of thought.

[29] See Sabra, "The Astronomical Origins of Ibn al-Haytham's Concept of Experiment," *Actes du XIIe congrès international d'histoire des sciences* 3a (1971): 133f; and Sabra, ed. and trans., *The Optics of Ibn al-Haytham* (London: The Warburg Institute, University of London, 1989), 2: 10–19.

[30] *The Grand Titration* (London: Allen and Unwin, 1969), p. 38 (hereafter cited as *GT*).

then everyone's proper social function would be clear and the social order would be restored.[31] One such effort resulted in a glossary of original meanings to which all scholarship was expected to conform.[32] This "back to the ancients" movement hoped "to restore the spirit of the ancient world and thereby to rehabilitate society."[33] When such evidential methods were applied to mathematics and astronomy (in reaction to the arrival of the Jesuits), the texts were exhumed in order to show "the depth and sophistication of native expertise in calendrical studies."[34] By 1750 this intellectual movement had turned into a rigid textualism: a "narrowly defined scholarly methodology had become an end in itself, narrow in interpretation and intolerant of the urge to generalize."[35] In a word, the various crises experienced by the literati of China, whether in the form of the fall of a dynasty or the incursion of foreign ideas from outside, were generally reacted to by going back to the classic and ignoring the new opportunities for change and development that were presented. Consequently, what appears to be a problem of language and conceptual expression may simply be a Chinese philosophical dislike of all foreign concepts and ideas.

The question of the ultimate influence of language and its stylistic uses remains indeterminate in the present context. For while it may be judged that Arabic was more facile than Chinese for scientific communication, in the end this was a matter of no consequence for the advancement of science. The fact is, as we have seen, that the Arabs laid the technical foundations for the scientific revolution but failed to initiate the revolution itself. This suggests that Bodde and others are probably on surer ground when they say that the Chinese language and traditional Chinese xenophobia, along with its internal standards of scholarship, combined to rigidly filter out foreign ideas.

Let us turn now to the ways in which these linguistic inclinations crystallized into stylistic forms that further magnified and possibly restricted Chinese modes of thought.

Chinese modes of thought

In the realm of conceptual thought Professor Bodde sees a reinforcing process whereby the linguistic forms of expression induced by literary Chinese

---

[31] Benjamin Elman, *From Philosophy to Philology: Intellectual and Social Aspects of Change in Late Imperial China* (Cambridge, Mass.: Harvard University Press, 1984), p. 45.

[32] Ibid., pp. 45f, 62.

[33] Ibid., p. 61.

[34] Ibid., p. 63.

[35] Nathan Sivin, "Wang Hsi-Shan," *DSB* 14: 163.

served to enhance and preserve the correlative or analogical mode of thinking that has so consistently been ascribed to Chinese thought. As we saw in the last chapter, the correlative mode of thinking begins with binary polarities and works from there to increasingly complex but balanced collectives of harmonies and antitheses. This reinforcing effect of linguistic usage and the analogical (correlative) mode of thinking is energized by the penchant of written Chinese to use parallel constructions with antonymies – paired opposites. For example, the English phrase "penny wise and pound foolish" is an illustration of the parallel construction that pairs opposites. Another such would be "easy come, easy go." And it is this stylistic thrust of written Chinese that in Bodde's view reinforces the Chinese tendency toward correlative thinking characterized by the constant use of paired dualities. This habit becomes the way through which the world is seen and described; it is highly constructive and largely devoid of empirical guidance. Because it persisted nearly exclusively in China through the end of the nineteenth century, it was, Derk Bodde contends, a powerful deterrent to the development of a scientific point of view.[36] One should notice, however, that Bodde repeatedly observes that while this point of view was especially strong among the literate elite, and thus affected philosophical thinking in the sciences, it was not as strong among the artisans and technologists – a claim that Needham does not appear to share.[37]

Thus the yang–yin polarity (with its attendant five elements [wu hsing]) has been fundamental to Chinese thinking for centuries. It operates on the principle that the world at all levels is a balanced set of paired forces, units, or elements. The most elementary of such comparisons might be light versus darkness, hot versus cold, or heaven versus earth. Within each of these dyads there are additional groupings of elements that are believed to share a primary quality of the polar category while simultaneously sharing an antonymous relationship with specific groups of elements in the category of the polar opposite.

For example, under the primal force *yang* there are the qualities of brightness, heat, dryness, hardness, and so on. In contrast, the *yin* contains such opposing qualities as darkness, cold, wetness, softness, and so on. A simple illustration of this balanced format is the following:[38]

---

[36] Bodde, *Chinese Thought*, p. 99; and cf. Needham, *SCC* 2: 285.

[37] See Bodde, *Chinese Thought*, pp. 121–2; and cf. Needham, *SCC* 4/2: 7.

[38] This I have taken from Bodde, *Chinese Thought*, p. 98, who also draws on A. C. Graham, *Yin-Yang and the Nature of Correlative Thinking* (Singapore: Institute of East Asian Philosophies, National University of Singapore, 1986), pp. 16–24, and especially Graham, *Disputers of the Tao* (La Salle, Ill.: Open Court Press, 1989), pp. 319–25.

Paradigm

		S	
		y	
1. Day		n	Night
		t	
2. Light		a	Darkness
		g	
		m	

Here the horizontal dimension represents the model or paradigmatic relationship and the vertical represents the collection of subsidiary qualities (syntagm) associated with the primal polarities. Many more elements could be added to each side of this metaphysical duality.

When the Chinese desire to create symmetry and centrality is set to work on all these categories, the result becomes a pictograph displaying a highly refined sense of harmony and balance. To most readers the astrological charts connecting the parts of the body and mind to locations and movements of the heavenly bodies is the most common kind of such visualization. But a variety of circles, hexagrams, and magic squares came to be used to represent these patterns of correlated qualities and forces. This reminds us, of course, that this analogical mode of thinking was not unique to the Chinese but existed worldwide. Indeed, some would say that this analogical mode of thinking is built into language use because of the tendency of the mind to think in binary polarities. This is the view of A. C. Graham who was inspired by the linguist Roman Jakobson, who also influenced Lévi-Strauss.[39] What is unique here, however, is the fact that the correlative mode persisted at the center of Chinese thinking into the twentieth century, as it was not displaced by mechanical or causal thinking.

There is another paramount property of correlative thinking – the numerological value of the elements. If the principles of symmetry and centrality are to be attained, then a precise number of units must be grouped together, and this leads to the accentuation of odd numbers. For example, the desire to attain centrality of expression in Chinese thought can be represented by the pairing of elements in such a fashion that a single grouping is always placed at the center of a linear or spatial sequence. In its most generic sense the principle of symmetry implies a sense of absolute orientation, that is, a sense of being related to the central point, the *axis mundi*.[40] But Bodde suggests that within these notions one can see another and more technical principle of

---

[39] See Graham, *Disputers of the Tao*, p. 320.
[40] Bodde, *Chinese Thought*, pp. 108 and 118–19.

centrality in operation. For example, aside from the paired dyad, one can imagine the aesthetic desire to pair all collections of units so that they are symmetrical or, more importantly, balanced around a central pair. Thus, the pairs AB/CD/EF represent a balanced pairing whereby the central pair is flanked on either side by single pairs of quantitatively equal units. A larger sequence of such pairs is AB/CD/EF/GH/IJ. Here the central pair (EF) is symmetrically balanced by two adjacent pairs of the same kind. Now it becomes evident why the pursuit of the principle of centrality places a high premium on odd numbers, for only such can achieve this kind of balanced symmetry.

If this principle is projected into space, then we have spatially balanced figures such as

$$
\begin{array}{ccc}
 & D & \\
A & B & C \\
 & E &
\end{array}
$$

Bodde notes that the requirements of this spatial symmetry are such that the numbers three, five, and nine, as well as higher odd numbers will work, but seven will not. This may account for the fact that the number seven has less significance in China than in the West.[41]

In short, Professor Bodde's analysis reveals an important connection between the stylistic habits of Chinese writing and the correlative-analogical mode of thought. We saw in the last chapter, moreover, that this style of evoking antithetical expressions was ensconced in the eight-legged essays of the official examinations as early as 1487. One might say that these instrumentalities of symbolic technology, along with the state apparatus, worked in tandem over the centuries to impede the development of scientific thought. Instead of moving toward mechanical and causal modes of thinking that recognized impersonal natural forces, the Chinese thrust has been ever toward creating a harmonious worldview that linked all forces and elements together in a man-centered cosmic harmony. In addition, changes within this harmony were characterized by only apparent phenomenal change, for all change according to traditional Chinese thought is but the recurrent interaction and flow of forces through cyclical (evolutive) phases.

Still another powerful tendency in Chinese thought is what Bodde and others call the technique of "scissors-and-paste" or "composition by compilation."[42] It was a technique that placed a high premium on classical and ancient allusions and, I believe, is related to the absence in Chinese philosophical thought of a genuine dialectic of disputation and a faith in reason.

[41] Ibid., p. 112.
[42] Bodde, *Chinese Thought*, pp. 82ff.

The technique of composition by compilation resulted in copying the works of others, virtually without comment, into an author's new work. In the case of historiography, it entailed a "verbatim reproduction of the records of earlier historians, no matter how extensive," but this was not regarded "as plagiarism, but rather the natural and reasonable process by which new histories of previously recorded events should be constructed."[43] The issue this technique raises is not that Chinese writers borrowed wholesale from the writings of their predecessors, but that this borrowing took place without the author's being aware that such a record is highly likely to contain contradictory assumptions and points of view, as well as contextually misplaced metaphors and allusions – all of which might confuse the reader.

Although not all Arabic historiography was of this nature, it did display an excessive penchant for producing commentaries on commentaries when Islamic civilization began to decline. Still, there may have been a greater inclination in classical Chinese scholarship for the scholar to be self-effacing than there was in either Arabic-Islamic civilization or the West. This is perhaps another expression of the Chinese conception of yieldingness, jang.[44] In contrast, there is a tradition in Arabic of execration poetry, the *hija'* tradition, which praises the self and attacks one's enemies.[45] There was also a deep-rooted tradition of forthright disputation in Arabic-Islamic culture.

The application of the scissors-and-paste method can be seen in many spheres of Chinese intellectual thought. Its extensive use suggests one of the detrimental consequences of the absence of a true dialectical method of argumentation in Chinese thought (on which more later). In the case of historical writing, as we have seen, the imperative was to extract verbatim reports from authors of the past and to compile them as a sort of impersonal history, "divorced from any kind of proprietorship by the author."[46] This technique was also widely employed in philosophy, medicine, and science.

For example, the greatest of the neo-Confucians, Chu Hsi (1130–1200), according to Derk Bodde, is most famous for his compilation of an authoritative canon of Confucian writings, not for his original sayings or systematic thought. He did not write a great systematic or original work, a summa, so that one has "to derive his system from a bewildering assortment of recorded

---

[43] Charles S. Gardner, *Chinese Traditional Historiography* (Cambridge, Mass.: Harvard University Press, 1938), as cited in Bodde, *Chinese Thought*, p. 83.

[44] Needham, *SCC* 2: 61f.

[45] See *EI²* 3, s.v. "Hija'." Evidence that this cultural form was still alive and in use during the recent Gulf War can be found in Ehud Ya'ari and Ina Friedman, "Curses in Verses," *The Atlantic Monthly*, Feb. 1991, pp. 22–6.

[46] Gardner, *Chinese Traditional Historiography*, p. 70, as cited in Bodde, *Chinese Thought*, p. 83.

sayings, commentaries on the classics, letters to friends and other scattered documents."⁴⁷ There are of course the "classified sayings of Master Chu," the *Chu-tzu yü lei*. But these are vernacular transcriptions of Chu's discussions with his disciples and, as such, are not of the same order as the systematic philosophical or theological writings of a Thomas Aquinas. In this regard, Chu Hsi is unlike the great philosophical thinkers of Islam and the West during the same period of time. That is, his writings have none of the disputatious and dialectical flavor so noticeable in the writings of al-Ghazali, Averroes, Abelard, and the milder Thomas Aquinas.

Similarly, the brilliant Chinese astronomer and mathematician Shen Kua (d. ca. 1095) left behind only a scattered set of writings that lack organization and theoretical acuteness. According to Nathan Sivin, "Notices of the highest originality stand cheek by jowl with trivial didacticism, court anecdotes, and ephemeral curiosities," providing little insight.⁴⁸ Donald Holzman likewise writes that Shen Kua "has nowhere organized his observations into anything like a general theory."⁴⁹ Moreover, as I pointed out in the preceding chapter, Western students of Chinese law have noted the lack of systematic treatments that would be equivalent to the great systematic works of European law such as Gratian's *Concordance of Discordant Canons* or later works similar to a "Gaius, a Cujas, a Pothier or a Gierke."⁵⁰

More importantly, Derk Bodde suggests that when the Chinese writers of the ancient and early imperial period compiled their anthologies by the scissors-and-paste method they did not even seem to be aware of the conflicts of meaning, the contextual displacement of allusions, and the loss of interpretative sense that arose from their pasting together of comments and remarks from diverse sources, writers, and periods of time.⁵¹ For them there was no awareness of – much less a pressing need to reconcile – the conflicting points of view or the contrasting claims to knowledge. Yet this awareness of sharply different interpretations – of the Bible, the church fathers, Aristotle, natural phenomena, and so forth – is what most characterizes European thought in the twelfth and thirteenth centuries. This is dramatically and classically illustrated by Abelard's *Sic et non*, but even more so by the great synthesis of opposing legal canons worked out by Gratian and the glossators. Furthermore, these Western writers were moved by a belief in the powers of reason to get at the truth and the need to pursue that end

---

⁴⁷ Bodde, *Chinese Thought*, p. 86.
⁴⁸ Sivin, "Shen Kua," *DSB* 12: 374.
⁴⁹ Donald Holzman, "Shen Kua and His Men-ch'i pi-t'an," *T'oung Pao* 46 (1958): 290, as cited in Bodde, *Chinese Thought*, p. 86 n99.
⁵⁰ Escarra, *Chinese Law*, p. 359, as cited in Needham, *SCC* 2: 524–5.
⁵¹ *Chinese Thought*, pp. 82–5.

aggressively. As we saw in Chapter 4, the development and use of the dialectical method in the West in the twelfth and thirteenth centuries resulted not only in the development of a science of law as well as a science of faith (that is, theology), but in a general breakthrough in the "logics of decision" whereby argumentative techniques were thought to yield new and imperative truths. Western thinkers believed that they had discovered a new universal method. This method was used not only in law and the sifting of sacred sources, that is, the Bible and the religious canons, but in the study of nature as well. In the works of Grosseteste, for example, "the object of inquiry was to provide 'demonstrated knowledge' (*scientia propter quid*), as distinct from bare empirical knowledge (*scientia quia*), of the facts. Demonstrated knowledge of a fact was had when it was deduced from a theory which related it to other facts and showed its causes."[52] But none of this thrust is evident among the medieval Chinese writers of the same period. Here, then, we can see the reinforcing effects of the cultural premium put on yielding to the priority of the classics, standardizing the meaning of all terms by philological reference to classical usage, and avoiding vigorous public debate. The result was the absence of daring innovation.

Joseph Needham, however, finds in the writings of the Taoists, especially the *Chuang tzu*, both a commitment to naturalistic inquiry and the rudiments of a dialectical method.[53] The passage he translates, however, does not present a convincing parallel with the dialectics of the Europeans, or the Greeks, for that matter. It seems rather closer to various forms of mystical enlightenment.[54] Indeed, Benjamin Schwartz sees in the long passage Needham translates from the *Chuang tzu*[55] not a commitment to positivist science, but rather "a positive injunction against seeking the underlying, unobserved causes of things which figure so largely in the book of Mo-tzu – particularly the dialectic chapters."[56]

As a method of logic, both this and that of the Mohists to which Needham directs our attention[57] can hardly be compared to the method of the European Scholastics, the method practiced by Abelard and his followers. A. C. Graham admitted that Mohist logic, above all the ethics of Mo-tzu, "is an achievement quite without parallel in Chinese philosophy, a highly ratio-

---

[52] A. C. Crombie, *Robert Grosseteste and the Origins of Experimental Science* (Oxford: At the Clarendon Press, 1953), pp. 290–1.

[53] Needham, *SCC* 2: 76–7.

[54] Benjamin Schwartz, *The World of Thought in Ancient China* (Cambridge, Mass.: Harvard University Press, 1985), p. 217.

[55] Needham, *SCC* 2: 40.

[56] Schwartz, *The World of Thought*, p. 221.

[57] Needham, *SCC* 2: 198–9.

nalized ethical system in which all key terms are defined."[58] It would be difficult, however, to point to any institutional location or, indeed, any informal school of scholars who practiced the method continuously down through the centuries. Hajime Nakamura argues that "the dialectic – the art of questioning and answering as a device for philosophical analysis – did not develop [in China] as it had in Greece."[59] There is not only the question of the technical aspects of such a method but also the question of its spirit. It was a method that, in its Western version, was intentionally designed to achieve progressively new ideas and relationships. In contrast, Chinese philosophy was especially hidebound and oriented toward preserving the past.[60] And this is just the other side of the story regarding the rise of the new evidential scholarship (k'ao-cheng) of the seventeenth and eighteenth centuries. This movement "from philosophy to philology" was mainly a great return to the unsullied past, to Confucian foundations untouched by neo-Confucian coloring, as Benjamin Elman demonstrates.[61]

We have now seen that the figures who are identified in the Sung and early Ming period as the leading intellectuals did not write original and systematic works in the fashion of such outstanding figures in Islam and the West as al-Ghazali, Averroes, Peter Abelard, Gratian, or Thomas Aquinas. Nor is there an example of a systematic natural philosopher of this period who produced either a progressive scientific work having the influence of, say, Avicenna's *Canon* or Ibn al-Haytham's *Optics* or a methodological work as advanced as that of Grosseteste.[62] In short, these habits of thought I have sketched and the lack of a tradition of disputation seem to have worked against the gestation of a rigorous naturalistic tradition that could call into question the traditional assumptions of Chinese thought and metaphysics. As Professor Bodde has put it, the absence in Chinese thought of an imperative to hypothesize, synthesize, and generalize prevented the evolution of a mode of inquiry "based on experimentation combined with observation," and, instead, Chinese thought remained committed to "uncritically accepted tradition."[63]

[58] As cited in Bodde, *Chinese Thought*, p. 96.
[59] See Hajime Nakamura, *Ways of Thinking of Eastern Peoples* rev. ed. (Honolulu: East West Center, 1964), p. 188.
[60] Ibid., chap. 18.
[61] Benjamin Elman, *From Philosophy to Philology*, chap. 2.
[62] Needham affirms that the theorizing of the neo-Confucians "about the acquisition of natural knowledge was not as advanced in its ways as that of the +13th century European scholars." *SCC* 3:163.
[63] Bodde, *Chinese Thought*, p. 88.

Institutional impediments and patterns of opportunity

We must turn our attention now to the social organization and institutional structures of medieval China. When we approach the question of why modern science did not develop in China from an institutional point of view, we see even more powerful inhibitions blocking the free and open pursuit of disinterested knowledge. It is these, I believe, that in the long run had the greatest impact on the development of science in China, just as we saw in the case of Arabic science. For even if the intellectual impediments Derk Bodde and others have enumerated did operate to the detriment of the rise of modern science in China, detailed explorations of typical styles and modes of thought always reveal exceptions or – positively stated – studious deviations that turn out to be intellectual innovations in the world of thought. The question then becomes, given the likelihood of innovative modes of thought, what is the chance that such innovations will find institutional support and be allowed to enter the world of public discourse? In China the chances were slim indeed.

The first of these impediments preventing the rise of free and open public discourse we saw in the preceding chapter. It is found in the simple lack of spheres of autonomy on any level. From a Confucian point of view, the natural order of things requires that there be social harmony sanctified by the mandate of heaven, and this could only be possessed by the rightly guided supreme ruler. Over time this autocratic manner, fused with Taoism, came to be called "the one right way."[64] From such a point of view, the idea of a Rechtsstaat, a legally ordered state, a constitutional order of man-made rules binding on all men (including the emperor) was unthinkable. It might be possible for the emperor to issue laws and sacred edicts to direct the conduct of his subjects, but he was above the law.[65] No separation of religion and state was possible, and no legal revolution such as the medieval European papal revolution which effected that separation ever took place in China. And thus, without the possibility of a constitutional order, a state ruled by law binding on all citizens (including the supreme ruler), there could be no authentic jurisdictions that granted legal autonomy to legitimate

---

[64] See the discussion in Harry White, "The Fate of Independent Thought in Traditional China," *Journal of Chinese Philosophy* 18 (1991): 53–72.

[65] This essential assumption is reported in many sources, but see Jack Dull, "The Evolution of Government in China," in *Heritage of China;* Hsiao Kung-ch'uan, *A History of Chinese Political Thought* (Princeton, N.J.: Princeton University Press, 1979); as well as Benjamin Schwartz, *The World of Thought.* The tradition of the emperor issuing sacred edicts and the districts sending notables to receive these messages is a more recent extension of the powers of the emperor, though rooted in ancient understandings. See J. Legge, "Imperial Confucianism," *The China Review* 6 (1877–8): 146–58.

social collectives – whether these be modified corporations, professional associations, or other legal entities such as trusts, the forerunners of corporations in Western law. Furthermore, without the separation of religion and state, there could be no philosophical and, hence, no legal distinction between the public and the private. For the foundation of that distinction resides in the distinction between ownership and jurisdiction: those who have jurisdiction – the lawful right to establish laws and to adjudicate conflicts – do not own the corporate assets, as they belong to the "whole body." Such a theory also entails the idea of delegated authority as well as rights to ownership, to rule making, and to representation (internally and externally): "What affects all must be decided by all" was the Roman legal maxim given new life by the European medievals. Of course, Chinese administrative officials had been delegated the authority – at the provincial, prefectural, and district levels – to act upon a variety of matters on behalf of the emperor, but one could not say that such officials had their own rights in opposition to the emperor. Nor can we attribute to them the power of legislation. The officials had duties and obligations and also privileges (such as the yin privileges), but these are not legal rights.

The absence of a separation between church and state (between the religious and the secular realms) and the absence of a legal theory spelling out the spheres of autonomy and self-regulation in traditional China had major social consequences. The first of these was the absence of cities and towns as autonomous legal units of self-governing citizens. Since this level of autonomy was lacking, the notion of citizens exercising self-government through legislation based on legitimate representation was missing – as were regionally autonomous courts of adjudication.[66] All such powers resided with the central government. Representative government, like constitutional order, was out of the question.

The second consequence was that institutions of higher learning, which would be equivalent to either the college in Islam or the university in the West, did not emerge in China. The various schools and academies (shu-yüan) of China were local organizational units that the imperial government tolerated, sometimes encouraged, but never gave any irrevocable powers. Nor did they have authoritative control over a standard curriculum. The evidence suggests that when the private academies transgressed the lines of orthodoxy, they were strongly reproved and persuaded to give up their privacy. "Implicitly, the way to correct the fault was to close the academy

---

[66] Of course this was noticed by Max Weber, but he did not see the connection between the legal theory of jurisdiction (or, more broadly, sovereignty) and the social order. See Weber, *The City*, and Weber, *The Religion of China*, pp. 13ff. Harold Berman has insightfully clarified these issues and criticized Weber in *Law and Revolution* (Cambridge, Mass.: Harvard University Press, 1983), chap. 12, especially pp. 392–9 and 399–403.

or bring it under state management. Privacy was a particular and vulnerable characteristic."[67] More to the point, the private came to be seen as selfish, and with the legists, as the illegal.[68] The academies (shu-yüan) were in that regard far less legally protected than the Islamic madrasas, for the latter were pious endowments and hence inalienable trusts under the law. No such status appears to have been available in China. Even the land on which the private academies were located was owned by the state and was only given conditionally.

On a higher level, the imperial academies (sometimes called universities in translation), such as the institutes and academies of the imperial government, were simply bureaucratic subdivisions of the administrative structure that could be expanded, reorganized, or abolished at a moment's notice, as they often were.[69] They had no charter, no legal standing, but only the weakest of customary force. As Needham generously put it, "the important point is that throughout the +1 millennium the conception of an institution of higher learning within the framework of the national bureaucracy was thoroughly rooted in Chinese culture."[70] Bluntly stated, there was no conception of a degree-granting institution of higher learning outside the national bureaucracy.[71] Even more remarkable is the fact that there was only one organizational unit of higher education in all of China (with some 120 million people) that had status enough to be (misleadingly) called a university. By way of contrast, Europe from the twelfth to the fourteenth centuries, with half the population of China, had at least eighty-nine universities, not to mention hundreds of colleges with more autonomy than existed anywhere in China.[72] It is true of course, as we saw above, that China had

---

[67] John Meskill, "Academies and Politics in the Ming Dynasty," in *Chinese Government in Ming Times*, ed. Charles O. Hucker (New York: Columbia University Press, 1968), p. 150.

[68] See John Watt, *The District Magistrate in Late Imperial China* (New York: Columbia University Press, 1972), chap. 11, pp. 162ff.

[69] Thomas Lee, reporting on the history of the imperial university, observes that it was granted independence with its own budget and buildings in 1044, but by 1045 it was forced to give up all thoughts of autonomy, lost its budget, and was forced into the quarters of a former military barracks. *Government Education and Examinations* (Hong Kong: Chinese University Press, 1985), pp. 63–4.

[70] Needham, "The Qualifying Examination," in *Clerks and Craftsmen in China and the West* (New York: Cambridge University Press, 1970), p. 383.

[71] It may also be noted that the Hanlin Academy during the Middle Ages was a loosely organized unit of scholars and was a policy-advising and edict-drafting unit that advised the emperor directly. But it had no teaching functions, and its research activities were confined to writing official histories and such. See Charles O. Hucker, *A Dictionary of Official Titles in Imperial China* (Stanford, Calif.: Stanford University Press, 1985) s.v., "han-lin yüan," no. 2154; and Adam Yuen-Chung Lui, *The Hanlin Academy, 1644–1850* (Hamden, Conn.: Archon Books, 1981).

[72] Hastings Rashdall, *The Universities of Europe*, new ed., ed. A. B. Emden and F. M. Powicke (Oxford: Oxford University Press, 1936), vol. 1, "Table of Universities," p. xxxiv.

many so-called private academies (shu-yüan) which were centers of learning devoted to a famous scholar and his disciples. Originally the purpose of these academies was to encourage self-cultivation in the Confucian tradition. During the Sung dynasty these academies became major centers for the dissemination of neo-Confucianism.[73] Later in Ming times they were frequently repressed by the government. Moreover, as John Meskill notes, "to describe an academy as independent was usually to condemn it."[74] As time wore on and the government's monopoly of the examination system became more firmly entrenched, the older model of teaching self-cultivation and enlightenment in the academies became subordinate to training for the government examination, above all for career advancement.[75] Apart from such considerations, there is no reason to doubt the existence of coteries of masters and disciples who continued to learn together and to pass on various forms of specialized information – both religious and ethical as well as scientific. Structurally this informal pattern of learning is not much different from that of the Arab Middle East that we discussed earlier, except that the Arab-Muslim scholars were the only officials qualified to grant the ijaza, the authorization to transmit knowledge.

It is for these reasons, then, that it has been said that "the Chinese sciences were not the basis of professions, nor even of coherent occupational groups."[76] In a word, China did not experience the legal and social revolution that the West experienced in the twelfth and thirteenth centuries, and as a result it developed no institutional locations – no institutionalized neutral spaces – that would allow autonomous self-regulation or could protect free thought from the incursions of the political and religious censors. This fact more than any other explains the retarded state of systematic scientific thought in traditional China.

A third consequence of the absence of a legal theory of corporate autonomy is the absence of professional associations or occupational guilds. As Nathan Sivin tells us, even as late as 1600, "there was no occupational group sufficiently autonomous or coherent to be called a 'profession.'"[77] The law

---

[73] John Meskill, "Academies," pp. 149ff; and Tilemann Grimm, "Academies and Urban Systems in Kwangtung," in *The City in Late Imperial China*, ed. G. William Skinner (Stanford, Calif.: Stanford University Press, 1977), especially pp. 476ff. For the generic meaning of "shu-yüan," see Hucker, *Dictionary*, s.v. "shu-yüan," no. 5471, p. 437.

[74] Meskill, "Academies," p. 152.

[75] Grimm, "Academies and Urban Systems," p. 477f.

[76] Sivin, "Science and Medicine in Imperial China – The State of the Field," *The Journal of Asia Studies* 47, no. 1 (1988): 54. Also see Golas, "Early Ch'ing Guilds," in *The City in Late Imperial China*, pp. 555ff.

[77] Sivin, "Why the Scientific Revolution," p. 545. It is to be noted that merchant guilds were not legally autonomous groups in this connection.

of China prohibited the emergence of all such groups with the consequence that the development of independent institutions of specialized knowledge could not take root: "No such institutions existed in China."[78] This stands in marked contrast to the emergence of autonomous professional guilds of physicians and surgeons in Europe in the thirteenth and fourteenth centuries, discussed in Chapter 5.

Nevertheless, we know that the natural sciences were studied and that the imperial government even occasionally encouraged this activity. The problem is that the status graduations in China were sharply and formally drawn, and the very idea of a scientist, or even the man of knowledge (insofar as this involved orthodox conceptions) was formally contained within that of the scholar-official. In his attempt to speak about the social position of the scientist, Needham couples this position with that of engineer. This attempt to link science and technology, as I suggested earlier, is highly misleading since it is only in the twentieth century that science and technology have generally been linked. Furthermore, in China the roles of scientist and engineer were even more sharply divided than in probably any other society because of the great gulf between the literate (officials) and the illiterate (commoners), which was made more rigid by the examination system. The category of knowledge worker was clearly composed of the two groups of high officials and the minor officials.[79] Since Needham's enterprise also concerns the history of technology and invention, he refers to three other groups: the commoners, the semi-servile groups, and the slaves. Although it may be true that "the greatest group of inventors is represented by commoners, craftsmen, and artisans"[80] (who were not court officials), there is no suggestion that they made any contributions to science per se. This is so for two reasons: science is, first of all, a set of arguments (or propositions) about the way the world is, and only those who are literate would have adequate access to such knowledge. Secondly, only highly literate individuals would be in a position to make a contribution to that ongoing and written debate about the nature of the world.[81]

---

[78] Sivin, "Why the Scientific Revolution," p. 545.
[79] Needham, *GT*, pp. 29ff.
[80] Ibid., p. 28.
[81] It should be noted that in sec. 27 of *Science and Civilisation in China*, "Mechanical Engineering," Needham attempts to link technology and science by referring to the former as "applied science," when of course the very existence of the science is in question. As a result his discussion is always qualified by the realization that the "scientific principles" are perhaps not "fully formulated." For example, he takes us into "the obscure expanse of the trades and husbandries" where there is a putative "application of scientific principles (whether or not always fully formulated)," *SCC* 4/2: 10. Consequently, Needham frequently concedes that Chinese artisans were remarkably good at carrying out "empirical procedures of which there

It was equally unlikely that even a gifted engineer would rise to a position of authority in the bureaucracy. Needham attributes this to the fact that "the real work [of engineering] was always done by illiterate or semi-literate artisans and master-craftsmen who could never rise across that sharp gap which separated them from the 'white-collar' literati in the office of the Ministry above."[82] It therefore remains the case that the individuals who pursued the orthodox sciences were mainly scattered government officials, including court physicians.[83]

We have seen that the civil service recruitment examination was a completely humanistic, literary, and poetic exercise, and that it contained nothing in the later Sung and Ming times that could be labeled scientific. Since passing the examinations became a self-contained activity, the system served mainly to reinforce both official and orthodox Chinese conceptions of history, ethics, and morality, while discouraging disinterested learning for its own sake. It is valid to say, therefore, that "the institution of the mandrinate had the effect of creaming off the best brains of the nation for more than 2000 years" and that it directed these minds away from scientific inquiry into the civil service.[84]

When we look at official bureaucracy itself, we encounter additional impediments to the free and unfettered pursuit of scientific knowledge. Here I refer to the elements of secrecy and excessive regulation in the study of astronomy and mathematics. Such secrecy obviously worked directly against the scientific norm of communalism, the free and open access to knowledge. Needham's account of the study of astronomy in China is littered with references to the security-minded manner and the semi-secrecy in which these disciplines were kept.

As we saw earlier, the study of astronomy in China was given a special place in the Bureau of Astronomy which was located in offices adjacent to the imperial city along with other administrative offices. Even this was an

---

was no scientific understanding," *SCC* 4/2: 47. Likewise when he discusses the work of Leonardo (whom he prefers to see as a protoscientist rather than a craftsman), Needham realizes that in Leonardo's scientific understanding and "hypothesis-making" skills "one may see . . . [a] relative theoretical backwardness," *SCC* 3: 160; and yet "what remarkable achievements may be effected without adequate scientific theory," ibid. Bodde (*Chinese Thought*, p. 233) remarks on the grave difficulty that artisans had in obtaining literacy and hence their inability to contribute to literate scientific discourse.

82 Needham, *GT*, p. 27.

83 We should not overlook the work of the alchemists who developed laboratory techniques. Yet just what their social location was remains unclear – as does their contribution to scientific thought generally. See N. Sivin, "Chinese Alchemy and the Manipulation of Time," *Isis* 67 (Dec. 1977): 513–26; as well as Needham (and Sivin), *SCC* 5/5, sec. 33, especially subsec. (h).

84 Needham, *GT*, p. 39.

advance from ancient China in which "astronomy was the secret science of priest-kings" and the observatory was "the emperor's ritual home."[85] As a result of Chinese cosmological beliefs, astronomical phenomena were taken to be the most visible and awesome signs of the harmonious state of the heavens. Since there was thought to be a correlation between the heavenly order and the political order, it behooved the emperor and his officials to take special notice of the cosmic realm. Moreover, just as the Muslims of the Middle East had their strong religious motives for studying the heavens – namely, to establish the correct times for prayer (five times daily) and the appropriate direction of prayer (toward Mecca, the qibla) – so too the Chinese had their religious reasons for diligently studying and mapping the patterns of the universe.

What the Chinese did, however, was to make this study a state secret and thereby drastically reduced the number of scholars who could, legitimately or otherwise, study astronomy. This restriction also greatly reduced the availability of the best and latest astronomical instruments and observational data. As Needham diplomatically put it, "From earliest times astronomy had benefitted from state support, but the semi-secrecy which it involved was to some extent a disadvantage."[86] This is to put the matter far too mildly, as Needham's account amply shows, for even Chinese historians were aware of the grave costs that this policy of secrecy exacted. One official historian wrote, "Astronomical instruments have been used from very ancient days, handed down from one dynasty to another, and closely guarded by official astronomers. Scholars have therefore had little opportunity to examine them."[87] Likewise, in the eleventh century the polymath Shen Kua wrote:

[between +1049 and +1053] the Ministry of Rites arranged for the examination-candidates to be asked to write essays on the instruments used for gaining knowledge of the heavens. But the scholars could only write confusedly about the celestial globe. However, as the examiners themselves knew nothing about the subject either, they passed them all with a higher class.[88]

This restriction of access to astronomical instruments and information during the Sung and Ming dynasties bordered on the paranoid. Because of the fear that astronomical observations could reveal disorder in the universe and thus could offer a naturalistic report on the sagacity of the conduct of the emperor, the staff of the Bureau of Astronomy "submitted confidential

---

[85] Needham, *SCC* 3: 189.
[86] Needham, *SCC* 3: 193.
[87] As cited in Needham, *SCC* 2: 193.
[88] Needham, *SCC* 3: 192.

reports to the Emperor whenever there were abnormalities."[89] To make sure that no confidential astronomical information could leak out of the bureau, "officers in the Bureau were not transferable to posts outside the Bureau; their children were not permitted to change to other professions."[90] It is question begging, therefore, to merely offer the opinion that "whether or not the best and greatest scientific achievements happen under such conditions is another question,"[91] when it would require a great stretch of the imagination to believe that they could. To suggest, moreover, that the study of astronomy "was possible during the Sung dynasty, at any rate . . . in scholarly families connected with the bureaucracy" is no recommendation at all.[92] The fact remains that virtually every move made by the astronomical staff had to be approved by the emperor before anything could be done, before modifications in instrumentation or traditional recording procedures could be put into effect. It is not surprising, therefore, that despite the existence of a bureau of astronomers staffed by superior Muslim astronomers (since 1368), Arab astronomy (based as it was on Euclid and Ptolemy) had no major impact on Chinese astronomy, so that three hundred years later when the Jesuits arrived in China, it appeared that Chinese astronomy had never had any contact with Euclid's geometry and Ptolemy's *Almagest*.[93] Moreover, contrary to Needham's arguments, more recent students of Chinese astronomy suggest that Chinese astronomy was perhaps not as advanced as Needham suggested and that "Chinese astronomers, many of them brilliant men by any standards, continued to think in flat-earth terms until the seventeenth century."[94]

If we consider the study of mathematics, in which the metaphysical implications of abstract thought may be less obvious to outsiders, and which may therefore give scholars more freedom of thought, we encounter an institutional structure equally detrimental to the advancement of science. In the first place, as we saw above, the institutional reward system established by imperial authorities in the T'ang and Sung eras was one that encouraged classical, literary, and historical studies. There was, all students of the examination system attest, nothing in the content of the civil service examination

---

[89] Ho Peng-yoke, "The Astronomical Bureau in Ming China," *Journal of Asian History* 3/4 (1969):137–53 at p. 142.

[90] Ibid., p. 144.

[91] Needham, *SCC* 3:193.

[92] Ibid.

[93] Needham's own efforts to understand the failure of Arab influences on Chinese astronomy and especially the "fit" between Arab-designed astronomical instruments and Chinese assumptions are detailed in *SCC* 3: 372–82.

[94] Christopher Cullen, "Joseph Needham on Chinese Astronomy," *Past and Present* no. 87 (1980): 39–53 at p. 42.

that could be called scientific. When an additional examination in mathematics was added during the T'ang dynasty, reports are that "nobody wanted to take it since it was not likely to lead to high advancement in the bureaucracy."[95] This led to Needham's lament that during the Sung era, "the greatest mathematical minds were now (with the exception of Shen Kua) mostly wandering plebeians or minor officials. Moreover, their attention was devoted less to calendrical work, and more to practical problems in which common people and technicians were likely to be interested."[96] It is remarkable that during the Sung flowering of mathematical thought the greatest mathematical innovators – for example, Ch'in Chiu-shao (ca. 1202–ca. 1261), Chu Shih-chieh (ca. 1280–1303), Li Ye (1178–1265), and Yang Hui (fl. ca. 1261–75) – were remote scholars unknown to each other.[97] Perhaps it was the breakdown in social order and imperial control in the transition from the T'ang to the Sung dynasty that provoked this flowering of thought. But rather quickly, it may be surmised, the widespread imposition of the examination system served to structure the world of learning so that aspiring scholars concentrated all their energies on memorizing the Confucian classics and writing stylized expositions of their meaning. Since all the rewards of the system were attached to passing the examinations, a culture of disinterested learning with its own norms, ideals, and standards of scholarship did not take root. All aspects of the learning process fell under the control of the central authority that created the examination questions, disseminated the standard works and commentaries to be studied, administered the examinations, corrected them, and rewarded the successful with official positions in the bureaucracy.

One can hardly imagine a more autocratic system designed to control the learning of a whole nation. It should not be forgotten that this monopoly of educational certificates was especially effective because there were no high positions of wealth and authority not controlled by the state. As Etienne Balazs pointed out in connection with "the rise of capitalism problem," the Chinese state in theory owned all the land and mineral wealth of the country, so that even mining operations – salt, iron, copper, and so forth – were operated as government monopolies supervised by state officials.[98] Likewise, all banking innovations such as letters of credit ("flying money"), long-distance facilitation of exchange, and so forth, were taken over as state

95 Needham, *SCC* 3: 192.
96 Needham, *SCC* 3: 42.
97 Needham, *SCC* 3: 41; and see Li Yan and Dù Shíràn, *Chinese Mathematics*, pp. 109–17.
98 Etienne Balazs, *Chinese Civilization and Bureaucracy* (New Haven, Conn.: Yale University Press, 1964), chap. 4, especially pp. 44–7.

monopolies.[99] There was no scope for entrepreneurial innovation, and thus disinterested learning, even for essentially entrepreneurial practice, was discouraged because these avenues of advancement were closed without state sponsorship. Beyond that, the Chinese status system continued to place a great premium on the holding of imperial office and hence the passing of the official examination (or the buying of an office). The status of a wealthy merchant consequently remained considerably below that of an officeholder and induced wealthy families to lavish their money on the support of traditional scholars and the training of their relatives and children to pass the official examinations – not the pursuit of science.[100]

In the centuries that followed, therefore, it is not surprising that of that spirit of "mathematics 'for the sake of mathematics' there was extremely little."[101] There was no incentive to study mathematics: there was only official discouragement in the sense that the rewards went to those who mastered the classical humanistic and ethical materials of the civil service examination, and thus there was a lack of interest in mathematical theory. There remains a final irony in this situation, and that is the general tradition of secrecy which also affected the study of mathematics. For, according to Needham, it is the extensive tradition of secrecy in China that "explains well enough why Matteo Ricci's mathematics books were confiscated when he was on his way to the capital in 1600."[102] At every turn there were great impediments to the disinterested pursuit of natural science and mathematics.

Given such a cultural context and the overwhelming structure of disincentives built into the examination system insofar as science was concerned, it might be anticipated that the Chinese would not match the Arabs in this area and that ultimately no scientific revolution would occur in China. The Sung renaissance in science and learning appears to have ended as abruptly as it began. The levels of economic output (as measured by tons of iron produced) dropped by 50 percent during the next three centuries, and even in the 1930s they were not as high as they had been in the late eleventh century in the Northern Sung.[103] The neo-Confucian orthodoxy that unfolded in the thirteenth century ushered Confucian scholars back into office, while

[99] Ibid., pp. 42–4.
[100] For aspects of this process among nineteenth-century salt merchants, see Ho Ping-ti, "The Salt Merchants of Yang-Chou: A Study of Commercial Capitalism in Eighteenth-Century China," *Harvard Journal of Asiatic Studies* 17 (1954): 130–68, especially 154ff.
[101] Needham, *SCC* 3: 153.
[102] Needham, *SCC* 3: 194.
[103] Hartwell, "A Revolution in Chinese Iron and Coal Industries in the Northern Sung, 960–1127 A.D.," *Journal of Asian Studies* 21, no. 2 (1962): 153–62, p. 154.

"mathematics was again confined to the back rooms of provincial yamens [government offices]," with a resultant decline in mathematics in China.[104] Perhaps Needham was speaking with hyperbole when he wrote that there was no one to tell the Jesuits when they arrived in the seventeenth century of the past glories of Chinese mathematics, yet other students of Chinese science and mathematics make similar statements. Lî Yan and Dù Shíràn report that the tremendous development of mathematics which had taken place in the Sung and Yüan dynasties had long since come to a halt, and in the Ming dynasty the great achievements in mathematics were poorly understood and in danger of being lost. By "the fifteenth century some Ming Dynasty mathematicians had virtually no understanding of the 'method of the celestial element' or the 'method of four unknowns.'"[105] Likewise, Nathan Sivin reports that by "1600 no one was able to fully comprehend the old numerical equations of higher-order, prototrigonometric approximations, applications of the method of finite differences, and other sophisticated techniques."[106]

It is at this point that Needham originally posed the great question: "What was it, then, that happened at the Renaissance in Europe whereby mathematical natural science came into being? And why did this not occur in China?"[107]

Conclusion

We can now see the contours of the answer to these two questions. Regarding the first – what happened in Europe – we see that it was not something that happened in the Renaissance of the fourteenth and fifteenth centuries, but something that happened much earlier in the twelfth and thirteenth centuries, something that did not happen in China. Yet there is a truth to Joseph Needham's claim that in viewing the set of changes that facilitated the birth of modern science in the West, "we seem to be in the presence of a kind of organic whole, a package of changes."[108] What happened in Europe was a social and legal revolution that radically transformed the nature of medieval society, in fact laying the foundations of modern society and civilization. Europe experienced a revolution that placed social life on an entirely new footing. From one point of view, it represented the grand fusion for the first time of Greek philosophy and science, Roman law, and Christian theology. All three structures were unique to the West – as there is nothing in Chinese

104 Needham, *SCC* 3: 153–4.
105 Lî Yan and Dù Shíràn, *Chinese Mathematics,* p. 175.
106 Sivin, "Wang Hsi-Shan," p. 161.
107 Needham, *SCC* 3: 154.
108 Needham, *GT,* p. 40.

thought equivalent to Greek philosophy as expressed by Aristotle, to Christian theology, or to the *corpus juris civilis*.

At the center of this revolution was a legal transformation that redefined the nature of social organization in all its realms – political, social, economic, religious, and intellectual. In the present context, the development of the law of corporations was foremost among these changes. At the time when the Christian church declared itself to be a corporation, a whole body for all legal purposes, it granted such status to a variety of other collectivities, such as residential communities, cities, towns, universities, and economic interest groups such as guilds. It also laid the legal foundations for professional associations. As a result, each of these collectivities of individuals was granted a legal autonomy to make its own internal laws and regulations, to own property, to sue and be sued, and to have legal representation before the king's court. In each of these cases local groups of citizens were granted local autonomy and legal jurisdiction that enabled them to work out in various public forums the best solutions that enlightened men of goodwill could fashion in the context of consensual decision making, that is, either majority vote or "the greater and sounder part." In creating many spheres of autonomy, the Western legal revolution created both neutral zones (universities) and a variety of public spaces in which various levels of open discussion and debate prevailed – including courts of law with regularized procedures, standards of due process, and the use of advocates to defend the rights of the accused.[109] It thereby made a huge contribution to the shaping of the modern culture of political institutions in which it is presumed that there are both collective and individual rights and interests which must be reconciled through open debate and representative delegation of authority. This revolution also sharply demarcated the religious domain – the moral and the ethical – from the secular state. Not least of all, these changes created both the legal and institutional foundations for the emergence of professional associations of physicians, lawyers, merchants, and, eventually, scientists. Moreover, scholars such as Harold Berman and Robert Lopez have pointed to a European commercial revolution during the twelfth and thirteenth centuries.[110]

Focusing on the world of learning, however, it was in the twelfth and thirteenth centuries in Europe that universities arose, establishing neutral zones of intellectual autonomy which allowed philosophers and scientists to pursue their agendas free from the dictates of the central state and the

---

[109] See Berman, *Law and Revolution*, pp. 250–3, 469ff, 423ff, and passim.
[110] Robert Lopez, *The Commercial Revolution of the Middle Ages, 950–1350* (New York: Cambridge University Press, 1977); and Berman, *Law and Revolution*, pp. 336–9 and 540–3.

religious authorities. The founders of the universities consolidated the curriculum around a basically scientific core of readings and lectures. This was embodied in the natural books of the new Aristotle that became known during those centuries. These books included Aristotle's *Physics, Meteorology, On Generation and Corruption, On the Soul, The Small Works on Natural Things,* and so forth.[111] Anyone who reads these works cannot but be impressed with the extraordinary concentration of energy on the naturalistic understanding of the world in all its dimensions that these works represent. They unmistakably enjoin the reader to adopt a naturalistic point of view and thereby to assume that the world is explainable in terms of fundamental elements, causal processes and principles, and dispassionate rational inquiry.

Not only was this new intellectual agenda thoroughly naturalistic, but it was made the core of an evolutionary course of study. Scholars were not only free to raise questions about it, they were taught how to raise questions and were even enjoined to dispute every aspect of it. In this process they asked whether the world had a beginning or had always been in existence; whether there were other worlds and, if so, whether the same physical laws would obtain in such worlds. In the domain concerning speculation about time, space, and motion, they asked questions about the existence of a vacuum and its properties. Could God cause the earth to be instantaneously accelerated in a straight line, and if he could, would this produce a vacuum?[112] These questions touched on every field – theological, medical, and scientific. It was just this set of naturalistic inquiries (including those about the heavens) that set the agenda of scientific inquiry for the next four hundred years in European universities. If Chinese philosophers and scholars-without-portfolio were asking these same questions, they could not do so in public forums – there were no established zones of officially sanctioned neutral space where such questions could be asked. Likewise, it was just this naturalistic agenda that was intentionally excluded from the Islamic madrasas.

To the second of Needham's questions posed above – why did the scientific revolution not occur in China? – we now see a large part of the answer in the deliberate displacement of naturalistic inquiries from the center of higher education in China and the failure of the Chinese legal and political systems to undergo the radical restructuring that took place in Europe in the twelfth and thirteenth centuries. As Needham put it, "The fact is that in the spontaneous autochthonous development of Chinese society no drastic

---

[111] The statute incorporating these works in the curriculum (and the schedule for their reading) at the University of Paris is translated in Edward Grant, ed., *A Source Book of Medieval Science* (Cambridge, Mass.: Harvard University Press, 1974), pp. 43f.

[112] See Grant, *A Source Book*, pp. 199ff.

change parallel to the Renaissance and the 'scientific revolution' in the West occurred at all."[113] More importantly, we see the failure of Chinese civilization to create neutral spheres of intellectual autonomy in which any intellectual agenda could be pursued independent of the interference of state authorities. In a similar vein, Nathan Sivin writes that "no institution had evolved through Chinese history to work out and resolve conflicts of political viewpoints."[114] That outcome was but the failure of the Chinese legal system to separate the sacred and the secular realms (the state and religion) and to evolve and develop legal conceptions beyond those of an inherently penal nature. What Chinese law lacked was a conception of law as an enabling set of structures and techniques – a set of objective structures that create zones of autonomy and independence as well as adequate procedures (due process) which facilitate the peaceful (and nonpenal) resolution of legitimate but conflicting interests and rights. The idea that law and legal structures only make for greater contentiousness was never supplanted in China by the conception that law and legal structures create neutral forums within which conflicts can be peacefully resolved without stigmatizing or punishing the participants.[115]

We glean further insight into the arresting of Chinese institutional autonomy when we learn that the first emperor of the Ming dynasty (1368–1644), Hung-wu (or T'ai-tsu [r. 1368–98]), having concluded that the students of the newly reopened imperial academy (kuo-tzu chien) were too unruly and undisciplined, appointed his young nephew as head of the institution.[116] Later in his reign T'ai-tsu feared that people and events had gotten out of hand and this caused him to issue a set of pronouncements. In the third of these proclamations (ca. 1386) there was a "list of 'bad' metropolitan degree holders," that is, chin-shih or "doctorates," along with the names of some students. "He prescribed the death penalty for sixty-eight metropolitan degree holders and two students; penal servitude for seventy degree holders and twelve students." The author of this account in the *Cambridge History of China* adds that these lists "must have discouraged men of learning."[117] Appended to the edict was a further reprimand. The emperor

---

[113] Needham, *GT*, p. 40.

[114] Sivin, "Shen Kua," *DSB* 14: 371.

[115] The view that traditional Chinese legal authorities were animated by a desire to stigmatize all those who came in contact with the legal system is expressed by Clarence Morris in *Law in Imperial China*, reprint (Philadelphia: University of Pennsylvania Press, 1973), p. 542, as well as Escarra, *Chinese Law*, trans. G. W. Browne (Harvard University Asian Research Center, 1961), p. 102.

[116] *Cambridge History of China*, ed. Fredrick W. Mote and Denis Twitchett (New York: Cambridge University Press, 1988), vol. 7, pt. 1, p. 122.

[117] Ibid., p. 154.

would put to death any man of talent who refused to serve the government when summoned. As he put it, "To the edges of the land, all are the king's subjects. . . . Literati in the realm who do not serve the ruler are estranged from teaching [of Confucius]. To execute them and confiscate the property of their families is not excessive.[118]

The trial and punishment of Galileo (confinement to his villa overlooking Florence) is nothing compared to this.

When we look at the larger picture of Chinese science, we can surely find technical lapses and inadequacies, as we can in any ongoing enterprise of progressive learning. The problem of Chinese science, however, was not fundamentally that it was technically flawed, but that Chinese authorities neither created nor tolerated independent institutions of higher learning within which disinterested scholars could pursue their insights. The same is of course true of Arabic-Islamic civilization, but there one finds far more individualism and public charity devoted to learning. Scholarship was frequently aided by wealthy patrons who gave covert shelter and support to scholarly inquiry in the natural sciences. Arabic-Islamic scholars were also greatly aided in their work by the existence of widely scattered but often lavishly appointed libraries open to all believers. There was a long-standing tradition that each mosque should maintain a library adjacent to it. Naturalistic works found their way into these libraries, and readers were free to read them at will. The prohibition on publicly teaching these works only drove the study of the natural sciences underground or, more correctly, into the homes and discussion centers (majlis) of private individuals. Nor, finally, did the Middle Eastern state manage to control either all the schools (through a uniform examination system) or all the important jobs that required intellectual skills. Physicians, as we saw in Chapter 5, were highly esteemed in the Middle East, and their intellectual tradition was one of rich philosophical learning. As a result, they were highly valued as both government officials and community leaders. This appears not to be true of Chinese physicians, and it undoubtedly relates to (but does not explain) the fact that a system of hospitals equivalent to those developed in the Middle East did not emerge in China.[119] Chinese physicians qua physicians could not rise to positions of power and authority in local government (as they could in the Middle East) because the local government was tightly controlled by the imperial system of district magistrates.

In China the disinterested pursuit of learning was generally acknowledged

[118] Ibid.
[119] Cf. Paul Unschuld, *Medicine in China: A History of Ideas,* especially p. 149, and Needham, "Medicine in Chinese Culture," chap. 14 in *Clerks and Craftsmen.*

to have been displaced by the desire to pass the examinations. Despite the existence of woodblock printing and the enormous multiplication of government-sponsored printed works, there does not appear to have been a rich tradition of maintaining public libraries and gathering important literary works. In fact, the Chinese repeatedly lost significant portions of their literary heritage. This seems to have been the case with regard to a treatise on medieval clockworks – as well as the actual technology – crafted by Su Sung (1020–1101). Despite the fact that the Chinese had invented a great mechanical clock – which was housed in a thirty- to forty-foot-tall tower in the eleventh century – when the Jesuits arrived in the seventeenth century, it appeared that the Chinese had not yet invented timekeeping devices with escapement mechanisms. This was so because the original clock had been disassembled and carried off, and the actual treatise by Su Sung describing the device had itself been lost for a long period of time.[120]

Thomas Lee has given us a description of the state of a county school library as reported by the great neo-Confucian philosopher Chu Hsi. At the county school of T'ung-an to which he had been assigned, there was a box of books, never catalogued, much less examined or studied, which had been received by the school eighty years before his arrival. Many of the books had been stolen, but otherwise they were strewn around in dust and consumed by insects.[121] Lee adds to this account that in general most library collections "were meager and some leading schools, such as the Prefectural School of Hangchow in the capital, did not even possess a library"[122] – this occurring at a time when block printing was in full bloom. In comparison to the Arabs, the Chinese were much less interested in libraries, while among the Europeans what books there were (fewer and hand-copied) were probably better taken care of.[123] The invention of block printing in China was undoubtedly a landmark achievement in the history of mankind, and the output of woodblock-printed works in China is indeed remarkable.[124] Nevertheless, the invention of printing in China did not lead to intellectual unrest, the promotion of national languages and identities, or a cultural and scientific revolution.[125] In terms of output, moreover, it is an open question whether China surpassed the Middle East during the latter's golden era (ca. 945–1300) when all forms of mechanical printing were rejected. For exam-

---

[120] Joseph Needham and Derek J. de Solla Price, *Heavenly Clockwork: The Great Astronomical Clocks of Medieval China*, 2d ed. (New York: Cambridge University Press, 1986).

[121] Thomas H. C. Lee, *Government Education*, p. 112 n33.

[122] Ibid.

[123] See John F. D'Amico, "Manuscripts," in *The Cambridge History of Renaissance Philosophy* (New York: Cambridge University Press, 1988), pp. 11–24.

[124] See Joseph Needham and Tsien Tseun-hsuin, *SCC* 5/1, especially pp. 159–83.

[125] Ibid., p. 382.

ple, the Peking library inherited the books of the former Chin, Sung, and Yüan dynasties, and in 1441 the collection consisted of "7,350 titles, in some 43,200 volumes (*tshe*) containing one million chüan [sections]."[126] In contrast, there are several reports of Arab libraries with more than 100,000 volumes. Although it is probably an exaggeration, the Marâgha observatory is said to have had 400,000 volumes.[127] The famous House of Wisdom of the Fatimids (tenth century), containing forty rooms of books, was estimated to have housed between 120,000 and 2 million books.[128] Books in the natural sciences alone were said to number 18,000.[129] One benefactor in Egypt is said to have given 100,000 volumes to the founding of the madrasa al-Fadiliya; another 100,000 given to the founding of the Qala'un hospital in Cairo.[130] Whatever the actual number of volumes may have been, there was a deep and long-standing tradition whereby each mosque maintained a library, and sizeable cities in the Middle East often had dozens of mosques.[131]

Finally, I must note that travel in the Middle East, as in Europe, was much more open and encouraged. In China the attitude toward travel was at best ambivalent. Many travel restrictions were imposed in keeping with the general cultural view that people should remain in their own towns and villages and that wandering scholars were to be both avoided and discouraged.[132] During Ming and Ch'ing times these restrictions were increased. The ultimate cultural impediment was the imposition of the *pao-chia* system that required every family to keep a list of occupants, visitors, and travelers, under pain of punishment for failure to do so. While the system was officially adopted at the inception of the Ch'ing dynasty (1644), its roots go back much further.[133]

The pursuit of the sciences in China was relegated to the periphery of intellectual endeavor. For that reason and those set out above, it is not puzzling that Chinese culture and civilization – unlike the case of Arabic science – did not give birth to modern science.

[126] Ibid., p. 175.
[127] Sayili, *The Observatory in Islam* (Ankara: Turkish Historical Society Series 7, no. 38, 1960), p. 194.
[128] Pedersen, *The Arabic Book* (Princeton, N.J.: Princeton University Press, 1984), pp. 118–19.
[129] Ibid., p. 116.
[130] Ibid., p. 119.
[131] For an overview of the early library tradition of the Middle East, see Ruth S. Mackensen, "Four Great Libraries of Medieval Baghdad," *Library Quarterly* 2 (1932): 279–99.
[132] White, "Fate of Independent Thought," pp. 68–9; and Chaffee, *The Thorny Gates of Learning in Sung China* (New York: Cambridge University Press, 1985), p. 33.
[133] Hsiao Kung-ch'uan, *Rural China: Imperial Control in the Nineteenth Century* (Seattle: University of Washington Press, 1960), p. 26; Ch'ü, *Local Government in China under the Ch'ing* (Cambridge, Mass.: Harvard University Press, 1962), pp. 2–4; and John R. Watt, *The District Magistrate*, pp. 145–50.

9

The rise of early modern science

Viewed from a comparative and civilizational point of view, the rise of modern science appears quite different than it does when seen exclusively as an intra-European movement. In the first instance, we realize that dedicated investigators of the processes of nature existed in other societies and civilizations around the globe. Over the course of time learned scholars exerted their utmost to fashion the technical tools and the explanatory devices needed to accomplish the task of mapping out and explaining all the realms of nature. What is perhaps most surprising is the fact that Arabic-Islamic culture and civilization had the most advanced science to be found in the world prior to the thirteenth and fourteenth centuries. In optics, astronomy, the mathematical disciplines of geometry and trigonometry, and medicine, its accomplishments outshone those of the West as well as China's. We also know that men of science in the Islamic world wrote treatises on experimental science (in optics, medicine, and astronomy) and that they applied these techniques to specific areas of inquiry, especially optics. Here one thinks of the research program designed to explain the rainbow and the controlled experiments performed to achieve that end. In addition, in the realm of experiment, one thinks also of efforts in medicine, pharmacology, and even astronomy.

It must be said, however, that these scientific activities were often scattered geographically, isolated in their influence, and conducted in semi-secrecy. The transmission of important correspondence and scientific treatises between investigators in distant places was often long delayed, incomplete, or even completely interrupted by local events and political upheavals. Still, the work went forward and, over the course of time, indispensable elements of scientific practice accumulated and became a unique heritage of human endeavor.

321

The Copernican revolution

If one takes the view of historians of science such as Edward Rosen, Herbert Butterfield, and notable others, the Copernican revolution was a major transformation in the Western conception of the universe and the individual's place in it. So viewed, the scientific revolution of the sixteenth and seventeenth centuries was a profound metaphysical revolution. At the same time, we recognize that this revolution erupted only in the West, not in Islam or China, and this fact compels many to ask what unique social and cultural factors existed in the West that enabled this great transformation to occur.

The fact that the Copernican shift was primarily a metaphysical transformation is even more dramatically accented by recalling some of the more impressive practices and achievements of Arabic science prior to the fifteenth century that we noted above. That is, in classical Arabic science we find elements of theoretical sophistication, exacting empirical observations, occasional uses of experimental techniques, and the use of highly advanced mathematical techniques – above all, the development of the so-called non-Ptolemaic planetary models of the Marâgha observatory in the thirteenth century. Considering such things, it is evident that the breakthrough to modern science, above all in astronomy, is not most usefully described either as the product of new observations or as a technical innovation within the narrow confines of mathematical astronomy. Indeed, it is now generally agreed that Copernicus's great new conception of the order of the universe was not built on any stunning new observations or new mathematical techniques that were not available to the Arabs. It was, rather, "a radical, purely intellectual shift,"[1] a sort of "transposition of the mind,"[2] which brought an old "bundle of data" into a new set of relationships. Furthermore, there is no doubt but that Copernicus borrowed heavily from the *Almagest* of Ptolemy, an action made easier by the advent of the printing press.

To say, however, that the revolutionary cosmological changes that Copernicus wrought "were unattended by mathematical complexity and quite independent of any new mathematical techniques of more than the most simple and unsophisticated variety"[3] is to give the wrong accent to the

[1] Robert S. Westman, "Proof, Poetics, and Patronage: Copernicus's Preface to *De revolutionibus*," in *Reappraisals of the Scientific Revolution*, ed. David C. Lindberg and Robert S. Westman (New York: Cambridge University Press, 1990), p. 170.

[2] Herbert Butterfield, *The Origins of Modern Science, 1300–1800*, rev. ed. (New York: Free Press, 1957), p. 13.

[3] Derek J. de Solla Price, "Contra-Copernicus: A Critical Re-estimation of the Mathematical Planetary Theory of Ptolemy, Copernicus, and Kepler," in *Critical Problems in the History of*

episode. For Copernicus's innovation was indeed radical at the time it was offered. Subjected to severe criticism on many grounds, Copernicus's physical description of the universe was indeed closer to the truth than the Ptolemaic earth-centered system. Moreover, those who followed Copernicus, for example, Kepler, Galileo, and later Newton, could not have achieved what they did without the switch from a geocentric to a heliocentric universe. Kepler, in particular, was indebted to Copernicus's idea that the sun was the center of the planetary system as then conceived, and without this idea, "without a sun-centered universe, the entire rationale of his book [*Mysterium cosmographicum* of 1596] would have collapsed."[4] In other words, the discovery of the elliptical orbit of the planet Mars and Kepler's cube root law were predicated on the Copernican hypothesis, though these discoveries were to come later (in 1609). In addition, in exposing the false preface attached to *De revolutionibus* by Osiander (which characterized the new sun-centered system as merely hypothetical), Kepler underscored his belief in the physical reality of the system, as well as Copernicus's own.

In short, Copernicus claimed a new reality – a new physical reality – on the basis of planetary models and observational data that were held in common with Arab astronomers. If those models and the observational data have been judged to be inadequate to support the new Copernican system,[5] we can see how radical and indeed courageous Copernicus was in setting forth his new astronomical system. The Copernican revolution was then a purely metaphysical leap that the Arabs were either unwilling or unable to make – despite their having had nearly two centuries of previous experience with the observational problems which the planetary models posed.

From a sociological point of view, the question is not whether Copernicus's theory was true or false or whether it was strongly or poorly supported by observational and logical considerations, but whether a set of cultural institutions, a modicum of neutral space, existed within which the merits of the new system could be debated without personal danger to those

Science, ed. M. Clagett (Madison, Wis.: University of Wisconsin Press, 1959), pp. 197–218 at p. 198.

4 Owen Gingerich, "From Copernicus to Kepler: Heliocentrism as Model and as Reality," *Proceedings of the American Philosophical Society* 117, no. 6 (1973): 513–22, at p. 520.

5 This was the position of many astronomers as well as the official position of the Catholic church in the time of Galileo. For a review of the facts of that situation, see Olaf Pedersen, "Galileo and the Council of Trent: The Galileo Affair Revisited," *Journal for the History of Astronomy* 14 (1984): 1–29. For the problem of the predictive inaccuracies of the Copernican system, see Price, "Contra-Copernicus," especially pp. 209ff. But also see Owen Gingerich, "Commentary: Remarks on Copernicus's Observations," in *The Copernican Achievement*, pp. 99–107. There it is suggested that "Copernicus was interested more in cosmology than in predictive accuracy," p. 107.

who defended it. The question is what kind of social and institutional supports existed that could provide at least a fair approximation of a dispassionate evaluation of a new cosmological system that could easily be declared heretical. For clearly enough, the Copernican system not only violated some of the principles of Aristotelian natural philosophy (for example, that a planetary body could have more than one motion, that is, a diurnal and a linear motion, and that astronomy as a discipline was subordinate to physics from which it took its first principles), but more importantly it violated the theological assumptions of Christian theology. According to the latter, the earth was the center of the universe and the Bible was the authoritative source declaring it so. Neither Copernicus nor his followers overlooked that fact but they worked to devise strategies which would skirt that authoritative source. Indeed, Copernicus's first and most ardent follower, Rheticus (1514–1574), wrote a treatise that attempted to reconcile the Scriptures and the new world system of Copernicus.[6] In a word, the clash between the new world system and the established theological views – an amalgam of Scripture and orthodox Aristotelianism – presented a formidable obstacle to the acceptance of the Copernican system and the conflict was bound to come to a head, either in the time of Copernicus or soon thereafter – as it did in the Galileo affair.

Moreover, from 1588 on there was still another rival to the Copernican system, namely, the geoheliocentric system of Tycho Brahe (1546–1601), which was published in his book *Concerning Recent Phenomena in the Celestial World* (1587). According to Brahe's model of the universe, the planets revolved around the sun, and the latter in turn revolved around the earth (Figure 9). Of course this system had its own technical problems, such as the fact that its planetary orbits required the intersection of the sun's orbit by those of Mars, Mercury, and Venus.[7] This was initially a serious defect since most astronomers continued to believe in the material reality of the celestial spheres that carried the planets in their orbits. As material entities they could not be interpenetrated by spheres carrying other planets. Tycho's observations (and those of others) of the comet of 1577, however, revealed that the comet had traversed a path which shot directly through "what Tycho and everyone else regarded as the Ptolemaic spheres of Mercury and Venus."[8] The implication was clear: "The very motion of the comets is the

---

6 See R. Hooykaas, "Rheticus's Lost Treatise on the Holy Scriptures and the Motion of the Earth," *Journal for the History of Astronomy* 15 (1984): 77–80.

7 For a good discussion of the Tychonic system and its development, see Victor E. Thoren, *The Lord of Uraniburg: A Biography of Tycho Brahe* (New York: Cambridge University Press, 1990), chap. 8.

8 Ibid., p. 257.

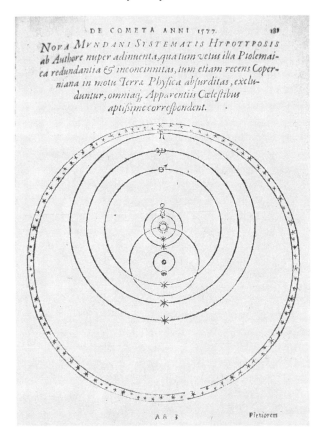

Figure 9. One of the greatest astronomers of the sixteenth century, Tycho Brahe (1546–1601) could not reconcile his theory and observations to the heliocentric Copernican system. In order to preserve the centrality of the earth in the universe, he devised a system in which the planets Mercury and Venus revolved around the sun, which in turn revolved around the earth. Likewise, the planets Mars, Jupiter, and Saturn continued to revolve in great circles about the earth placed at the center of the universe. (Photograph courtesy of Owen Gingerich and the Harvard College Library.)

strongest argument that the planetary spheres cannot be solid bodies."[9] That good news saved Tycho's geoheliocentric system from certain rejection, as there was now no reason why the orbit of Mars (or Mercury and Venus) could not intersect that of the sun (Figure 10). Furthermore, a great strength

[9] These are the words of the German astronomer Christoph Rothmann, as cited in ibid., p. 258–9.

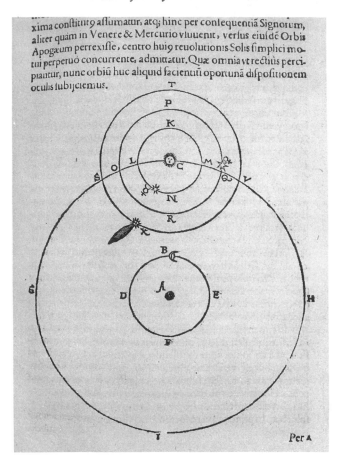

ximaconftituto aflumatur, atq; hinc per conlequentiā Signorum, aliter quàm in Venere & Mercurio viuuenit, verlus eiuldē Orbis Apogæum perrexifle, centro huio reuolutionis Solis fimplici motui perpetuó concurrente, admittatur. Quæ omnia vt rectiùs percipiantur, nunc orbiū huc aliquid facientiti oportunā dilpofitionem oculis lubijciemus.

Figure 10. This diagram of Tycho Brahe's cosmology shows the circular orbits of Mercury (*LKMN*) and Venus (*OPQR*). It also depicts the comet of 1577 (at point *x*) in orbit around the sun. Only later were astronomers to realize that the comet's path through the heavens must have pierced the spheres to which the stars and planets were thought to be attached. (Photograph courtesy of Owen Gingerich, and the Harvard College Library.)

of the Tychonic system in the eyes of many was the fact that Tycho rejected the diurnal movement of the earth and preserved the centrality of the earth in the universe.

We see then that the sixteenth century was a time of astronomical innovations which implied radically different arrangements of the cosmos. Although mathematical astronomers were not generally acceded the right to make claims about the actual physical shape of the universe, it is clear that

Copernicus, Kepler, Tycho, and the Jesuit Christoph Clavius did believe in a realist interpretation of the universe and that not all systems of cosmology – the Ptolemaic, the Copernican, and the Tychonic – could be true. It is justly said, therefore, that "Copernicus is really the initiator of a very basic attitude which came to be held in some form or other by most of the great figures in the Scientific Revolution – namely, that fundamental principles in the form of hypotheses or assumptions about the universe must be physically true, and incapable of being otherwise."[10] It is well to remind ourselves, therefore, of the following two perspectives articulated by Benjamin Nelson:

(1) The pioneers [of sixteenth- and seventeenth-century philosophy and science] *had to fight* in the name of certitude and truth. They did not really have the option of remaining ensconced inside their disciplines by the acknowledgement of the merely "hypothetical" character of their experimental and theological views.

(2) To be an innovating physicist or philosopher in bygone days meant to risk becoming embroiled in dangerous conflicts with theological authorities, perhaps to place one's life in the balance for the sake of an idea. Had the pioneers not risked everything in struggles *against* fictionalism and probabilism, today's physicists would not have been as free as they now are to champion fictionalist and probabilist positions.[11]

If we are to appreciate the climate of receptivity and the institutional resources of intellectual defense that could be mustered in the West to aid innovators in such a struggle (and the absence of anything like it in the world of Islam and China during the same period of time), we must return again to the European medieval legal, social, and institutional revolution. For it was that which transformed the nature of learning and propelled the universities into the center of the religious, metaphysical, and scientific debates that continue down to the present. In order to fully comprehend the theoretical issues that lie behind this great switch in the civilizational center of authority of the West, we need to consider the institutionalization problem, that is, the institutionalization of science problem, which has generally been associated with the social and intellectual changes which took place in the sixteenth and seventeenth centuries in England. These considerations in turn require some additional comments regarding the "Merton thesis," which asserts that there was a marked increase in scientific activity in seventeenth-century England and that Puritanism was a spur to this new movement – surely an unobjectionable thesis so stated. Likewise, questions are raised regarding Joseph

[10] Edward Grant, "Late Medieval Thought, Copernicus, and the Scientific Revolution," *Journal of the History of Ideas* 23 (1962): 197.

[11] Nelson, "The Early Modern Revolution in Science and Philosophy," in *On the Roads to Modernity*, ed. Toby E. Huff (Totowa, N.J.: Rowman and Littlefield, 1981), pp. 125–6.

Ben-David's assertion that the study of science was peripheral in the medieval university.

The institutionalization problem

In its most unornamented form, the thesis connecting Puritanism and the rise of modern science is the suggestive thesis that during the latter half of the seventeenth century there was a pronounced shift in intellectual interests in cultivated circles in England that expressed itself as a rise in the study and pursuit of science and technology. Although Robert Merton did separate science from technology conceptually, his tabulations sometimes lead to combining in one trend the separate trends of both scientific discoveries and technological inventions. Merton claimed that among the several growing social currents of the time (that is, political, economic, military, and utilitarian), a complex of values variously defined as the *Puritan ethos* or the *Protestant ethic* was a spur to this development. At many points in his classic study, moreover, Merton affirmed that his interest was in the rise of a new scientific movement, "a well defined social movement," which clearly had become prominent in the latter half of the seventeenth century.[12] His interest was in what appeared to be a "new fashionable" interest in, even "publicistic" promotion of, science.[13] So stated, the thesis is not objectionable, for Merton's own analysis of the shifts in occupational pursuits and intellectual orientations gave ample support to these aspects of his thesis.

It must also be said, however, that Merton's larger quarry was a thesis about the interconnections between the rise of a relatively new social activity, that is, the public, open, and enthusiastic pursuit of science as a socially valued activity, and the social and cultural values which supported that activity as an ongoing avocation. This part of the thesis concerns, on the one hand, the sources of the support for this newly enhanced evaluation of science and, on the other, the process of institutionalization whereby science became an institutionally autonomous enterprise. It is in the unfolding of this part of the thesis that Merton's analysis ran into some difficulties – difficulties caused by the failure of historians to grasp the logic of Merton's analysis and by certain ambiguities in Merton's own claims.

It should also be noted that the analysis of this problem of the rise of modern science entails some purely theoretical issues which rise or fall on the

---

[12] Robert K. Merton, *Science, Technology, and Society in Seventeenth-Century England*, reprint (New York, Harper & Row, [1938] 1970), pp. 43 and 27–8, 95–6 (hereafter cited as *STS*).

[13] Merton, *STS*, pp. 28 and 96; and cf. Gary Abraham, "Misunderstanding the Merton Thesis," *Isis* 74 (1983): 368–87, p. 372.

wings of our conceptual language. That is to say, theoretical problems have to be discovered before they can be solved. As Merton acknowledged, "I did not [then] recognize that theoretical problems in sociology as in other disciplines had to be *invented* before they could be *solved*."[14] In terms of sociological theory the rise of modern science is now clearly described as a problem of institutionalization, whereas Merton's own language of the late 1930s rarely uses this locution at all. There is one paragraph containing two references to science as having become institutionalized,[15] but Merton admits that this scarcely qualifies as a full assault on the problem of articulating the norms of science and the process of their institutionalization in the society at large. Instead of referring to the accomplished fact of the institutionalization of science or the process of institutionalizing science, Merton talks more frequently of "the genesis and development of" science,[16] the fact that "science had definitely been elevated to a place of high regard in the social system,"[17] and the fact that science and its pursuit were not "accredited and organized."[18] Indeed, when Merton published the thesis of his monograph as a separate paper in 1936 under the title "Puritanism, Pietism, and Science," it was through the use of the terminology of the American social theorist Talcott Parsons, that is, "value-integration," that he linked the values of the two domains of religion and science.[19] In Merton's new preface to the 1970 edition of *Science, Technology, and Society in Seventeenth-Century England*, however, it is clear not only that he was dealing with the interrelations between different social institutions (for example, religion, economy, and science), but that one main set of issues in his thesis was "the interrelations between Puritanism and the institutionalization of science."[20] The theoretical vocabulary of sociology is such, as Gary Abraham has pointed out, that even sociologists are prone to use terms inconsistently, and historians are even more inclined to use the term *institution* to refer to a single organization rather than to the much broader and more deeply rooted societal process.[21]

---

[14] R. K. Merton, "STS: Foreshadowings of an Evolving Research Program in the Sociology of Science," in *Puritanism and the Rise of Modern Science*, ed. I. B. Cohen et al. (New Brunswick, N.J.: Rutgers University Press, 1990), pp. 334–71, at p. 337.

[15] Merton, *STS*, p. 83.

[16] Ibid., p. xxxi.

[17] Ibid., p. 28.

[18] Ibid., p. 55.

[19] Reprinted in *Social Theory and Social Structure*, enlarged ed. (New York: Free Press, 1968), pp. 628–60, at pp. 641f. One might also note that in the index to this expanded edition of Merton's famous theory text, there is no entry for the idea of the institutionalization of science, though five of Merton's classic papers in the sociology of science are included in the volume, including the "ethos of science" paper.

[20] Merton, *STS*, p. xi.

[21] Abraham, "Misunderstanding the Merton Thesis," pp. 374f.

If one were to set out the idea of an *institution* in the sociological sense, the following would be implied:

First the patterns of behavior which are regulated by institutions ("institutionalized") deal with some perennial basic problems of any society. Second, institutions involve the regulation of behavior of individuals in society according to some definite, continuous, and organized patterns. Finally, these patterns involve a definite normative ordering and regulation; that is, regulation is upheld by norms and sanctions which are legitimized by these norms.[22]

From such a point of view, an institution in a strict sociological sense is not simply an organization but rather an institutional complex of patterned behavior that is generalized throughout a society. At an incipient stage of development a new set of values might be realized in only one organization, but if they do not transcend that organization to permeate the other institutions of society, such patterns of behavior are not expressions of the institutional foundations of the society. On the other hand, and what has often not been noticed, most social institutions rest on an implicit set of legal (and sometimes religious) authorizations, authorizations that grant legitimacy to rights of jurisdiction, ownership, representation, and of course communication. But neither sociologists nor historians have given much thought to these deeper issues, for they appear most obvious only in civilizational contexts, where one does not find all the legal assumptions characteristic of the West.

It was the contribution of Joseph Ben-David to fashion what he called an institutional approach to the sociology of science with an emphasis on the concept of a role (which is always embedded in a social institution). To say that some activity or, perhaps better, some social function has been institutionalized is to presuppose the following:

(1) the acceptance in a society of a certain activity as an important social function valued for its own sake; (2) the existence of norms that regulate conduct in the given field of activity in a manner consistent with the realization of its aims and autonomy from other activities; and finally (3) some adaptation of social norms in other fields of activity to the norms of the given activity. A social institution is an activity that has been so institutionalized.[23]

---

[22] S. N. Eisenstadt, "Social Institutions: The Concept," *International Encyclopedia of the Social Sciences* 14: 409–421, at p. 409a. It is interesting to compare these sociological assumptions about "institutions" with those of an economist; see Douglass C. North, *Institutions, Institutional Change, and Economic Performance* (New York: Cambridge University Press, 1990).

[23] Ben-David, *The Scientist's Role in Society* (Englewood Cliffs, N.J.: Prentice-Hall, 1971), p. 75.

As we saw in Chapter 1, for Ben-David the problem of the rise of modern science was reduced to the rise of a new social role (that of the scientist), by which he referred to "the pattern of behaviors, sentiments, and motives conceived by people as a unit of social interaction with a distinct function of its own and considered as appropriate in given situations."[24] But just as with the problem of the rise of modern science in Merton's framework, there is still in Ben-David's formulation a question of the rise of new social values and their institutionalization in the social order. "Therefore, the emergence of the scientific role was connected to changes in the normative patterns ('institutions') regulating cultural activities."[25] So formulated, the problem seems essentially the same as in Merton's much earlier formulation, except that Ben-David has radically foreshortened the role of values in science and sociocultural process. For Ben-David adds a positivistic note by asserting that "in the case of the scientific role, that change in values meant the acceptance of the search for truth through logic and experiment as a worthwhile intellectual pursuit."[26] As we saw earlier, this narrow formulation leaves out the greater part of those values that Merton later called "the ethos of science" as well as those Thomas Kuhn refers to as those "commitments without which no man is a scientist,"[27] elements evidently much broader than those delimited by Ben-David. Worse, Ben-David has eviscerated Merton's major thesis of the essential connection between the values of science per se and those of the surrounding culture. Without this connection, Merton went to great pains to argue, science as a persistent endeavor will falter. "The persistent development of physical science occurs," Merton said in 1938, "only in societies of a definite order, subject to a peculiar complex of tacit presuppositions and institutional constraints."[28] It is of "no small moment" to ascertain the origin, nature, and function of those "cultural values which underlie the large-scale pursuit of science,"[29] and without identifying those underlying values and their sources, one could hardly be credited with explaining the rise of modern science, much less its institutionalization. But perhaps most important of all, Merton placed some stress on the fact that "separate institutional spheres are only partially autonomous, not completely so" and that "it is only after a typically prolonged development that social institu-

---

[24] Ibid., p. 17.
[25] Ibid.
[26] Ibid.
[27] Kuhn, *The Structure of Scientific Revolutions*, 2d enlarged ed. (Chicago: University of Chicago Press, 1970), p. 42.
[28] Merton, *STS*, p. 225.
[29] Ibid., and p. xxxi.

tions, including the institutions of science, acquire a significant degree of autonomy."[30]

At this point we can see that the question of the institutionalization of science hinges on two fundamental issues: (1) what is the "it" (science in this case) that is being institutionalized; and (2) what are the appropriate indicators of the fact of institutionalization? There can be little doubt that Robert Merton over the course of his career recognized the many dimensions of this problem and attempted to deal with them. Still, it comes as a surprise that Merton's early thesis, as expressed in the monograph, did not employ the concept of the ethos of science much less focus attention on the institutionalization of that ethos in seventeenth-century England. There are some oblique anticipations of the idea of an ethos, as when Merton, discussing science as a social activity, affirms that the continuance of science "presupposes disinterestedness, integrity, and honesty of the scientists and is thus oriented toward moral norms."[31] The study of the rise and institutionalization of these norms, however, does not figure prominently in Merton's research agenda of the 1930s. Nor, we may add, did Joseph Ben-David pursue this course in his major work some thirty-three years later. In retrospect, it appears that although Merton did go forward in the conceptual elaboration of the many elements connoted by the term science, he did not attempt to bring his mature insights together to resolve the many issues at hand – probably because such an effort would have entailed an even larger comparative and historical inquiry such as the present one. Such a study would also have shown that many of Merton's attempted contrasts between the value commitments of medieval intellectuals and his seventeenth-century English intellectual elite were not as sharp as some of his remarks suggested.

Nevertheless, Merton's theoretical analysis did go forward so that he was able to articulate the ethos of science as a major ingredient of the scientific enterprise. This was done in his famous article originally titled "Science and Technology and the Democratic Order," published in 1942.[32] We have seen that in Merton's doctoral thesis he was concerned primarily with the social and cultural sources of the scientific movement, which was evidenced by the shifts in public and private attention to scientific thought and inquiry. But now, in 1942, Merton articulated the many other dimensions of science. The term science, he observed, is commonly used to denote the following: "(1) a set of characteristic methods by means of which knowledge is certified; (2) a stock of accumulated knowledge stemming from the application of these

---

30 Ibid., p. x.
31 Ibid., p. 225.
32 Reprinted in *The Sociology of Science: Theoretical and Empirical Investigations* (Chicago: University of Chicago Press, 1973), as chap. 13, "The Normative Structure of Science."

methods; (3) a set of cultural values and mores governing the activities termed scientific; or (4) any combination of these."[33] From a sociological point of view, attention is rightly directed toward the cultural structure of science, the mores with which the methodological canons are hedged about. In a word, the "methodological canons are often both technical expedients and moral compulsives," but they are hedged about, or institutionalized, in a larger cultural context that gives them validity.[34] Here then we come to the ethos of science, those "affectively tinged" norms that are "expressed in the form of prescriptions, proscriptions, preferences, and permissions."[35] These, as we have seen, are the norms of universalism, communalism, disinterestedness, and organized skepticism. If these norms were taken to be the *sociological* core of the scientific enterprise, then it would be a major endeavor to trace out their historical sources and rise to prominence, but this was not attempted by Merton, not least of all because he had not – in the 1930s – worked out a conception of science as a social institution.

It must be said, nevertheless, that Merton did attempt "to find the specific sources of this newly expressed vitality in science" in the cultural values of the Puritan ambience. And in the process of making his points as sharply and clearly as possible, he slipped into the habit of drawing strong contrasts between the value commitments of the Puritan-inspired enthusiasts of science in seventeenth-century England and their seemingly reactionary medieval counterparts. This took the form of suggesting that the leading intellects of the pre-Reformation were inclined to mysticism, that they believed that scientific inquiry took the life out of reality and that they failed to answer "the ultimate questions," leading on balance to a retardation of science.[36] Indeed, in some passages Merton implies that the enthusiasm for scientific discovery and innovation "would have been unthinkable in the medieval period, save as referring, at best, to the intellectual amalgam of science and theology presented by an Aquinas."[37] High-ranking officials are cited to the effect that "all the worldly sciences are absurdities and fooleries."[38] Merton further suggested that, in the medieval worldview, "to regard with high esteem scientific discoveries attained empirically and without reference to Scriptural or other sacred authority would have been almost as heretical as making the discoveries themselves."[39] To be sure, Merton did not paint the

---

[33] Ibid., p. 268.
[34] Ibid., pp. 268–9.
[35] Ibid., p. 269.
[36] Merton, *STS*, p. 74.
[37] Ibid., p. 76.
[38] Ibid., p. 77, referring to Peter Damian, chancellor to Pope Gregory VII.
[39] Ibid., pp. 76–7.

period all black, and he fully acknowledged that questions of internal history must be taken into account, such as the fact that "a fixed order must prevail in the appearance of scientific discoveries; each discovery must await certain prerequisite developments."[40] Still, the message was clear: the medieval, pre-Reformation period held little stock in the virtues of science and the progress of knowledge, but, in the seventeenth century, "the social values inherent in the Puritan ethos were such as to lead to an approbation of science because of a basically utilitarian orientation, couched in religious terms furthered by religious authority."[41] And, if this were true, then indeed it was some amalgam of Puritanism and utilitarianism from whence the positive evaluation of science arose and became quite possible in the seventeenth century.

There is an implication throughout *Science, Technology, and Society* that the universities were backwaters of intellectual life and at best served to retard the development and growth of science. Here again Merton's account is a moderated version of a far more radical view put forward by the English reformers themselves, who exaggerated the nature of the new learning, the new experimental philosophy, the new astronomy, and so forth.[42] Merton's temperate view is simply that "the universities remained largely outside the stream of scientific development during this period."[43] And that there was "slow development of the sciences in the universities during this period."[44] Consequently, while there were internal shifts in the universities, shifts resulting from the establishment of new chairs in mathematics and astronomy, at best the role of the universities in the rise of modern science was reluctant, if not reactionary. As further evidence of this recalcitrant mentality of the universities, Merton tells us that "until about 1630 official university statutes declare that Bachelors and Masters of Arts who do not faithfully follow Aristotle were liable to a fine of five shillings for every point of divergence, or for every fault committed against the *Organon*."[45] That the universities might be the principal center of scientific learning was clearly not within Merton's ken and, of course, not within Ben-David's either. Surely what is needed here is a reappraisal of the continuity as well as the discontinuity of intellectual thought and institution building between the Middle Ages and the seventeenth century.

[40] Ibid.
[41] Ibid., p. 79.
[42] See John Gascoigne, "A Reappraisal of the Role of the Universities in the Scientific Revolution," in *Reappraisals of the Scientific Revolution*, pp. 207–60 at p. 20f.
[43] Merton, *STS*, p. 28f.
[44] Ibid., p. 31.
[45] Ibid., p. 229.

Science, learning, and the medieval revolution

As I have pointed out in earlier discussions, the medieval West experienced a profound social, intellectual, and legal revolution that dramatically altered the nature of social relationships (see the conclusion to Chapter 8). The legal revolution created a variety of new forms of social relatedness, group and social agency, and new domains of political and intellectual autonomy. From the point of view of the rise of early modern science, the most significant event was the legal breakthrough that allowed the formation of autonomous institutions of higher learning, namely, the studium generale and the university.

As the universities developed their curricula in the twelfth and thirteenth centuries, they moved increasingly toward a core of readings and lectures that were basically scientific. The most emblematic sign of this scientific orientation was the existence of the naturalistic writings of Aristotle at the heart of the curriculum. As we have seen, these books included Aristotle's *Physics, Meteorology, On Generation and Corruption, On the Soul, The Small Works on Natural Things,* and others.[46] Anyone who reads these works or compares them with the philosophical writings of China cannot fail to see the uniqueness of the Aristotelian emphasis on explaining the natural world in terms of fundamental elements, causal processes, and rational inquiry. It was this educational agenda that formed the core of the arts curriculum through which all students passed on their way to study in the three higher faculties of the universities, that is, law, theology, and medicine.[47] And it was this intellectual organization of the universities into four faculties (arts, law, theology, and medicine) that was still in place at the time of Copernicus, Galileo, and Kepler.

In carrying out all of these reforms, the European medievals, fully cognizant or not, created autonomous, self-governing institutions of high learning and then they imported into them a methodologically powerful and metaphysically rich cosmology that directly challenged and contradicted many aspects of the traditional Christian worldview. Instead of holding these foreign sciences at arm's length, they made them an integral part of the official and public discourse of higher learning. By importing, indeed, ingesting the corpus of the new Aristotle and its rigorous methods of argumentation and inquiry, the intellectual elite of medieval Europe established an impersonal

---

[46] The statute incorporating these works in the curriculum (and the schedule for their reading) of the University of Paris is translated in Grant, *A Source Book in Medieval Science* (Cambridge, Mass.: Harvard University Press, 1974), pp. 43f.

[47] See Pearl Kibre and Nancy Siraisi, "The Institutional Setting: The Universities," in *Science in the Middle Ages,* especially pp. 126ff.

intellectual agenda whose ultimate purpose was to describe and explain the world in its entirety in terms of causal processes and mechanisms. This disinterested agenda was no longer a private, personal, or idiosyncratic pre-occupation, but a shared set of texts, questions, commentaries, and in some cases, centuries-old expositions of unsolved physical and metaphysical questions that set the highest standards of intellectual inquiry. By incorporating the natural books of Aristotle in the curriculum of the medieval universities, a disinterested agenda of naturalistic inquiry had been institutionalized. It was institutionalized as a curriculum, a course of study, in which logic and the exact sciences, especially at Paris and Oxford, took pride of place. It was asserted by all that logic should be the first of the seven liberal arts, since it (in the words of Hugh of St. Victor) "provides ways for distinguishing between modes of argument and the trains of reasoning themselves . . . it teaches the nature of words and concepts, without both of which no treatise of philosophy can be explained rationally."[48]

As a set of intellectual puzzles, this new corpus was a research agenda for the elite of the university. This is evident in Aristotle's *Physics* where he enunciates the naturalistic framework and makes it evident that the highest form of knowledge is based on "principles, causes, or elements" and that it is through acquaintance with these that knowledge and understanding are attained. "For we do not think that we know a thing until we are acquainted with its primary causes or first principles," Aristotle wrote, "and have carried our analysis as far as its elements. Plainly . . . in the science of nature too our first task will be to try to determine what relates to its principles."[49] It was just such principles, moreover, that moved Galileo four hundred years later when he wrote in his *First Letter on Sunspots* (1612) that he wanted to solve "the greatest and most admirable problem there is, the true constitution of the universe. For such a constitution exists, and exists in only one, true, real way, that could not be otherwise."[50]

From this point of view, the organized skepticism that we associate with modern, present-day views of things has a long history in the West, and it begins not later than the biblical criticism of the twelfth and thirteenth centuries when the superiority of rational demonstration to biblical literalism was asserted by the *moderni* of the schools and universities. As Tina Stiefel put it, "the most daring of all intellectual enterprises" of this time was

---

[48] As cited in ibid., p. 127.
[49] *The Complete Works of Aristotle*, rev. Oxford trans., ed. Jonathan Barnes (Princeton, N.J.: Princeton University Press, 1984) 1:315.
[50] As cited in A. C. Crombie, "Sources of Galileo's Early Natural Philosophy," in *Reason, Experiment, and Mysticism in the Scientific Revolution*, ed. R. Bonelli and William Shea (New York: Science History Publications, 1975), pp. 157–74 at p. 158.

the work of a few scholars, including William of Conches, Thierry of Chartres, and Adelard of Bath, who set about establishing the methodological foundations of a new natural science. One foundation for this mode of procedure is to be found in the European medieval belief that man is a rational creature, one possessed of reason and conscience, and by virtue of these capacities is capable of understanding and deciphering the secrets of nature, with or without the aid of Scripture.[51] Similarly, the medieval Europeans frequently deployed the metaphors of the "world machine" (*machina mundi*) and the "Book of Nature,"[52] two devices giving pattern and intelligibility to the study of nature. Both ideas were integral to the teachings of the medievals (as in the writings of Grosseteste and Sacrobosco), and this shows again how deeply the metaphysical and religious roots of scientific culture are imbedded in the history of the West.

When all of these elements were finally assimilated into the discourse of the universities by the end of the thirteenth century, along with the formal elements of the Aristotelian corpus, a powerful, methodologically sophisticated, intellectual framework for the study of nature had been institutionalized. They had been made part of the normative pattern of higher learning by being made a regular part of university instruction, frequently with regularly scheduled times of reading and discussion.

The new Aristotle presented a most formidable challenge to Christian theology, and yet it was incorporated into the new curriculum. The new Aristotle presented a new and imposing edifice of scientific and secular learning and it was accompanied by Arabic commentaries. Together these two rational currents – the Platonism of the twelfth century and the new Aristotle – constituted the foundation of an unending disinterested agenda of inquiry ensconced in the institutions of higher learning in Europe. As Professor Grant puts it, "For the first time in the history of Latin Christendom, a comprehensive body of secular learning, rich in metaphysics, methodology, and reasoned argumentation, posed a threat to theology and its traditional interpretation."[53] It should also be noted that the philosophical

---

[51] Tina Stiefel, " 'Impious Men': Twelfth-Century Attempts to Apply Dialectic to the World of Nature," in *Science and Technology in Medieval Society,* ed. Pamela Long (New York: New York Academy of Sciences, 1985), pp. 187–8. Also see Stiefel, *The Intellectual Revolution in Twelfth-Century Europe* (New York: St. Martin's Press, 1985), especially chaps. 1–3.

[52] Nelson, *On the Roads to Modernity,* chap. 9. Also see Lynn White, Jr., *Machina Ex Deo* (Cambridge, Mass.: MIT Press, 1968), pp. 100–1; and White, *Medieval Technology and Social Change* (New York: Oxford University Press, 1962), pp. 125 and 174 n5.

[53] Edward Grant, "Science and Theology in the Middle Ages," in *God and Nature: Historical Essays on the Encounter Between Christianity and Science,* ed. David C. Lindberg and Ronald L. Numbers (Berkeley and Los Angeles: University of California Press, 1986), pp. 49–75 at p. 52.

justifications of the natural study of the world (whether Platonist or Aristotelian) were far more sophisticated and powerful than their counterparts to be found in thirteenth-century China where the neo-Confucians spoke rather simply of "the investigation of the nature of things."[54] For not only did the Chinese inquiry pertain mainly to the human and moral domain, but Chinese philosophy lacked the rigorous logic of proof to be found in Aristotle as well as Euclidean mathematical proof. So well recognized was this subversive dimension of Greek philosophy, moreover, the Arabs kept it out of the colleges and sequestered in private homes and carefully controlled intimate discussion groups.

In the West, the adoption of this whole metaphysics created an intellectual space within which men could entertain all sorts of questions about the constitution of the world. By establishing a public examination system within the university, controlled by the scholars as a faculty, the Europeans broke from the Arabic-Islamic tradition of learning from masters who validated the learning of a single text. The Europeans vested intellectual authority for dispute resolution in the collective wisdom of scholars, not a state-sponsored bureaucracy as in China.

By creating public forums for lectures – that is, both legitimately authorized classroom discussion and public lectures – the university system throughout Europe took a step toward allowing and encouraging universal participation in scholarly discussion. Although the oral examination is perhaps not as objective and impersonal as the written exam, European scholars did take a major step toward establishing impersonal standards within the university by giving to a body of scholars the authority to collectively examine all new scholars. Likewise, as a mechanism of peer review, the process of public declamation is not as efficient or impartial as the mechanism of the written communication, the book and the scientific journal. These additions to the institutional structure of science had to wait for the arrival of the printing press in the mid-fifteenth century. Once that technology arrived, Europeans, as opposed to their counterparts in the Middle East and China, quickly pushed the new technology into the service of scientific publication.[55] Such a move clearly represents a major step toward the imper-

---

54 See Needham, *SCC* 2: 455ff, and Chan Wing-tsit, *Chu Hsi: Life and Thought* (New York: St. Martin's Press, 1987), pp. 136–7, 44.

55 For the importance of the printing press in the advance of scientific discourse, see Elizabeth Eisenstein, *The Printing Press as an Agent of Change* (New York: Cambridge University Press, 1979), vol. 2, chap. 6, "Technical Literature Goes to Press: Some New Trends in Scientific Writing and Research." Of course, it goes without saying in the present context that the effects of the printing press are almost wholly dependent on the cultural ambience in which they are located, for neither the Chinese nor the Arabs allowed the free use of the printing press, nor did the appearance of movable type in China lead to a cultural renaissance

sonal review process of evaluating philosophical and scientific arguments
and suggests that the social activity of advancing scientific knowledge was
well underway before the seventeenth century.

Furthermore, we know that the allure of disinterested inquiry quickly
propelled scholars into conflict with the vested interests of the traditional
religionists. Efforts were made to condemn certain arguments and assumptions that appeared to place limits on God's powers. Most notably, this took
the form of the condemnation of 219 suspect ideas by the Bishop of Paris in
1277. The condemnation, however, did not put an end to philosophical
inquiry but spurred philosophers to conduct a great variety of thought
experiments, to imagine the impossible in the service of reconciling Aristotelian thought with Christian theology. This meant imagining non-
Aristotelian possibilities, with the result that such speculations paved the
way for the overthrow of the Aristotelian worldview in the sixteenth and
seventeenth centuries.[56] In effect, the philosophers of the universities asserted their right to continue their inquiries on a number of grounds, not
least of all, those of pursuing truth for its own sake. Legitimation for this
kind of inquiry was to be found not only in Aristotle and his commentators
but in the Bible itself. For it was written, "And ye shall know the truth, and
the truth shall make you free."[57] All in all, there were multiple sources
(philosophical and religious) in Europe that served to create a new foundation for the study of the natural world and for challenging the Scriptures as
the sole authority regarding the world.

In brief, this probing philosophical and scientific curriculum based on the
work of Aristotle was the main thrust of European universities for more
than four hundred years, from 1200 to 1650. During that period all those
who trained for the master of arts studied this curriculum, and it served
thereby to shape and instill a major commitment to the norms of disinterestedness and organized skepticism that are at the heart of modern science. In the seventeenth century, there was a dramatic shift toward the

---

and innovation such as occurred in Europe. Among others, also see A. Demeerseman, "Un
étape décisive de la culture et de la psychologie sociale islamique: Les données de la controverse autour du problème de l'Imprimerie," *Institut des Belles Lettres Arabes* 16 (1953): 347–
89, 17 (1954): 1–48 and 113–40; and the article on "*matba'a*" [printing] in *EI*² 6: 779–803; as
well as Needham and Tsien Tsuen-hsuin, *SCC* 5/1.

[56] See Grant, "Science and Theology," pp. 55–9; and "The Condemnation of 1277, God's
Absolute Power, and Physical Thought in the Late Middle Ages," *Viator* 10 (1979): 211–44,
among others. While this thesis is consistent with Duhem's insistence on the importance of
the medieval precursors of Galileo and modern science, it does not agree with Pierre Duhem's assigning of 1277 as the "birthday" of modern science. For further discussion of Pierre
Duhem's claim that the Condemnation of 1277 marks the birth of modern science, see
Nelson, *On the Roads to Modernity*, pp. 126–8.

[57] John 8:32.

greater use of empirical (as opposed to logical and mathematical) techniques of research, but the naturalistic agenda with its highly developed forms of argumentation had long been established.

On the other hand, in the madrasas of the Islamic world there was nothing like a standard curriculum, because there was no faculty. Since the pattern of learning and certification was based solely on the individual scholar and the students' erratic choices of the masters they would follow, there was no coherent curriculum. The study of any particular subject was largely the result of happenstance and the personal preference of the scholarly master. To grasp the nature of the rejection of the philosophical and scientific path of learning introduced into the European universities, one has only to read al-Ghazali's vociferous efforts to reject the deterministic arguments of the philosophers, that is, the Greek and Greek-trained Muslim Aristotelians. Clearly Greek natural philosophy was widely recognized in the Islamic world as a metaphysical system opposed to the indeterminism of the Islamic worldview. Moreover, in the Islamic world the great center of research with its observatory at Marâgha was completely gone from the scene by the first decade of the fourteenth century. Even its buildings were soon to disappear without a trace. With no further communication from it after 1304–5, Marâgha had a short life span of forty-five to fifty-five years.[58] It was thus but another ephemeral attempt to institutionalize naturalistic inquiry in Islamic culture. Other observatories were indeed built, but none of them had anything like the intellectual significance or longevity of Marâgha, short as its life was. Sadly, no educational institutions were established in the Islamic world that were dedicated to the pursuit of naturalistic knowledge until the twentieth century.

In the case of China the national focus of education was on the moral and humanistic classics of China's ancient past. These included nothing that could be called scientific. Studying for the official examinations based on this material was analogous to studying and memorizing all the books of the Bible along with officially sanctioned commentaries. For the purpose of recruiting mathematicians and astronomers into the state bureaucracy, however, special examinations were held periodically, but this system did not lead to the institution of a special curriculum or autonomous specialized academies. Those who would qualify for these examinations would be drawn either from those who belonged to families with traditions of special access to training in state-sponsored scientific research or from those who, as lesser officials of the state, had rare access to the mathematicians and astronomers of the imperial service.

[58] Aydin Sayili, *The Observatory in Islam* (Ankara: The Turkish Historical Society Series 7, no. 38, 1960), p. 213.

It is paradoxical, therefore, that sociologists and even many historians of science have neglected the central role that the universities of the West played in the rise of modern science. For a dispassionate examination of the educational backgrounds of major scientists from the fifteenth to the seventeenth century shows that the vast majority of them were in fact university educated. As John Gascoigne has shown, "Something like 87% of the European scientists born between 1450 and 1650 [who were] thought worthy of inclusion in the *Dictionary of Scientific Biography* were university educated."[59] More importantly, "A large proportion of this group was not only university educated but held career posts in a university." For the period 1450–1650 this was 45 percent, and for 1450–1550, it was 51 percent.[60] If one speaks of particular scientists, then one must immediately acknowledge that Copernicus, Galileo, Tycho Brahe, Kepler, and Newton were all extraordinary products of the apparently procrustean and allegedly scholastic universities of Europe.[61] In short, sociological and historical accounts of the role of the university as an institutional locus for science and as an incubator of scientific thought and argument have been vastly understated. While the university has always been reluctant to give up its (defective) assumptions, as more recent appraisals of the trajectory of intellectual discourse in the universities have shown, the universities were highly instrumental in disseminating many new intellectual currents in scientific thought, and, most important of all, they were the primary locations of severe criticism of both old and new ideas.[62] It remains the case that it was only in the West that the scientific revolution took place, and the existence of the university with its uniquely scientific and philosophical curriculum made a major contribution to that outcome.

The revolution in authority and astronomy

If we return now to the advent of the Copernican system of the universe, it is evident that the struggle over this new system was more than a narrow scientific dispute and that a successful adoption of the system – whether in its initial form as stated in *De revolutionibus* or in the corrected form provided by Kepler's discoveries – required a major civilizational debate. In the

[59] Gascoigne, "A Reappraisal," p. 208; and his table 5.1.
[60] Ibid.
[61] For more on this, see Gascoigne, "A Reappraisal." That Kepler worked outside the university was largely a matter of his personal choice and the possibility of even greater freedom under royal patronage. But that does not negate the sociological fact that scientific education since the medieval period had been and still was dominated by university faculties.
[62] See Charles Schmitt, "Toward a Reassessment of Renaissance Aristotelianism," *History of Science* 11 (1973): 159–93.

final analysis, as Benjamin Nelson put it, "the fundamental issue at stake in the struggle over the Copernican hypothesis was not whether or not the particular theory had or had not been established, but whether in the last analysis the decision regarding truth or certitude could be claimed by anyone who was not an officially authorized interpreter of revelation."[63] In somewhat more technical terms, the central issue raised by the new Copernican world system was the right of "the mathematical astronomer to make claims in natural philosophy."[64]

According to the traditional division of the sciences that prevailed in the universities, mathematical astronomy was meant to be subordinate to physics, that is, to natural philosophy. While natural philosophers had the right to speak about natural reality and its workings, the mathematical astronomer could only devise predictive calculating devices that would describe the position and motion of the heavenly bodies. The mathematical astronomer was not otherwise capable of offering true descriptions of the universe. Since the time of Eudoxus (ca. 400–ca. 350 B.C.) the planetary system had been supposed to be embedded in a set of spheres that carried the heavenly bodies in perfect circular motions around the center of the universe, which was thought to be occupied by the earth. Though Ptolemy's *Almagest* did not give his planetary models such a coherent system, his collection of models was generally placed within the geocentric conception with the further authority of Aristotelian assumptions.

As we saw earlier, the medieval social revolution was accompanied by a revolution in educational thought and organization signified by the rise of the universities. Alongside the new Aristotle in the curriculum there also emerged in the twelfth and thirteenth centuries a body of scientific knowledge that has been called the *corpus astronomicus*.[65] This corpus included standard texts, scientific instruments, and collections of data, that is, tables of astronomical observations, which allowed the determination of local time as well as the prediction of astronomical events such as eclipses and conjunctions of planetary bodies. Among the most important scientific instruments introduced into the West about this time were the astrolabe – a hand-held observational device that allowed the determination of time, day or night (see Figure 11) – the abacus, and the armillary sphere. There were also other astronomical instruments fashioned by Europeans during the medieval pe-

---

[63] Nelson, "The Early Modern Revolution," in *On the Roads to Modernity*, p. 133.

[64] Robert S. Westman, "The Astronomer's Role in the Sixteenth Century: A Preliminary Study," *History of Science* 18 (1980): 105–47 at p. 126.

[65] Olaf Pedersen, "Astronomy," in *Science in the Middle Ages*, pp. 315ff; and John North, "The Medieval Background to Copernicus," in *Vistas in Astronomy*, vol. 17, ed. Arthur Beer and K. Aa. Strand (New York: Pergamon Press, 1975), pp. 3–16, especially 8ff.

Figure 11. Although the astrolabe appears to be of Greek origin, the Arabs in the medieval period perfected its design and use. A flexible instrument that had many purposes in astronomy and surveying, it was also and most importantly used for determining local time at day or night. The first reported astronomical use of the astrolabe in the West was in October 1092. The astrolabe depicted here was made in Seville, Spain, in 1222/3 by Muhammad b. Fattuh al-Khamai'ri. Later it was transported to northern Europe, where it was fitted with a Latinate rete, which appears to be of sixteenth-century Flemish design. (Photograph courtesy of the Time Museum, Rockford, Ill. Catalog no. 3407.)

riod.[66] It is surely symptomatic of the deep interest in astronomy and naturalistic inquiry among medieval Christian scholars that these instruments were introduced as teaching aids by Gerbert of Aurillac (ca. 945–1003), a man who was later to become Pope Sylvester II.[67] It should also be noted that while Hindu-Arabic numerals became available to Europeans (in Spain) in the tenth century (ca. 960), it was not until the thirteenth and fourteenth centuries that the new "arithmetic mentality" emerged.[68] So strange was this new system of counting and computation, moreover, that the Europeans created literally dozens of sets of Arabic numerals as they embarked upon the process of adopting them as a universal system of numeration.[69] By 1200, the system, with clear examples, had been explained in a number of handbooks for commercial and other uses of it.

In brief, between the eleventh and fourteenth centuries, a new set of universal mathematical symbols and a corpus of manuals, texts, and other documents were assembled in the West for the purpose of university instruction in astronomy. This assemblage included instructional texts on mathematics, geometry, planetary theory, and the practical art of devising calendars. Cosmology per se belonged to the discipline of natural philosophy rather than mathematical astronomy, and it was thus taught independently.[70]

Because the *Almagest* of Ptolemy was the most sophisticated astronomical treatise in existence at that time, and because it was only introduced to the West in Latin translation in 1160 and 1175, it was too advanced for instructional purposes. To bridge this gap, European scholars developed their own textbooks, which provided a nontechnical and more accessible path to the most difficult problems in mathematical astronomy. By far the most popular of such works were three treatises by Sacrobosco (d. ca. 1256), an Englishman who taught at Paris from about 1230 to about 1256. The first of these was his manual on arithmetic, which borrowed extensively from the ninth-century Arab mathematician al-Khwarizmi.[71] The second work in this compilation was a work called *On the Sphere*, which provided a nonmathematical introduction to the elements of astronomy.[72] This work on the sphere continued to be enormously popular down to the time of Galileo. It was

---

[66] See North, "The Medieval Background," pp. 9–10.

[67] Ibid., p. 309. And See David C. Lindberg, "The Transmission of Greek and Arabic Learning to the West," in *Science in the Middle Ages*, pp. 52–90, especially 60–1. For more on Gerbert's role in the transmission process, see Alexander Murray, *Reason and Society in the Middle Ages* (Oxford: At the Clarendon Press, 1978), pp. 163ff.

[68] Alexander Murray, *Reason and Society*, chap. 7.

[69] Ibid., p. 168.

[70] Edward Grant, "Cosmology," in *Science in the Middle Ages*, pp. 264–302.

[71] For a selection from this treatise, see Grant, *A Source Book*, pp. 94–101.

[72] Ibid., pp. 442–51.

among the earliest scientific books published by Erhalt Ratdolt following the prospectus of Regiomontanus with the aid of the new print technology (in 1482 and 1485).[73] The third part of this instructional material was another treatise by Sacrobosco, in this case, devoted to the art of reckoning time.[74] This was supplemented by a work by Robert Grosseteste called *Calendar.*[75]

In addition to these classroom materials, the medieval astronomers also worked with large sets of astronomical observations compiled into tables. These tables, usually derived from the Arab zij tables, were occasionally revised, as they were under King Alphonse X of Spain in the last quarter of the thirteenth century. This set of tables is commonly referred to as the Alphonsine Tables, though they were extensively revised by John of Saxony about 1325 and were still in use in Copernicus's time.

While these tables and works provided an essential foundation for astronomical training, they lacked the theoretical foundation for mathematical astronomy that could only be provided by the *Almagest* or some superior work, which was not to appear until the time of Copernicus. To bridge this gap between routine instruction and the most advanced aspects of mathematical astronomy, a work was introduced called *The Theory of the Planets.*[76] It was this work that became the standard treatise on planetary theory from the early fourteenth century through the sixteenth. Although the author of this work was unknown (it is sometimes ascribed to Gerard of Cremona), it is evident that many of the arguments adduced in the work were derived from Ptolemy and the *Almagest.* It was a work admirably suited to its purpose and was far more popular than the *Almagest* itself, and in the view of many historians of science it had a justly deserved popularity.[77] Accordingly, it was this work and not the *Almagest* that was most well known to astronomers right up to the time of Copernicus. It has been suggested, furthermore, that when Copernicus wrote the first published account of his heliocentric theory (before 1514, and perhaps as early as 1511–3),[78] he did not have access to the *Almagest* but possibly to various epitomes of it, such as Georg Peuerbach's *Theoricae novae planetarum* (1454) and the *Epitome of the Almagest* of Peuerbach and Regiomontanus (1496).[79]

---

[73] Owen Gingerich, "Copernicus and the Impact of Printing," in *Copernicus Yesterday and Today, Vistas in Astronomy,* vol. 17, p. 203.
[74] Pedersen, "Astronomy," p. 315.
[75] Ibid.
[76] See Grant, *A Source Book,* pp. 451–65; and Pedersen, "Astronomy," pp. 316ff.
[77] Pedersen, p. 316.
[78] The second date is that suggested by Edward Rosen, *Three Copernican Treatises,* 3d ed. (New York, Octagon Book, 1971), p. 345.
[79] Noel Swerdlow, "The Derivation and First Draft of Copernicus's Planetary Theory: A

In short, the European medievals introduced a continuous tradition of astronomical teaching and research into the universities of the West. This discipline was of course stimulated by new translations of important astronomical and mathematical works from Greek and Arabic sources, and it thus built on old foundations. Unlike in the Arabic-Islamic world, however, this new technical learning was built into the main centers of higher learning, namely, the universities. Hence students from the thirteenth century onward "were taught spherical astronomy and planetary theory; they were provided with calendars and tables which enabled them to calculate the positions of the heavenly bodies and to predict particular phenomena, such as conjunctions and eclipses; and they were taught how to construct instruments for both observation and computation."[80] The study of astronomy had been institutionalized in the universities of Europe.

At the philosophical center of this study of astronomy, there remained a fundamental problem that was both an interdisciplinary debate and a genuine scientific problem regarding the constitution of the universe. It was also a theological problem. It was created by the contrasting assumptions of Aristotelian cosmology and physics and those of Ptolemaic mathematical astronomy. As we noted earlier, the Aristotelian universe centered on the earth, above which there were the regions of water, air, and fire, followed by the spheres of the moon and the superior planets (Figure 12). The superior planets were each attached to a giant sphere that moved (as required by Aristotle's physics) in uniform circular motion (and in the required directions) about the center of the universe, which was the earth.

Mathematical astronomers, on the other hand, were concerned with the task of predicting the motions of the heavenly bodies using whatever mathematical devices they could devise. These included representational devices that clearly did not conform to uniform and circular motion based on concentric circles. Instead they involved eccentric circles and circles within a circle (or epicycles). It was apparent (as all the great Arab astronomers had painfully noted) that Ptolemaic astronomy violated the major tenets of Aristotelian physics and, therefore, it could not be a true representation of the universe. It was happily the case, however, that to the degree that astronomical predictions worked at all, they did not depend upon either Aristotelian cosmology or the cosmological implications of the calculating devices of Ptolemaic astronomy. In other words, these were merely calculating de-

Translation of the Commentariolus with Commentary," *Proceedings of the American Philosophical Society* 117 (1973): 425f.
[80] Pedersen, "Astronomy," p. 320.

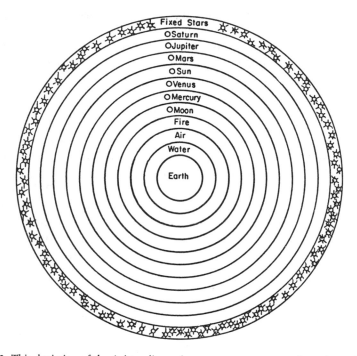

Figure 12. This depiction of the Aristotelian universe was very common throughout the medieval period. According to Aristotelian cosmology, there was a qualitative distinction between the heavenly spheres above the moon and the domain below. Only in the sublunar domain could one find the elements of earth, water, air, and fire.

vices.[81] There were calendrical problems, of course, as well as inexplicable observations of the superior planets and wandering stars, but these were just unsolved problems. The times of day could be calculated with the use of an astrolabe and standard tables, and so the mathematical assumptions and devices of the astronomers were useful devices. Aristotelian physics reigned supreme, and astronomers did their work but could not make claims about the actual shape and constitution of the world – that was the task of physicists (or, better, natural philosophers).

It is this context (among others) in which the great struggle over the Copernican hypothesis ought to be seen. For the question at hand was, as

[81] Specialists will recognize here a great debate carried on in philosophy and the history of science regarding the Duhem thesis. It does not serve our purpose to enter the technical literature of this debate, but see Duhem, *To Save the Phenomena*, trans. E. Doland and C. Maschler (Chicago: University of Chicago Press, 1969).

we have seen, twofold: (1) could anyone outside Christian theology speak authoritatively regarding the constitution of the world, and (2) could astronomers in particular make claims about the physical arrangement of the universe? These were the mighty issues over which the great debate raged for a little over a century.

It was just this intellectual situation that Copernicus inherited when he entered the scene at the University of Cracow in 1491, and a little over a hundred years after the death of Ibn al-Shatir. Copernicus is thus a prime example of a scholar trained in the European university system, having first studied at the University of Cracow and then at Bologna and Padua. When in Bologna he studied law, but while enrolled in Padua he studied medicine before earning a doctorate in law at Ferrara in 1503.[82] At Bologna he also served as assistant to the astronomer Dominico Maria da Novara as the two of them made astronomical observations.[83] Although Copernicus later studied astronomy on his own and in his spare time, there is strong indication that Copernicus's interest in astronomy and its problems goes back to his early studies in the liberal arts at Cracow. For Cracow was clearly an important center of astronomical education during the fifteenth century, though probably not the international center suggested by some.[84] Likewise, the man who is most credited with the establishment of the Copernican system as a mathematical system centered on the sun, namely, Johannes Kepler, was given his most valuable lessons in Copernican astronomy at the University of Tübingen under the tutelage of Michael Maestlin (1550–1631). In the introduction to *Mysterium cosmographicum*, Kepler reported that "on the basis of Maestlin's lectures and his own reflections, he gradually compiled a list of superiorities of Copernicus over Ptolemy from the mathematical point of view."[85] Although the principal thesis of the *Mysterium* was mistaken, the idea that astronomers should seek actual physical explanations of the universe (and with the sun as the center) was a clear injunction for others and led Kepler himself to the discovery of his three laws of planetary motion.

---

[82] Edward Rosen, "Copernicus," *DSB* 3: 401ff.

[83] See Rheticus, "Narratio Primus," in Rosen, *Three Copernican Treatises*, p. 111.

[84] See Paul W. Knoll, "The Arts Faculty at the University of Cracow at the End of the Fifteenth Century," in *The Copernican Achievement*, ed. Robert S. Westman (Berkeley and Los Angeles: University of California Press, 1975), pp. 137–56. Knoll cites the claim of L. Birkenmaier that Cracow "emerged as the international center of astronomical education at the end of the middle ages" (p. 146), but this does not appear to have been established in the eyes of historians of science.

[85] Eric J. Aiton, introduction to Kepler, *Mysterium cosmographicum: The Secret of the Universe*, trans. Alistair M. Duncan (New York: Abaris Books, 1981), p. 17, as cited in Edward Rosen, "Kepler's Early Writings," *Journal for the History of Ideas* 46 (1985): 450.

There is no denying the fact that the scientific revolution centered in astronomy was a product of the unique configuration of the Western university and its course of study. As Paul Knoll strikingly put it, the early "medieval universities were not places; they were states of being, paper corporations whose reality rested not in some physical locus, but in privileges, in disputations, in persons."[86] Accordingly, the revolution in astronomy and authority was the outcome of the research and scholarly debate of individuals trained in that tradition, thrashing out the greatest metaphysical problems of that time. Conversely, there was no equivalent institutional setting in which such an open, public debate could occur in either Islam or China.

Although it has traditionally been said that the adoption of the Copernican hypothesis was slow and gradual, it can also be said that within a decade or so of its publication, Copernicus's great work was being read and studied widely throughout Europe. As Professor Owen Gingerich's efforts to conduct a census of the first and second editions of *De revolutionibus* show, the volume was often read with pencil in hand.[87] Furthermore, in at least some German universities in the 1570s, at Wittenberg University, for example, students taking the natural sciences were exposed to Copernicus's ideas, while a master's candidate at the same university "would not only encounter Copernican planetary models and values in his textbooks but he would be openly encouraged to read *De revolutionibus*."[88]

When the central issues for the debate became truly focused, they concerned, on the one hand, theological questions about the nature of the world as seen through the Scriptures and, on the other, a disciplinary debate about whether or not mathematical astronomers could speak to physical questions about the constitution of the universe and the nature of physical reality. Perhaps it was Kepler, writing as a philosopher of science, rather than the combative but theologically informed Galileo, who most clearly articulated the central issue insofar as it concerned the role of the astronomer. In his posthumous *Apologia*, a defense of Tycho Brahe's originality against the claims of Nicolas Ursus, Kepler challenged Ursus. According to Ursus (who, like Brahe, was one of Kepler's predecessors as imperial mathematician in Prague), hypotheses are mere fabrications for aiding the prediction of planetary movements. The duty of astronomers is but to predict the future motions of celestial bodies. According to Kepler, however:

---

[86] Knoll, "The Arts Faculty," p. 140.
[87] Owen Gingerich, "The Great Copernicus Chase," *American Scholar* 49 (1979): 81–8.
[88] Robert Westman, "Three Responses to the Copernican Theory," in *The Copernican Achievement*, pp. 285–345 at p. 286.

What [Ursus] says here is not without qualification true. For even though what he mentions is the primary duty of an astronomer, the astronomer ought not to be excluded from the community of philosophers who inquire into the nature of things. He who predicts as accurately as possible the movements and positions of the stars performs well the duty of the astronomer. But he who employs true opinions about the form of the universe does it better, and is held worthy of greater praise. The former, to be sure, draws conclusions that are true as far as what is observed is concerned; the latter not only does justice in his conclusions to what is seen, but also, as I explained above, in order to draw his conclusions embraces the innermost form of nature.[89]

Clearly Kepler believed that the astronomer could speak to questions of the "innermost form of nature," as he had attempted in his *Mysterium cosmographicum* and as he did in his *New Astronomy* of 1609. As for the suggestion that astronomers were mere observers, Kepler asserted that "in all acquisition of knowledge it happens that, starting out from the things which impinge on the senses, we are carried by the operation of the mind to higher things which cannot be grasped by any sharpness of the senses. The same thing happens in the business of astronomy."[90]

It is in this context that we should not forget what some would call the "constellation of rhetorical possibilities"[91] which were available to Copernicus when he presented *On the Revolutions of the Heavenly Spheres* to the world in 1543. These included both the scientific and the humanistic audiences that often resided together in the papal court. As Robert S. Westman has shown, Copernicus was steeped in the humanist tradition and used many rhetorical flourishes to commend the study of the heavens as the most divine, perfect, and consummate undertaking of the learned scholar.[92] On the other hand, he could and did boldly appeal to what he took to be the unique power of mathematical arguments (and mathematicians) to arrive at certain knowledge about the world.[93] It was in this context that he wrote that "mathematics is written for mathematicians" and his book was meant for

[89] N. Jardine, *The Birth of History and Philosophy of Science: Kepler's "A Defense of Tycho Against Ursus" with Essays on Its Provenance and Significance* (New York: Cambridge University Press, 1984), pp. 144f. I have used Robert Westman's translation of this passage. See Westman, "The Astronomer's Role," p. 126. For the compellingly told story behind Kepler's writing of his *Apologia*, see Edward Rosen, *Three Imperial Mathematicians: Kepler Between Tycho Brahe and Ursus* (New York: Abaris Books, 1985).

[90] Jardine, *The Birth*, p. 144.

[91] Westman, "Proof, Poetics, and Patronage," p. 174.

[92] Westman, "The Astronomer's Role," p. 109; and "Proof, Poetics, and Patronage," pp. 176ff.

[93] Westman, "Proof, Poetics, and Patronage," p. 181; and Westman, "The Copernicans and the Churches," in *God and Nature*, p. 80.

them.[94] By writing his book with this intention, he tipped the balance within the universe of learned discourse toward the ascendancy of mathematics and away from physics (natural philosophy).

Most importantly, Copernicus chose to address his work to an ecclesiastical audience. This he did in two ways. First, he affixed to *De revolutionibus* the 1536 letter from Cardinal Nicholas Schönberg that alluded to Copernicus's heliocentric thesis and enthusiastically called upon him to publish his book. Secondly, by dedicating the preface of the work to Pope Paul III and citing the previous support of his work by the bishop of Kulm, he called upon papal authority for an impartial hearing of his work. Copernicus also referred to the probable existence of "triflers who though wholly ignorant of mathematics abrogate the right" of mathematical astronomers to make judgments about the universe, "because of some passages in Scripture wrongly twisted to their purpose,"[95] and dismissed them. When he did that, notably as a church canon, it surely seems that Copernicus believed he had the intellectual space to freely present his new world system.[96] While Copernicus was faced with potential critics in many quarters, he also found support for his endeavors in many others, not least of all in the papal court itself. Despite the church's reactionary stance in the time of Galileo, it has been noted by Islamic scholars, that "while in Europe the clergy felt the need of studying Greek learning and, as a result scientists were produced from among their ranks, in Islam religious authorities rather sought to discourage independent individual ventures into philosophical and scientific study. This was certainly a great handicap and an irreparable loss to the pursuit of science in Islam."[97] This was so because they were the most literate group of individuals and thus the most capable of delving into scientific and philosophical knowledge, but they were ideologically opposed to such a course of inquiry. In contrast, the European tradition was one in which the study of natural sciences was not only encouraged but indulged in by men of the cloth.

It should now be evident that the study and pursuit of scientific inquiry by the time of Copernicus was deeply entrenched in the universities and received widespread support from the highest religious authorities. As an

---

[94] Copernicus, *On the Revolutions of the Heavenly Spheres*, trans. by A. M. Duncan (New York: Barnes and Noble Books, 1976), p. 27.

[95] Ibid., p. 26.

[96] Of course, it might be said that the actions of the Council of Trent (1546) shortly after Copernicus's death altered the situation as regards the interpretation of Scripture. See Olaf Pedersen, "Galileo and the Council of Trent," pp. 15ff.

[97] Sayili, *The Observatory*, p. 413.

institutionalized endeavor, it had all the backing that law and tradition could provide and it was taught and practiced in all the countries of Western Europe. Accordingly, it is well to reiterate the conclusion of Olaf Pedersen in his review of the trial of Galileo: "The assumption of an essential incompatibility between science and Christianity cannot stand up to an historical test, and . . . the notion of an unchanging hostile attitude towards science on the part of the Church has to be abandoned."[98]

Whatever the attitude adopted by church officials from this time forward, with the advent of the printing press in the 1450s, an indomitable mode of communication had been put at the disposal of all scientifically inclined individuals. The history of early modern scientific printing shows, I believe, that despite the need to get official approval and the existence of censorship, scholars and scientists were able to get their works published in one city or another over the objections of overly zealous officials. This was especially true in the case of Galileo, who used Dutch publishers for some of his more controversial works.

Although the Copernican model of the universe had not been fully embraced by the majority of academic astronomers by the time of Galileo, this was due as much to a lack of supporting evidence as to reactionary habits among qualified European astronomers. Indeed, Professor Westman suggests that in the mid-sixteenth century it was rare indeed to find "an astronomer who rejected outright all features of Copernicus's heliocentric theory."[99] Perhaps the most compelling empirical evidence in support of the Copernican thesis was the fact that the length of yearly periods of the superior planets corresponded very nicely with their distance from the sun. This fact was jubilantly recited by Copernicus's early disciple Rheticus, as well as by Kepler somewhat later. Yet other astronomers failed to find this evidence compelling.[100]

Kepler and Galileo were keenly aware that they had no "indubitable proofs" that the Copernican system was true, though Galileo pretended otherwise. Indeed, when Kepler asked Galileo to send him such proofs (as Galileo claimed to have in his letter to Kepler in 1596 acknowledging the publication of the *Mysterium cosmographicum*), Galileo terminated the correspondence.[101] Thereafter Galileo failed to use Kepler's actual achievements, especially those of the *New Astronomy*, to support his own case.

Furthermore, it is evident that the study, teaching, and attempted refuta-

---

[98] Pedersen, "Galileo and the Council of Trent," p. 9.
[99] Westman, "The Astronomer's Role," p. 106.
[100] See Robert Westman's discussion of this in the case of the Wittenberg Interpretation, especially the case of Praetorius (1537–1616), in "Three Responses," pp. 296–303.
[101] Pedersen, "Galileo and the Council of Trent," p. 5.

tion of the ideas of Copernicus in major (but not all) universities went forward, largely unhampered, from about the second decade after the latter's death until 1616 – about three-quarters of a century.[102] The eruption of the Galileo affair is thus an anomaly that occurred because of a variety of personal motives, personal vendettas, hubris, and not a little malfeasance. It is so important precisely because it clearly violates all of our Western notions about the freedom of the individual and his right to seek and speak the truth, even if it violates cherished religious notions or political interests. As an event in the history of Western civilization, the Galileo affair symbolizes the clash of individual rights and presumably ignorant but powerful authorities. On a lower level concerned with individual actions, personal motivations, and vested interests, it is probably less glamorous and far more untidy. For example, there is little doubt that Galileo was a bold and provocative intellect who preferred frontal assaults to quiet diplomacy. Father James Brodrick's characterization of him as "the wrangler"[103] is probably quite near to the truth.

Conversely, it is now evident that there were indeed malevolent individuals (the "Liga") joined in consort together, who, through their contacts with church officials, sought to attack and harm Galileo in any way they could.[104] There were also those, however misguided, who legitimately feared the loss of their biblical faith if the conjectured Copernican system were accepted. Cardinal Bellarmine was one such who clearly held fast to the standard and literal interpretation of the Scriptures, even while supervising Galileo's 1616 sentence to desist from expounding the Copernican thesis as a true description of the universe. Bellarmine was no novice in astronomy, as he had taught it as a young Jesuit at Bologna. In reply to the query of Cardinal Foscarini regarding the potential conflict between the Copernican thesis and selected passages of Scripture, Bellarmine wrote in April 1615 that "there is no danger in [speaking suppositionally] that, by assuming the earth moves and the sun stands still, one saves all the appearances better than by postulating eccentrics and epicycles; and that is sufficient for the mathematician."[105] But of course it was a different matter "to affirm that in reality the sun is at the center of the world and only turns on itself without moving

---

[102] See Westman, "Three Responses" and "The Melanchthon Circle," *Isis* 66 (1975): 165–93; and for the circulation of copies of *De revolutionibus*, see Gingerich, "The Great Copernicus Chase."

[103] James Brodrick, *Galileo: The Man and His Work and Misfortunes* (New York: Harper and Row, 1964), chap. 2.

[104] Pedersen, "Galileo and the Council of Trent," pp. 6–8 and 26 n17.

[105] Bellarmine, in Maurice Finocchiaro, *The Galileo Affair: A Documentary History* (Berkeley and Los Angeles: University of California Press, 1989), p. 67.

from east to west" and that is "a dangerous thing."[106] The crux of the problem was scientifically more problematic:

Third, I say that if there were a true demonstration that the sun is at the center of the world and the earth in the third heaven, and that the sun does not circle the earth but the earth circles the sun, then one would have to proceed with great care in explaining the Scriptures that appear contrary, and say rather that we do not understand them than that what is demonstrated is false. But I will not believe that there is such a demonstration, until it is shown me.[107]

Galileo was in no position to provide such a demonstrative proof. Given that indubitable proofs did not exist in support of Copernicus and many first-rate astronomers also doubted parts of the technical details of the system, it is easy to understand the temptations of religious authorities to overreact to Galileo's bold, public, and often inflammatory declarations that Scripture and Aristotle were now dead insofar as science was concerned. It should also be noted, however, that Cardinal Bellarmine (like many of his religious peers) was in fact a scriptural fundamentalist, as James Brodrick points out in the revised edition of Bellarmine's biography.[108] At the very time that he was involved with the supervision of the 1616 proceedings against Galileo, Bellarmine had written a spiritual work called *De ascentione mentis in Deum*. In it he recited passages from the Bible and marveled at the naturalistic implications of the biblical language suggesting that the sun "does run its course." If the circumference of the earth is twenty thousand miles – an estimate common in those times and probably derived from the Arab astronomer al-Farghani (d. after 861) – "then it needs must follow that the Sun every houre runneth many thousands of miles."[109] He goes on:

I myself being once desirous to know in what space of time the Sun set at sea, at the beginning thereof I began to recite the Psalm *Miserere*, and scarce had read it twice over before the Sun was wholly set. It must needs be, therefore, that the Sun in that short time in which the Psalm *Miserere* was read twice over, did run much more than the space of 7,000 miles. Who would believe this unless certain reason demonstrate it?[110]

The Copernican system was an assault on accepted theological understandings in many circles, and even though a person might have astronomical knowledge, it still clashed with deeply held religious sentiments.

[106] Ibid., p. 68.
[107] Ibid.
[108] James Brodrick, *Robert Bellarmine, Saint and Scholar* (Westminster, Md.: Newman, 1961), p. 365, as cited in Nelson, *On the Roads to Modernity*, p. 143.
[109] Bellarmine in Brodrick, *Robert Bellarmine*, p. 335, as cited in Nelson, *On the Roads*, p. 144.
[110] Ibid., p. 145.

Granted all that, the Galileo affair was probably not the greatest challenge to the pursuit of science and natural philosophy during the rise of early modern science. That honor goes to the condemnation of 219 suspect propositions at the University of Paris in 1277, which I mentioned earlier. It was there and then that the battle was joined between the tenets of Christianity and the forces of reason and logic brought into the universities by the new Aristotle. As we saw earlier, despite the official nature of the ruling by the bishop of Paris and its sweeping condemnation of naturalistic assumptions and modes of thought – some of which ensnared even Thomas Aquinas – the ban had little effect other than to encourage a variety of new thought experiments that eventually aided the overthrow of unsupportable Aristotelian ideas. By the first quarter of the fourteenth century, the ban had been lifted and the University of Paris returned to its original mission.

Accordingly, the Galileo affair, coming nearly 350 years later, did not involve anything like the assault on the university that the 1277 condemnation did. While Galileo was put under house arrest and confined to his villa in Arceti, in point of fact the revolution went forward. It went forward in the sense that the agenda of scientific inquiry ensconced in the medieval universities went forward; the Copernican thesis was examined in every possible way, inside universities and out; and print technology served to continue the flow of scientific information all over Europe. The church attempted to restrict scientific discourse to the issuance of merely probable and hypothetical conjectures in regard to the constitution of the universe, but its attempt was totally ineffectual as either a matter of practice or a matter of theology. Moreover, unlike the situation in China or in Islam, there was in effect no possible way in which a centralized authority, religious or otherwise, could stamp out either the new theoretical ideas of Copernicus or the religious, legal, and philosophical groundings of them that were embedded in the major institutions of Western civilization. Even seventy-five years earlier as Copernicus lay on his deathbed, *De revolutionibus* went into the hands of the printer and thus copies of it immediately got into the hands of scholars all over Europe and England.[111] While in Galileo's time the censors might try to correct suspect passages in a book, as they did in the cases of Galileo's *Dialogue*,[112] the nexus of institutional support for open

---

[111] Gingerich, "Copernicus's *De revolutionibus:* An Example of Renaissance Scientific Printing," in *Print Culture in the Renaissance*, ed. Gerald P. Tyson (Newark: University of Delaware Press, 1986), pp. 55–73; Gingerich, "Copernicus and the Impact of Printing," pp. 201–18; and Gingerich, "The Great Copernicus Chase."

[112] See P. F. Grindler, "Venice, Science, and the Index of Prohibited Books," in *The Nature of Scientific Discovery*, ed. Owen Gingerich (Washington, D.C.: The Smithsonian Institution Press, 1975), pp. 335–47.

inquiry was too widespread, too deeply entrenched, and too multidimensional to control.[113]

The triumph of the Copernican revolution was thus a vindication of the efficacy of the institutional structures that had been put in place to encourage, protect, and preserve spheres of neutral space within which offensive, revolutionary, and even heretical ideas could be openly debated. That remains the legacy of the medieval revolution and the West in general. But we should not imagine that scholars then were as free as we are today to entertain all sorts of libertine ideas. There were indeed doctrinal boundaries – within the university and without – and these could not be crossed with impunity, but they rarely concerned strictly scientific issues. Robert Westman, for example, mentions the case of the sixteenth-century German astronomer Caspar Peucer, son-in-law of the educational reformer Philipp Melanchthon, who was imprisoned, not for his scientific views, but because of his religious views. For being a "crypto-Calvinist" he was imprisoned for twelve years.[114] When weighing the relative degree of freedom enjoyed by medieval and early modern scholars in Europe, one should keep in mind (for our purposes) the contrast with the situations in Islam and China.

The sixteenth century in Europe witnessed a triple revolution: a revolution in cosmology, a revolution in the disciplinary balance of the sciences (with astronomers using physical arguments to support the reality of their constructions), and a revolution in church authority – that is, the Reformation. The scientific revolution was undoubtedly helped along by the Reformation, but insofar as it concerns the revolution in astronomy and cosmology, that revolution had already been sparked by Copernicus and executed by Galileo, Kepler, and their successors.

It is undoubtedly true that the far-reaching reform of the German educational system begun in 1545 by Philipp Melanchthon (1497–1560), the close associate of Martin Luther, elevated the study of mathematics and astronomy within the German universities.[115] It is also true that this reform, spread by the disciples of Melanchthon and their pupils, institutionalized a Copernican interpretation within the German universities and thus gave the study of Copernicus its most concerted audience in Europe. Yet, as Robert

---

[113] Professor Owen Gingerich found that of the thirty copies of the second edition of *De revolutionibus* in Italian libraries, 60 percent had been censored by crossing out offending passages. But of the first-edition copies, only 14 percent had been censored ("The Great Copernicus Chase," p. 87).

[114] Westman, "The Melanchthon Circle," p. 178.

[115] See ibid., as well as Westman, "The Copernicans and the Churches," pp. 82–9; and for the Jesuit colleges and universities, pp. 93–5. On the Jesuit Collegio Romano, see William Wallace, *Galileo and His Sources* (Princeton, N.J.: Princeton University Press, 1984).

Westman has shown in great detail, the exponents of the so-called Wittenberg interpretation did not accept the heliocentric cosmology of Copernicus, and Melanchthon himself rejected the idea that the earth itself moves.[116] By and large, they ignored the new cosmology and the evidence of the ordering of the planets as data in support of the heliocentric thesis. This was not true of Rheticus, but for most of the others until Maestlin it was. The Wittenberg interpretation sanitized the Copernican system and made it easier to accept within the university (Protestant or Catholic), and only toward the end of the sixteenth century did scholars like Maestlin and his pupil Kepler begin to appreciate the evidence and arguments for a realist interpretation of the Copernican system. Finally, while Melanchthon was a close associate of Luther and thus championed the Lutheran religious reform, his university education obviously had taken place in a pre-Reformation setting, a setting imbued with Aristotle's scientific spirit and which had likewise shaped Copernicus's outlook.

There is much to be said for the Weberian account that attributes a new spirit to economy, science, and society after the Reformers. For after the Reformation all of the fundamental terms of intellectual life and discourse – not just those relating to science – were to be radically transformed. This included new stresses on and new connotations attached to such terms as the light of reason, conscience, the world machine, the Book of Nature, as well as causes, experience, and experiment. All of these were to undergo reconstruction, not only in the thought of the Reformers, but in other places as well.[117] Insofar as the reform pertains to science, the change can be seen in the textbook tradition of the universities in the first half of the seventeenth century. Based on an analysis of a sample of twenty textbooks written by scholars in Germany, France, Holland, Denmark, Italy, and England, Patricia Reif noted a pronounced characteristic of these works and their image of science:

What we find completely absent from any of the textbooks is the Baconian vision and Galilean practice of using knowledge to get control over the forces of nature. There is absolutely no notion of putting knowledge to work, or manipulating natural things to discover their capacities, in short, of a truly practical science. Practical purposes seem to be considered utterly irrelevant to natural philosophy.[118]

---

[116] Westman, "The Copernicans and the Churches," p. 83. On Christopher Clavius's contribution to the rise of the study of mathematics in Jesuit circles, see Frederick A. Homann, "Christopher Clavius and the Renaissance of Euclidean Geometry," *Archivum Historicum Societatis Jesu* 52 (1983): 233–46.

[117] As contributions to new structures of consciousness, many of these themes have been explored by Benjamin Nelson; see *On the Roads to Modernity*, especially chaps. 3, 4, and 12.

[118] Patricia Reif, "The Textbook Tradition in Natural Philosophy," *Journal of the History of Ideas* 30 (1969): 17–32, at 22.

The impact of the Reformation mentality on the thought and practice of science in the seventeenth century and thereafter was real enough. Robert Merton was aware of some of the semantic shifts that took place, as when he noted that the medieval and the seventeenth-century concepts of reason and rationalism were different. "For the Puritans," he says, "reason takes on a new connotation, the rational consideration of data. Logic is reduced to a subsidiary role."[119] One may agree with Merton that "the rise of science which antedated the Reformation or developed quite independently of it does not negate the significance of ascetic Protestantism in this respect."[120] As Thomas Kuhn put it, as a thesis about the scientific movement, and not the revolution in science, the appeal of Merton's thesis is "vastly larger," especially if it is focused on the "Baconian sciences" of magnetism, electricity, chemistry, geology, and so on.[121] For many of these, the experimental sciences, "the universities had no place before the last half of the nineteenth" century.[122]

For these reasons, we should not forget that the so-called role of the scientist is really a role-set, an array of associated roles which usually include that of a college or university professor, a teacher of students, a member of a disciplinary department, a researcher, a writer and author, and, quite possibly, a gatekeeper who referees knowledge claims produced by other scientists. It is of course the very interdependence of these various roles that creates both the integrity and the autonomy of the scientist. Likewise, it is the far-reaching complexity of this role-set that prevents science from ever becoming completely autonomous. It is far too much to have expected all of this to have come into existence either in the twelfth and thirteenth centuries, or even in the seventeenth. The rise of modern science was clearly a long and protracted event, and its institutionalization in all its dimensions in new professional associations, in institutes, and in polytechnics has continued into the twentieth century, not least of all with the advent of the think tank and similar associational devices in the 1940s.

In a word, the domains of cultural activity "are only partially autonomous, not completely so."[123] To say that science at any point in time has been institutionalized (and hence gained a significant measure of autonomy) is only to say that it has been publicly and officially sanctioned, but that

[119] Merton, *STS*, p. 71.
[120] Ibid., p. 136.
[121] Kuhn, "Mathematical vs. Experimental Traditions," *Journal of Interdisciplinary History* 7, no. 1 (1976): 26.
[122] Ibid., p. 19. And see E. Mendelsohn, "The Emergence of Science as a Profession in Nineteenth-Century Europe," in *The Management of Science*, ed. Karl Hill (Boston: Beacon Press, 1964), pp. 3–48.
[123] Merton, *STS*, p. x.

official support can be withdrawn, thereby calling everything into doubt. When we bring the legal foundations of social institutions into the picture, we see that the withdrawal of public and official support from science, for example, cannot be accomplished without due process of law and that the foundations of Western social institutions (and especially science) are far deeper than is generally acknowledged. That is the genius of Western institution building, and has been so since the Middle Ages, though none of this is to say that such mechanisms are without flaw. On the other hand, the existence of such legal protections does not automatically banish forever the kind of misguided religious and political forces that erupted in the Galileo affair. The existence of a legal substructure supporting social institutions only means that the playing field has been leveled so that the forces of reason have a nearly equal chance of winning the day.

Epilogue: science and civilizations
East and West

The civilizations of Islam, China, and the West over the course of their histories have been inspired by contrasting images of reason, rationality, and the man of knowledge. In the world of Islam, we have seen both highly committed natural philosophers and deeply inspired religious souls who could not reconcile the demands of logic with the sensibilities of the religious life. From the outset, devout Muslims were inclined to think that all wisdom was contained in the Quran and thus all true sciences must also be found there. Such was the origin of the idea of Prophetic medicine, that is, medical knowledge derived from the sacred teachings of the Prophet Muhammad.

For a time the scientists and natural philosophers of Islam managed to carry on their work, reaching great heights of technical knowledge and even laying the foundations for later scientific development in the West. But while the Greek modes of reason and logic were well known to the enlightened men of medicine, philosophy, and science, they were held at bay by the religious authorities of Islam, so that in the long run, no social institutions were founded that could protect and support freethinking, a term commonly denoting heresy. For a large number of Muslims living in Islamic countries today, modern science is perceived as un-Islamic, and those who embrace it are thought to have taken the first (and fatal) step toward impiety. Among Islamic fundamentalists, these sentiments are all the stronger, reinforced by ideological rhetoric identifying science with decadence and the evils of the West. A recent assessment of attitudes toward science and scientific education in Muslim countries is gloomy indeed.[1]

---

[1] See Pervez Hoodbhoy, *Islam and Science: Religious Orthodoxy and the Battle for Rationality*, with a foreword by Mohammed Abdus Salam (London: Zed Books Ltd, 1990), especially

In China the image of the man of knowledge was first and foremost that of an enlightened individual who was also morally committed to the traditional ways. The epitome of the learned man was one who had mastered the Confucian classics and who, through long and arduous study, understood the place of man in a harmonious cosmos. It was he who could advise the emperor on statecraft and moral affairs and, being thus enlightened, could follow a course that would avert natural catastrophes and social unrest. It was toward the end of understanding the ebbs and flows of man and nature – the organic harmony of man and nature – that the man of learning directed his being. The center of attention in this framework was man and the social order – the microcosm – rather than nature and the macrocosm. The classic ways of knowing were not predicated on reason and logic but on listening with a sixth sense. Above all, learning entailed the mastery of the Confucian classics. This does not mean that there was no room for the techniques of the exact sciences, for practical men of industry, or for the pursuit of science but that these ideals were subordinate to the classical forms of being.

It has been suggested that in the seventeenth century, the Jesuit-sponsored importation of new geocentric and trigonometric approaches to cosmological questions in China "precipitated what can only be called a scientific revolution."[2] The same author readily admits that this transformation "did not lead to the fundamental changes in thought and society" that were experienced in the West. Instead, the leaders of this movement "felt a responsibility for strengthening and perpetuating traditional ideas."[3] In other words, China did not experience a true scientific revolution but only a minor shift in mathematical astronomy. The Chinese clung fast to the old correlative modes of thinking, and intellectual authority was retained by a tiny elite of state officials. Educational institutions that resembled those of the West did not appear, and there were no breakthroughs in the logics of action and decision, to use the phrase of Benjamin Nelson. Up until the present, China has had three major encounters with Western science: that of

---

chap. 4. The virtue of this book is its description of Islamic attitudes toward science and scientific education, especially in contemporary Pakistan. There the Islamization project launched by the Zia ul-Haq government wreaked great havoc on education in general and science education in particular. The author of this book has been courageous in detailing this sad state of affairs and in publicly protesting the absurdities of "Islamic science" (see his appendix, "They Call It Islamic Science"). But otherwise the book must be used with a good deal of caution as it contains a number of howlers, such as the allegation that Max Weber thought that modern science arose in the West because of the genetic superiority of Europeans (p. 20). Likewise, his description of medieval European history uses old canards regarding the church's attitude toward science.

[2] Nathan Sivin, "Science and Medicine in Chinese History," in *Heritage of China*, ed. Paul S. Ropp (Berkeley and Los Angeles: University of California Press, 1990), p. 192.

[3] Ibid., p. 193.

the seventeenth century sponsored by the Jesuits; that of the late nineteenth century forced upon China by the British; and the freely chosen embrace of "Mr. Science," which took place following the Revolution of 1911. Yet in each of these episodes, the traditional ways of China reasserted themselves.[4]

From our point of view, the modern scientific revolution was a social and intellectual revolution that at once reorganized the scheme of natural knowledge and validated a new set of conceptions of man and his cognitive capacities. The forms of reason and rationality that had been fused out of the encounter of Greek philosophy, Roman law, and Christian theology laid a foundation for believing in the essential rationality of man and nature. More importantly, however, this new metaphysical synthesis found an institutional home in the cultural and legal structures of medieval society. Together they laid the foundations validating the existence of neutral institutional spaces within which intellects could pursue their guiding lights and ask the most probing questions. The religious, legal, and philosophical presuppositions built into medieval state and society created the foundations of modernity that have continued to spread around the world. Whether that will continue to occur remains to be seen. For the prevailing images of reason, rationality, and legality around the world are perhaps not yet fully consonant with each other. Moreover, the contrasting images of the scientist and the man of knowledge mentioned above are still very much with us in the different parts of the world.

It must be said, however, that science is now fully institutionalized in the West and the desire to pursue scientific inquiry ought to be judged universal, despite the existence of many political and ideological obstacles around the globe. Although we have just begun to study the spread of science as an indigenous activity around the world, a sizable body of literature on this topic has begun to emerge.[5] Today there are dozens of developing countries

---

[4] D. W. Y. Kwok, *Scientism in Chinese Thought 1900–1950* (New Haven, Conn.: Yale University Press, 1965), pp. 137ff; as well as Merle Goldman and Denis Fred Simon, "Introduction: The Onset of China's New Technological Revolution," in *Science and Technology in Post-Mao China*, pp. 4–6; and Richard Baum, "Science and Culture in Contemporary China," *Asian Survey* 22, no. 12 (1982): 1166–86.

[5] For a sampling of this literature, see George Basalla, "The Spread of Western Science," *Science* 156 (1967): 611–22; J. Davidson Frame, Francis Naren, and Mark Carpenter, "The Distribution of World Science," *Social Studies of Science* 7 (1979): 501–16; Eugene Garfield, "Mapping Science in the Third World, Part 1," *Current Contents*, no. 34 (1983): 254–63; Garfield, "Third World Research, Part 2," *Current Contents*, no. 34 (1983): 264–75; S. Arunachalam and K. C. Garg, "A Small Country in a World of Big Science: A Preliminary Bibliometric Study of Science in Singapore," *Scientometrics* 8 (1985): 301–13; J. Irvine and B. R. Martin, "International Comparisons of Scientific Performance Revisited," *Scientometrics* 15 (1989):

that have their own science journals and whose nationals seek desperately to enter the international world of scientific dialogue. One of the consequences of this is that the United States and Europe are currently reaping the benefit of the influx of large numbers of young foreign students seeking to study science. We saw earlier (in Chapter 7 note 48) that today students from China are by far the largest group of foreign students in the United States. Among the graduate students from China (which accounts for 82 percent of Chinese foreign students), 33 percent are in the life and physical sciences or in what is now called science technology. Except for Japan and Taiwan, these numbers of foreign exchange students are more than 30 percent higher than for all other countries. What the implications of this development are for the future of the world remains to be seen.[6]

In the world of Islam, as in China, the central issue is whether there can be free thought and, above all, criticism of all forms of the status quo, which is at the heart of the scientific enterprise. Today as we approach the end of the twentieth century, there is great anticipation that China will emerge as an economic giant, but much less anticipation that Chinese authorities (or orthodox Islamic authorities) will greatly expand the spheres of free discussion, democratic process, and voluntary action. If we cast this discussion on the level of individual motivation and the rise of individualism – a phenomenon that occurred much later than the era discussed in this book – we again find strong contrasts between China and the West. Reflecting on this problem, William de Bary found an almost endless list of elements in China regarding the historical absence of a Western-style individualism. These include

the extreme weakness of the middle class, the nondevelopment of a vigorous capitalism, the absence of a church which fought for its rights against the state, or of competing religions which sought to defend the freedom of conscience against arbitrary authority; the lack of university centers of academic freedom . . . the want of a free press supported by an educated middle class.[7]

369–92; and Institute for Scientific Information, "No Slippage Yet in Strength of U.S. Science," *Science Watch* 2, no. 1 (1991): 1–2.

[6] One of the best sets of speculations about this outcome is that by Shigeru Nakayama, "The Shifting Center of Science," *Interdisciplinary Science Reviews* 16 (1991): 82–8. But also see the concluding chapter by Richard P. Suttmeier, "Science, Technology, and China's Political Future – A Framework for Analysis," in *Science and Technology in Post-Mao China*, ed. Denis Fred Simon and Merle Goldman (Cambridge, Mass.: The Council on East Asian Studies at Harvard University, 1989), pp. 375–96.

[7] William de Bary, "Individualism and Humanitarianism in Late Ming Thought," in *Self and Society in Ming Thought*, ed. William Theodore de Bary (New York: Columbia University Press, 1970), pp. 145–247 at p. 220. For a more contemporaneous assessment of the problems

To enter fully into the global world of scientific discourse, China must undergo major social and cultural changes to remedy these deficits. Accomplishing that task would represent the most profound intellectual and institutional revolution in the history of China. While that is a tall order, objective observers must be far more optimistic about China's (and Asia's) contribution to modern science in the twenty-first century than that of the world of Islam. The fate of science and education in the Middle East as a result of Islamic revolution is dramatically revealed by Iran's plummeting position in terms of students sent to study in the United States. From being ranked first in 1979, it has now plunged to fifteenth.[8] The overall effect of the Islamic revolution of the Middle East on education may be gauged by the fact that in 1979 foreign students from the Middle East studying in the United States constituted above 29 percent of all foreign students, whereas in 1990 they had dropped to about 8 percent.[9] Conversely, foreign students from Asia increased from 29 to nearly 59 percent of all foreign students in 1991–2.[10]

A major question facing developing countries today is not whether they will accept the results of natural science but rather whether their governing elites will grant autonomy to their own aspiring scientists to freely pursue their insights in the world of learning. If that should occur, the next question would be whether the developing countries would allow scientists – social and natural – to objectively describe the social world and publicly report their results, above all, when those results cast political authorities in a poor light. That challenge to authority has always been at the heart of the scientific enterprise. Creating the cultural and institutional conditions which permit that pursuit of the life of the mind cannot be a matter of indifference for those who do not yet enjoy it. If those conditions are not met elsewhere, the flow of scientific talent to the West, to the United States in particular, will continue unabated.

of individualism in China, see Lucian W. Pye, "The State and the Individual: An Overview Interpretation," *China Review* 127 (1990): 443–66.

[8] See M. Zikopoulos, ed., *Open Doors, 1990–91* (New York: Institute of International Education, 1991), Table 2.4, p. 21.

[9] Ibid., Table 2.1, p. 16.

[10] Institute of International Education, *Open Doors, 1991–92*, as reported in *The Chronicle of Higher Education*, Nov. 25, 1992, p. A29.

Selected bibliography

Abel, Armand, 1957. "La place des sciences occultes dans la décadence." In *Classicisme et déclin culturel dans l'histoire de l'Islam,* edited by R. Brunschvig and G. E. Von Grunebaum, pp. 291–311. Paris: G.-P. Maisonneuve et Larose.

Abelson, Paul. 1965. *The Seven Liberal Arts.* New York: Russel and Russell.

Abraham, Gary. 1983. "Misunderstanding the Merton Thesis." *Isis* 74:368–87.

Alabaster, Ernest. 1899. *Notes and Commentaries on Chinese Criminal Law and Cognate Topics, with Special Relation to Ruling Cases, Together with a Brief Excursus on the Law of Property.* London: Luzac.

Anawati, George. 1970. "Science." In *The Cambridge History of Islam,* vol. 2, edited by P. M. Holt, pp. 741–80. New York: Cambridge University Press.

Anderson, J. N. D. 1976. *Law and Reform in the Muslim World.* London: Athlone Press.

Appignanesi, Lisa, and Sara Maitland, eds. 1990. *The Rushdie File.* Syracuse, N.Y.: Syracuse University Press.

Ash'ari. 1983. "The Elucidation of Islam's Foundation." In *The Islamic World,* edited by William H. McNeill and Marilyn R. Waldman, pp. 152–66. Chicago: University of Chicago Press.

Averroes. 1954. *Tahafut al-Tahafut* (The Incoherence of the Incoherence). Translated with introduction by Simon Van den Bergh. 2 vols. London: Luzac.

Baer, Gabriel. 1970. "Guilds in Middle Eastern History." In *Studies in the Economic History of the Middle East,* edited by M. A. Cook, pp. 11–30. London: Oxford University Press.

Balazs, Etienne. 1964. *Chinese Civilization and Bureaucracy.* New Haven, Conn.: Yale University Press.

Bassiouni, M. Cherif, ed. 1982. *The Islamic Criminal Justice System.* New York: Oceana.

Baylor, Michael. 1977. *Action and Person: Conscience in Late Scholasticism and the Young Luther.* Leiden: E. J. Brill.

Beaujouan, Guy. 1982. "The Transformation of the Quadrivium." In *Renaissance*

365

and Renewal in the Twelfth Century, edited by Robert Benson and Giles Constable, pp. 463–87. Cambridge, Mass.: Harvard University Press.

Ben-David, Joseph. 1965. "The Scientific Role: The Conditions of Its Establishment in Europe." *Minerva* 4, no. 1, 15–54.

1971. *The Scientist's Role in Society.* Englewood Cliffs, N.J.: Prentice-Hall.

Ben-David, Joseph, and Abraham Zloczower. 1962. "Universities and Academic Systems in Modern Societies." *European Journal of Sociology* 3:45–84.

Benson, Robert, and Giles Constable, eds. 1984. *Renaissance and Renewal in the Twelfth Century.* Cambridge, Mass.: Harvard University Press.

Berman, Harold. 1983. *Law and Revolution: The Formation of the Western Legal Tradition.* Cambridge, Mass.: Harvard University Press.

Bodde, Derk. 1957. "Evidence for 'Laws of Nature' in Chinese Thought." *Harvard Journal of Asiatic Studies* 20:709–27.

1979. "Chinese 'Laws of Nature': A Reconsideration." *Harvard Journal of Asiatic Studies* 39:139–55.

1981 (1963). "Basic Concepts of Chinese Law: The Genesis and Evolution of Legal Thought in Traditional China." In *Essays on Chinese Civilization,* 2d ed., pp. 171–94. Princeton, N.J.: Princeton University Press.

1991. *Chinese Thought, Science, and Society: The Intellectual and Social Background of Science and Technology in Pre-Modern China.* Honolulu: University of Hawaii Press.

Bodde, Derk, and Clarence Morris. 1973. *Law in Imperial China.* Reprint. Philadelphia: University of Pennsylvania Press.

Bol, Peter. 1989. "Chu Hsi's Redefinition of Literati Learning." In *Neo-Confucian Education: The Formative Period,* edited by William de Bary and John Chaffee, pp. 151–85. Berkeley and Los Angeles: University of California Press.

Browne, Edward, G. 1962. *Arabian Medicine.* New York: Cambridge University Press.

Brunschvig, R., and G. E. Von Grunebaum, eds. 1957. *Classicisme et déclin culturel dans l'histoire de l'Islam.* Paris: G.-P. Maisonneuve et Larose.

Bullough, Vern. 1966. *The Development of Medicine as a Profession.* New York: Hafner.

Bürgel, J. Christoph. 1976. "Secular and Religious Features of Medieval Arabic Medicine." In *Asian Medical Systems: A Comparative Study,* edited by Charles Leslie, pp. 44–62. Berkeley and Los Angeles: University of California Press.

Butterfield, Herbert. 1957. *The Origins of Modern Science, 1300–1800.* Rev. ed. New York: Free Press.

Cahen, Claude, 1971. "Kasb." In *Encyclopedia of Islam,* 2d ed., vol. 4, pp. 690–2. Leiden: E. J. Brill.

Cahen, Claude, and M. Talbi. 1971. "Hisba." In *Encyclopedia of Islam,* 2d ed., vol. 3, pp. 485–9. Leiden: E. J. Brill.

Carra de Vaux, Baron. 1937. "Astronomy and Mathematics." In *The Legacy of Islam,* 1st ed., edited by T. Arnold and A. Guillaume, pp. 376–97. Oxford: Oxford University Press.

Carter, Thomas F. 1955. *The Invention of Printing in China and Its Spread Westward.* Rev. ed. by L. C. Goodrich. New York: Ronald Press.

Castro, Americo. 1971. *The Spaniards.* Berkeley and Los Angeles: University of California Press.

Catton, Henry. 1955. "The Law of Waqf." In *Law in the Middle East,* edited by M. Khadduri and H. Liebesny, pp. 203–22. Washington, D.C.: The Middle East Institute.

Chaffee, John W. 1985. *The Thorny Gates of Learning in Sung China: A Social History of Examinations.* New York: Cambridge University Press.

Chan Wing-tsit. 1987. *Chu Hsi: Life and Thought.* New York: St. Martin's Press.

Chejne, Anwar. 1974. *Muslim Spain: Its History and Culture.* Minneapolis: University of Minnesota Press.

Chenu, M.-D. 1969. *L'éveil de la conscience dans la civilisation médiévale.* Paris: J. Vrin.

   1968. *Nature, Man, and Society in the Twelfth Century.* Selected, edited, and translated by Jerome Taylor and Lester K. Little. Chicago: University of Chicago Press.

Ch'ü, T'ung-tsu. 1962. *Local Government in China under the Ch'ing.* Cambridge, Mass.: Harvard University Press.

Clagett, Marshall, Gaines Post, and R. Reynolds, eds. 1966. *Twelfth-Century Europe and the Foundations of Modern Society.* Madison, Wis.: University of Wisconsin Press.

Cobban, A. B. 1975. *The Medieval Universities: Their Organization and Development.* London: Methuen.

Cohen, I. B. 1985. *Revolution in Science.* Cambridge, Mass.: Harvard University Press.

Conrad, Lawrence. 1985. "The Social Structure of Medicine in Medieval Islam." *The Society for the Social History of Medicine Bulletin* 37:11–15.

Coulson, N. J. 1957. "The State and the Individual in Islamic Law." *International and Comparative Law Quarterly* 6:49–60.

   1964. *A History of Islamic Law.* Edinburgh: Edinburgh University Press.

   1969. *Conflicts and Tensions in Islamic Jurisprudence.* Chicago: University of Chicago Press.

Courtenay, William J. 1974. "Nominalism in Late Medieval Religion." In *The Pursuit of Holiness in Late Medieval and Renaissance Religion,* edited by Charles Trinkaus and Heiko Oberman, pp. 26–59. Leiden: E. J. Brill.

Crombie, A. C. 1952. "Avicenna's Influence on the Medieval Scientific Tradition." In *Avicenna,* edited by G. Wickens, pp. 84–107. London: Luzac.

   1953. *Robert Grosseteste and the Origins of Experimental Science, 1100–1700.* Oxford: At the Clarendon Press.

   1955. "Grosseteste's Position in the History of Science." In *Robert Grosseteste, Scholar and Bishop,* edited by D. A. Callus, pp. 98–120. Oxford: Oxford University Press.

   1959a. *Medieval and Early Modern Science.* Rev. ed. 2 vols. New York: Doubleday.

1959b. "The Significance of Medieval Discussions of Scientific Method for the Scientific Revolution." In *Critical Problems in the History of Science*, edited by Marshall Clagett, pp. 79–102. Madison, Wis.: University of Wisconsin Press.

1975. "Sources of Galileo's Early Natural Philosophy." In *Reason, Experiment, and Mysticism in the Scientific Revolution*, edited by R. Bonelli and William Shea, pp. 157–74. New York: Science History Publications.

1988. "Designed in the Mind: Western Visions of Science, Nature, and Humankind." *History of Science* 26:1–12.

D'Arcy, Eric. 1961. *Conscience and Its Right to Freedom*. New York: Sheed and Ward.

Dawson, John. 1978. *The Oracles of the Law*. Reprint. Westport, Conn.: Greenwood Press.

de Bary, William Theodore. 1957. "Chinese Despotism and the Confucian Ideal: A Seventeenth-Century View." In *Chinese Thought and Institutions*, edited by John K. Fairbank, pp. 163–203. Chicago: University of Chicago Press.

1970. "Individualism and Humanitarianism in Late Ming Thought." In *Self and Society in Ming Thought*, edited by William de Bary, pp. 145–247. New York: Columbia University Press.

1981. *Neo-Confucian Orthodoxy and the Learning of the Mind-and-Heart*. New York: Columbia University Press.

1983. *The Liberal Tradition in China*. New York: Columbia University Press.

ed. 1975. *The Unfolding of Neo-Confucianism*. New York: Columbia University Press.

de Bary, William Theodore, and John W. Chaffee, eds. 1989. *Neo-Confucian Education: The Formative Period*. Berkeley and Los Angeles: University of California Press.

Demaitre, Luke. 1975. "Theory and Practice in Medical Education at the University of Montpellier in the Thirteenth and Fourteenth Centuries." *Journal of the History of Medicine* 30:103–23.

Demeerseman, A. 1953–54. "Un étape décisive de la culture et de la psychologie sociale islamique: Les données de la controverse autour du problème de l'Imprimerie." *Institut des Belles Lettres Arabes* 16:347–89; 17:1–48 and 113–40.

Dictionary of Scientific Biography. 14 vols. With Supplements. New York: Charles Scribner's.

Dijksterhuis, E. J. 1961. *The Mechanization of the World Picture*. London: Oxford University Press.

D'Irsay, Stephen. 1933. *Histoire des universités, française et etrangère*. 3 vols. Paris: J. Vrin.

Dols, Michael. 1984. "Introduction" to *Medieval Islamic Medicine: Ibn Ridwan's Treatise "On the Prevention of Bodily Ills in Egypt,"* pp. 3–73. Berkeley and Los Angeles: University of California Press.

1987. "The Origins of the Islamic Hospital: Myth and Reality." *Bulletin of the History of Medicine* 61:367–90.

Drake, Stillman, ed. and trans. 1957. *Discoveries and Opinions of Galileo.* New York: Doubleday.

1970. "Early Science and the Printed Book: The Spread of Science Beyond the Universities." *Renaissance and Reformation* 6:43–52.

Dull, Jack C. 1990. "The Evolution of Government in China." In *Heritage of China,* edited by Paul S. Ropp, pp. 55–85. Berkeley and Los Angeles: University of California Press.

Eisenstein, Elizabeth. 1979. *The Printing Press as an Agent of Change: Communications and Cultural Transformation in Early Modern Europe.* 2 vols. New York: Cambridge University Press.

Elgood, Cyril. 1951. *Medical History of Persia and the Eastern Caliphate.* Cambridge: Cambridge University Press.

Elman, Benjamin. 1984. *From Philosophy to Philology: Intellectual and Social Aspects of Change in Late Imperial China.* Cambridge, Mass.: Harvard University Press.

Elvin, Mark. 1973. *The Pattern of China's Past.* Stanford, Calif.: Stanford University Press.

Encyclopedia of Islam. 2d ed. 1960–. Leiden: E. J. Brill.

Escarra, Jean. 1933. "Chinese Law." *Encyclopedia of the Social Sciences* 5:249–54.

1961. *Chinese Law.* Translated by Gertrude W. Browne. W. P. A. University of Washington; photo-mechanical reproduction by Harvard University, Harvard University Asian Research Center.

Fairbank, John K., ed. 1957. *Chinese Thought and Institutions.* Chicago: University of Chicago Press.

Fakhry, Majid. 1958. *Islamic Occasionalism and Its Critique by Averroes and Aquinas.* London: Allen and Unwin.

Feuerwerker, Albert. 1990. "Chinese Economic History in Comparative Perspective." In *Heritage of China,* edited by Paul S. Ropp, pp. 224–41. Berkeley and Los Angeles: University of California Press.

Fingarette, Herbert. 1972. *Confucius: The Sacred as Secular.* New York: Harper and Row.

Finocchiaro, Maurice. 1989. *The Galileo Affair: A Documentary History.* Berkeley and Los Angeles: University of California Press.

Frank, Richard. 1971. "Some Fundamental Assumptions of the Basra School of Mu'tazila." *Studia Islamica* 33:5–18.

Fung Yu-lan. 1968. *A History of Chinese Philosophy.* 2 vols. Translated by Derk Bodde. Princeton, N.J.: Princeton University Press.

Gardet, L. 1971. "Kasb." In *Encyclopedia of Islam,* 2d ed., vol. 3, pp. 692–4. Leiden: E. J. Brill.

Gardet, L., and M. M. Anawati. 1970. *Introduction à la Théologie Musulmane.* 2d ed. Paris: J. Vrin.

Gascoigne, John. 1990. "A Reappraisal of the Role of the Universities in the Scientific Revolution." In *Reappraisals of the Scientific Revolution,* edited by David

C. Lindberg and Ronald L. Numbers, pp. 207–60. New York: Cambridge University Press.

Gernet, Jacques. 1982. *A History of Chinese Civilization.* Cambridge: Cambridge University Press.

al-Ghazali. 1962. *Book of Fear and Hope.* Translated by William McKane. Leiden: E. J. Brill.

Gibb, H. A. R. 1947. *Modern Trends in Islam.* Chicago: University of Chicago Press.

    and Harold Bowen. 1965. *Islamic Society and the West.* vol 1. Reprint. Oxford: Oxford University Press.

Gingerich, Owen. 1973. "From Copernicus to Kepler: Heliocentrism as Model and as Reality." *Proceedings of the American Philosophical Society* 117, no. 6, 513–22.

    1975. "Copernicus and the Impact of Printing." In *Copernicus Yesterday and Today, Vistas in Astronomy,* vol. 17, edited by Arthur Beer and K. Aa. Strand, pp. 201–18. New York: Pergamon Press.

    1975. "Commentary: Remarks on Copernicus's Observations." In *The Copernican Achievement,* edited by Robert S. Westman, pp. 99–107. Berkeley and Los Angeles: University of California Press.

    1979. "The Great Copernicus Chase." *The American Scholar* 49:81–8.

Glick, Thomas F. 1979. *Islamic and Christian Spain in the Early Middle Ages.* Princeton, N.J.: Princeton University Press.

Goitein, S. D. 1963. "The Medical Profession in the Light of the Cairo Geniza Documents." *Hebrew Union College Annual* 34:177ff.

    1968–71. *A Mediterranean Society.* 2 vols. Berkeley and Los Angeles: University of California Press.

Goldstein, Bernard. 1972. "Theory and Observation in Medieval Astronomy." *Isis* 63:39–47.

Goldziher, Ignaz. 1981 (1916). "The Attitude of Orthodox Islam Toward the Ancient Sciences." In *Studies in Islam,* edited by Merlin Swartz, pp. 185–215. New York: Oxford University Press.

Goodman, Leonard E. 1978. "Did al-Ghazâli Deny Causality?" *Studia Islamica* 42:83–120.

Graham, A. C. 1973. "China, Europe, and the Origins of Modern Science: Needham's *Grand Titration.*" In *Chinese Science: Explorations of an Ancient Tradition,* edited by Shigeru Nakayama and Nathan Sivin, pp. 45–69. Cambridge, Mass.: MIT Press.

    1978. *Later Mohist Logic, Ethics, and Science.* Hong Kong: Chinese University Press.

    1986. *Yin-Yang and the Nature of Correlative Thinking.* Singapore: Institute of East Asian Philosophies, National University of Singapore.

    1989. *Disputers of the Tao.* La Salle, Ill.: Open Court Press.

Grant, Edward. 1962. "Late Medieval Thought, Copernicus, and the Scientific Revolution." *Journal of the History of Ideas* 23:197–220.

1978. "Cosmology." In *Science in the Middle Ages*, edited by David C. Lindberg, pp. 265–302. Chicago: University of Chicago Press.

1979. "The Condemnation of 1277, God's Absolute Power, and Physical Thought in the Late Middle Ages." *Viator* 10:211–44.

1984. "Science and the Medieval University." In *Rebirth, Reform, Resilience: Universities in Transition, 1300–1700*, edited by James M. Kittelson and Pamela J. Transue, pp. 68–102. Columbus, Ohio: Ohio State University Press.

1986. "Science and Theology in the Middle Ages." In *God and Nature: Historical Essays on the Encounter Between Christianity and Science*, edited by David C. Lindberg and Ronald L. Numbers, pp. 49–75. Berkeley and Los Angeles: University of California Press.

ed., 1974. *A Source Book of Medieval Science*. Cambridge, Mass.: Harvard University Press.

Grimm, Tilemann. 1968. "Ming Education Intendants." In *Chinese Government in Ming Times*, edited by Charles O. Hucker, pp. 127–47. New York: Columbia University Press.

1977. "Academies and Urban Systems in Kwangtung." In *The City in Late Imperial China*, edited by G. William Skinner, pp. 475–98. Stanford, Calif.: Stanford University Press.

Grindler, Paul F. 1975. "Venice, Science, and the Index of Prohibited Books." In *The Nature of Scientific Discovery*, edited by Owen Gingerich, pp. 335–47. Washington, D.C.: The Smithsonian Institution Press.

Hamarneh, Sami. 1970. "Medical Education and Practice in Medieval Islam." In *The History of Medical Education*, cited by C. D. O'Malley, pp. 39–71. Berkeley and Los Angeles: University of California Press.

1971a. "Arabic Medicine and Its Impact on Teaching and Practice of the Healing Arts in the West." *Oriente e Occidente* 13:395–426.

1971b. "The Physician and the Health Professions in Medieval Islam." *Bulletin of the New York Academy of Sciences* 47, no. 9, pp. 1088–110.

Haren, Michael 1985. *Medieval Thought: The Western Intellectual Tradition from Antiquity to the Thirteenth Century*. New York: St. Martin's Press.

Hartner, Willy. 1957. "Quand et comment s'est arrêté l'essor de la culture scientifique dans l'Islam?" In *Classicisme et déclin culturel dans l'histoire de l'Islam*, edited by R. Brunschvig and G. E. von Grunebaum, pp. 319–37. Paris: G.-P. Maisonneuve et Larose.

Hartner, Willy, and Matthias Schramm. 1963. "Al-Biruni and the Theory of the Solar Apogee: An Example of Originality in Arabic Science." In *Scientific Change*, edited by A. C. Crombie, pp. 206–18. New York: Basic Books.

1973. "Copernicus, the Man, the Work, and His Achievement." *Proceedings of the American Philosophical Society* 117, no. 6, 413–22.

Hartwell, Robert M. 1962. "A Revolution in Chinese Iron and Coal Industries in the Northern Sung, 960–1127 A.D." *Journal of Asian Studies* 21, no. 2, 153–62.

1966. "Markets, Technology, and the Structure of Enterprise in the Development of the Eleventh- and Twelfth-Century Chinese Iron and Steel Industry." *The Journal of Economic History* 26:29–58.

1967. "A Cycle of Economic Change in Imperial China: Coal and Iron in Northeast China, 750–1350." *Journal of the Economic and Social History of the Orient* 10:102–59.

1971. "Historical Analogism, Public Policy, and Social Science in Eleventh- and Twelfth-Century China." *American Historical Review* 76:670–727.

1982. "Demographic, Political, and Social Transformation in China, 750–1550." *Harvard Journal of Asiatic Studies* 42:365–445.

Haskins, Charles. 1928. *Studies in the History of Medieval Science.* Cambridge, Mass.: Harvard University Press.

1957 (1927). *The Renaissance of the Twelfth Century.* New York: Meridian.

Heinen, Anton M. 1982. *Islamic Cosmology: A Study of As-Suyuti's "al-Hay'a as-saniya fi l-hay'a as-sunniya."* With critical edition, translation, and commentary. Beirut: Franz Steiner Verlag.

Henderson, John B. 1984. *The Development and Decline of Chinese Cosmology.* New York: Columbia University Press.

Ho Peng-yoke. 1969. "The Astronomical Bureau of Ming China." *Journal of Asian History* 3–4:137–53.

1977. *Modern Scholarship on the History of Chinese Astronomy.* Canberra: Faculty of Asian Studies, The Australian National University.

Ho Ping-ti. 1967. *The Ladder of Success: Aspects of Mobility in China 1368–1911.* Rev. ed. New York: Columbia University Press.

Hodgson, Marshall G. S. 1974. *The Venture of Islam.* 3 vols. Chicago: University of Chicago Press.

Hoodbhoy, Pervez. 1990. *Islam and Science: Religious Orthodoxy and the Battle for Rationality.* With a foreword by Mohammed Abdus Salam. London: Zed Books Ltd.

Hooykaas, R. 1984. "Rheticus's Lost Treatise on the Holy Scriptures and the Motion of the Earth." *Journal for the History of Astronomy* 15:77–80.

Hourani, Albert H., and S. M. Stern, eds. 1970. *The Islamic City.* Philadelphia: University of Pennsylvania Press.

Hourani, George. 1971. *Islamic Rationalism: The Ethics of 'Abd al-Jabbar.* Oxford: The Clarendon Press.

Hourani, George, ed. and trans. 1976. *Averroes: On the Harmony of Religion and Philosophy.* Reprint. London: Luzac.

Hucker, Charles O. 1966. *The Censorial System of Ming China.* Stanford, Calif.: Stanford University Press.

1978a. *China to 1850: A Short History.* Stanford, Calif.: Stanford University Press.

1978b. *The Ming Dynasty: Its Origins and Evolving Institutions.* Ann Arbor: University of Michigan Center for Chinese Studies.

1985. *A Dictionary of Official Titles in Imperial China.* Stanford, Calif.: Stanford University Press.

Huff, Toby E. 1984. *Max Weber and the Methodology of the Social Sciences.* New Brunswick, N.J.: Transaction Books.

1989. "On Weber, Law, and Universalism: Some Preliminary Considerations." *Comparative Civilizations Review* no. 21 (1989):47–79.

Iskandar, A. Z., and R. Arnaldez. 1975. "Ibn Rushd." In *Dictionary of Scientific Biography,* vol. 12, 1–9.

Jardine, Nicholas, 1984. *The Birth of History and Philosophy of Science: Kepler's "A Defense of Tycho Against Ursus" with Essays on Its Provenance and Significance.* New York: Cambridge University Press.

Johnson, David. 1985. "The City-God Cults of T'ang and Sung China." *Harvard Journal of Asiatic Studies* 45:363–457.

Kantorowizc, Hermann. 1939. "The Quaestiones Disputatae of the Glossators." *Tijdschrift voor Rechtgeschedenis/Solidus Revue d'Histoire du droit* 16:1–67.

1966. "Kingship under the Impact of Scientific Jurisprudence." In *Twelfth-Century Europe and the Foundations of Modern Society,* edited by M. Clagett, G. Post, and R. Reynolds, pp. 89–114. Madison, Wis.: University of Wisconsin Press.

Kennedy, E. S. 1956. "A Survey of Islamic Astronomical Tables." *Transactions of the American Philosophical Society* 46, pt. 2 (new series): 123–77.

1966. "Late Medieval Planetary Theory." *Isis* 57:365–78.

1970. "The Arabic Heritage in the Exact Sciences," *Al-Abhath* 23:327–344.

1975. "The Exact Sciences [The Period of the Arab Invasion . . .]." *The Cambridge History of Iran* 4:378–95. Cambridge: Cambridge University Press.

1983. "The History of Trigonometry: An Overview." In *Studies in the Islamic Exact Sciences,* edited by E. S. Kennedy et al., pp. 3–29. Beirut: American University Beirut Press.

1986. "The Exact Sciences in Timurid Iran." *The Cambridge History of Iran* 6:568–80. Cambridge: Cambridge University Press.

Kennedy, E. S., and Victor Roberts. 1959. "The Planetary Theory of Ibn al-Shâtir." *Isis* 50:227–35.

Khadduri, Majid. 1966. *War and Peace in the Law of Islam.* Reprint. Baltimore, Md.: Johns Hopkins University Press.

1972. *Major Middle Eastern Problems in International Law.* Washington, D.C.: American Enterprise Institute for Public Policy.

1979. "The *Maslaha* (Public Interest) And 'Illa (Cause) in Islamic Law." *New York University Journal of International Law and Politics* 12:213–17.

1984. *The Islamic Conception of Justice.* Baltimore, Md.: Johns Hopkins University Press.

trans. 1966. *The Islamic Law of Nations: Shaybani's Siyar.* Baltimore, Md.: Johns Hopkins University Press.

Khadduri, Majid, and Herbert Liebesny, eds. 1955. *Law in the Middle East.* Washington, D.C.: The Middle East Institute.

Kibre, Pearl. 1974. *Scholarly Privileges in the Middle Ages.* Cambridge, Mass.: Medieval Academy of America.

　　1979. "Arts and Medicine in the Universities of the Later Middle Ages." In *The Universities in the Late Middle Ages*, edited by Jacques Paquet and J. Ijsewign, pp. 213–27. Louvain: Louvain University Press.

Kibre, Pearl, and Nancy Siraisi. 1978. "The Institutional Setting: The Universities." In *Science in the Middle Ages*, edited by David C. Lindberg, pp. 120–44. Chicago: University of Chicago Press.

King, David A. 1975a. "On the Astronomical Tables of the Islamic Middle Ages." *Colloquia Copernicana* 3:36–56.

　　1975b. "Ibn al-Shatir." In *Dictionary of Scientific Biography*, vol. 12, 357–64.

　　1983. "The Astronomy of the Mamluks." *Isis* 74, no. 274, 531–55.

　　1985. "The Sacred Direction in Islam: A Study of the Interaction of Religion and Science in the Middle Ages." *Interdisciplinary Science Reviews* 10, no. 4, 315–28.

Kirk, K. E. 1927. *Conscience and Its Problems: An Introduction to Casuistry.* London: Longmans, Green.

Kittelson, James, and Pamela Transue, eds. 1984. *Rebirth, Reform, and Resilience: Universities in Transition, 1350–1770.* Columbus: Ohio State University Press.

Klibansky, Raymond. 1966. "The School of Chartres." In *Twelfth-Century Europe and the Foundations of Modern Society*, edited by M. Clagett, G. Post, and R. Reynolds, pp. 3–15. Madison, Wis.: University of Wisconsin Press.

Kneale, William, and Martha Kneale. 1962. *The Development of Logic.* Oxford: Oxford University Press.

Knowles, David. 1962. *The Evolution of Medieval Thought.* New York: Viking.

Koester, Helmut. 1968. "Nomos and Physeôs: The Concept of Natural Law in Greek Thought." In *Religions of Antiquity: Essays in Memory of E. R. Goodenough*, edited by Jacob Neusner, pp. 521–41. Leiden: E. J. Brill.

Kogan, Barry S. 1985. *Averroes and the Metaphysics of Causation.* Albany, N.Y.: SUNY Press.

Kuhn, Thomas. 1957. *The Copernican Revolution.* New York: Vintage.

　　1970. *The Structure of Scientific Revolutions.* 2d enlarged ed. Chicago: University of Chicago Press.

　　1972. "The Growth of Science: Reflections on Ben-David's 'Scientist's Role.'" *Minerva* 10, no. 1, 166–78.

　　1976. "Mathematical vs. Experimental Traditions in the Development of Physical Science." *Journal of Interdisciplinary History* 7, no. 1, 1–31.

Kuttner, Stephen. 1982. "The Revival of Jurisprudence." In *Renaissance and Revival in the Twelfth Century*, edited by Robert Benson and Giles Constable, pp. 299–323. Cambridge, Mass.: Harvard University Press.

Kwok, D. W. Y. 1965. *Scientism in Chinese Thought, 1900–1950.* New Haven, Conn.: Yale University Press.

Lapidus, Ira. 1967. *Muslim Cities of the Later Middle Ages.* Cambridge, Mass.: Harvard University Press.

1988. *A History of Islamic Societies.* New York: Cambridge University Press.

Lea, Henry. 1968. *A History of Auricular Confession and Indulgences.* 3 vols. Reprint. New York: Greenwood Press.

Leaman, Oliver. 1985. *An Introduction to Islamic Philosophy.* Cambridge: Cambridge University Press.

Lee, Thomas H. C. 1985. *Government Education and Examinations in Sung China.* Hong Kong: Chinese University Press.

1989. "Sung Schools and Education before Chu Hsi." In *Neo-Confucian Education: The Formative Period,* edited by William de Bary and John Chaffee, pp. 105–36. Berkeley and Los Angeles: University of California Press.

Leff, Gordon. 1968. *Paris and Oxford in the Thirteenth and Fourteenth Centuries.* New York: John Wiley.

Legge, James, trans. 1960. *The Chinese Classics.* 5 vols. Reprint. Hong Kong: Hong Kong University Press.

Leiser, Gary. 1983. "Medical Education in Islamic Lands from the Seventh to the Fourteenth Centuries." *The Journal of the History of Medicine and Allied Sciences* 38:48–75.

Lemay, Richard. 1962. *Abu Ma'shar and Latin Aristotelianism in the Twelfth Century.* Oriental Series no. 38. Beirut: American University of Beirut Press.

Levey, Martin. 1973. *Early Islamic Pharmacology.* Leiden: E. J. Brill.

Lewis, Bernard. 1953. "Some Observations on the Significance of Heresy in the History of Islam." *Studia Islamica* 1:43–63.

Liebesny, Herbert. 1955. "The Development of Western Judicial Privileges." In *Law in the Middle East,* edited by M. Khadduri and Herbert Liebesny, pp. 309–33. Washington, D.C.: The Middle East Institute.

1985/6. "English Common Law and Islamic Law in the Middle East and South Asia: Religious Influences and Secularization." *Cleveland State Law Review* 34:19–33.

Liebesny, Herbert, ed. 1975. *The Law of the Near and Middle East.* Albany, New York: SUNY Press.

Lindberg, David C. 1971. "Lines of Influence in Thirteenth Century Optics: Bacon, Witelo, and Pecham." *Speculum* 46:66–83.

1978. "The Transmission of Greek and Arabic Learning to the West." In *Science in the Middle Ages,* edited by David C. Lindberg, pp. 52–90. Chicago: University of Chicago Press.

ed. 1978. *Science in the Middle Ages.* Chicago: University of Chicago Press.

Lippman, Matthew, Sean McConville, and Mordachai Yerushalmi. 1988. *Islamic Criminal Law and Procedure.* New York: Praeger.

Livingston, John W. 1971. "Ibn Qayyim al-Jawziyyah: A Fourteenth-Century Defense Against Astrological Divination and Alchemical Transmutation." *Journal of the American Oriental Society* 91:96–103.

Lloyd, G. E. R. 1975. "Greek Cosmologies." In *Ancient Cosmologies,* edited by C. Blacker and M. Loewe, pp. 198–224. London: Allen and Unwin.

Lo, Winston W. 1987. *An Introduction to the Civil Service in Sung China.* Honolulu: University of Hawaii Press.

Lopez, Robert. 1977. *The Commercial Revolution of the Middle Ages, 950–1350.* Reprint. New York: Cambridge University Press.

Luscombe, D. E. 1982. "Natural Morality and Natural Law." In *Cambridge History of Later Medieval Philosophy*, pp. 705–19. New York: Cambridge University Press.

1976. *The School of Peter Abelard.* Oxford: Oxford University Press.

McCarthy, Richard J., ed. and trans. 1953. *The Theology of Ash'ari.* Beirut: Imprimerie Catholique.

McEvory, James. 1982. *The Philosophy of Robert Grosseteste.* Oxford: The Clarendon Press.

McIlwain, Charles. 1947. *Constitutionalism, Ancient and Modern.* Rev. ed. Ithaca, N.Y.: Cornell University Press.

Mackensen, Ruth S. 1932. "Four Great Libraries of Medieval Baghdad." *Library Quarterly* 2:279–99.

1934/35. "Background of the History of Moslem Libraries." *American Journal of Semitic Languages and Literature* 51:114–25; 52:22–33; and 104–10.

1935/36. "Arabic Books and Libraries in the Umaiyad Period." *American Journal of Semitic Languages and Literature* 52:245–53; 54:41–61.

McLaughlin, Mary M. 1977. *Intellectual Freedom and Its Limits in the Twelfth and Thirteenth Centuries.* Reprint. New York: Arno Press.

McNeill, John. 1964. *A History of the Cure of Souls.* New York: Harper.

McNeill, William H., and Marilyn W. Waldman, eds. 1983. *The Islamic World.* Chicago: University of Chicago Press.

Mahdi, Muhsin. 1974. "Islamic Theology and Philosophy." In *Encyclopedia Britannica*, vol. 9, 1012–25.

1970. "Language and Logic in Classical Islam." In *Logic in Classical Islamic Cultures*, edited by G. E. Von Grunebaum, pp. 51–83. Wiesbaden: Otto Harrassowitz.

Maimonides, Moses. 1963. *The Guide of the Perplexed.* 2 vols. Edited and translated by Shlomo Pines. Chicago: University of Chicago Press.

Makdisi, George. 1961. "Muslim Institutions of Learning in Eleventh Century Baghdad." *Bulletin of the School of Oriental and African Studies* 24:1–56.

1970. "Madrasah and University in the Middle Ages." *Studia Islamica* 32:255–64.

1974. "The Scholastic Method in Medieval Education: An Inquiry into Its Origins in Law and Theology." *Speculum* 49:640–61.

1980. "On the Origin and Development of the College in Islam and the West." In *Islam and the Medieval West*, edited by Khalil I. Seeman, pp. 26–49. Albany, N.Y.: SUNY Press.

1981. *The Rise of Colleges: Institutions of Learning in Islam and the West.* Edinburgh: Edinburgh University Press.

1984. "The Guilds of Law in Medieval Legal History: An Inquiry into the Origins of the Inns of Court." *Zeitschrift für Geschichte der Arabisch-Islamischen Wissenschaften* 1:233–52.

ed. and trans. 1962. *Ibn Qadama's Censure of Speculative Theology*. London: E. J. W. Memorial Series, Luzac.

Mamura, Michael. 1965. "Ghazali and Demonstrative Science." *Journal of the History of Philosophy* 3:183–204.

1968. "Causation in Islamic Thought." *Dictionary of the History of Ideas* 1:286–9. New York: Scribners.

1975. "Ghazali's Attitude Toward the Secular Sciences." In *Essays on Islamic Philosophy and Science*, edited by G. Hourani, pp. 100–11. Albany, N.Y.: SUNY Press.

Margoliouth, D. S. 1905. "The Discussion Between Abû Bishr Mattâ and Abâ Sa'id al-Sîrafî on the Merits of Logic and Grammar." *Journal of the Royal Asiatic Society* 79–129.

Mendelsohn, Everett. 1964. "The Emergence of Science as a Profession in Nineteenth Century Europe." In *The Management of Science*, edited by Karl Hill, pp. 3–48. Boston: Beacon Press.

Merton, Robert K. 1968. *Social Theory and Social Structure*. Enlarged ed. New York: Free Press.

1970. *Science, Technology, and Society in Seventeenth-Century England*. Reprint. New York: Harper and Row.

1973. "The Normative Structure of Science." In *The Sociology of Science: Theoretical and Empirical Investigations*, edited by Norman Storer. Chicago: University of Chicago Press.

1989. "Le molteplici origini e il carattere epiceno del termine inglese *Scientist*." In *Scientisa: L'immagine e il mondo. 80th Anniversario della rivista*, pp. 279–93. Commune di Milano.

1990. "STS: Foreshadowings of an Evolving Research Program in the Sociology of Science." In *Puritanism and the Rise of Modern Science: The Merton Thesis*, edited by I. B. Cohen (with the assistance of K. E. Duffin and Stuart Strickland), pp. 334–371. New Brunswick, N.J.: Rutgers University Press.

Meskill, John. 1968. "Academies and Politics in the Ming Dynasty." In *Chinese Government in Ming Times*, edited by Charles O. Hucker, pp. 149–74. New York: Columbia University Press.

Meyerhof, Max. 1931. "Science and Medicine." In *The Legacy of Islam*. 1st ed., edited by T. Arnold and A. Guillaume, pp. 311–56. London: Oxford University Press.

1933. "Thirty-Three Clinical Observations by Rhazes." *Isis* 23:321–55.

1935. "Ibn An-Nafis (Thirteenth Century) and His Theory of the Lesser Circulation." *Isis* 22:100–20.

1944. "La surveillance des professions médicales et para-médicales chez les Arabs," *Bulletin de l'Institut d'Egypt* 26:119–34.

Michaud-Quantin, Pierre, 1970. *Universitas: Expressions du mouvement communautaire dans le moyen-âge Latin*. Paris: J. Vrin.

Moody, Ernest. 1957. "Galileo and Avempace: Dynamics of the Leaning Tower Experiments." In *Roots of Scientific Thought: A Cultural Perspective*, edited by Philip P. Wiener and A. Noland, pp. 176–206. New York: Basic Books.

1966. "Galileo and His Precursors." In *Galileo Reappraised,* edited by Carlo Golino, pp. 23–43. Berkeley and Los Angeles: University of California Press.

Mottahedeh, Roy. 1970. *Loyalty and Leadership in an Early Islamic Society.* Princeton, N.J.: Princeton University Press.

1985. *The Mantle of the Prophet.* New York: Simon and Schuster.

Murdoch, John. 1971. "Euclid: Transmission of the Elements." In *Dictionary of Scientific Biography,* vol. 4, 443–65.

1975. "From Social to Intellectual Factors: An Aspect of the Unitary Character of Late Medieval Learning." In *The Cultural Context of Medieval Learning,* edited by John Murdoch and Edith Sylla, pp. 271–338. Boston: Reidel.

Murdoch, John, and Edith Sylla. 1978. "The Science of Motion." In *Science in the Middle Ages,* pp. 206–64. Chicago: University of Chicago Press.

Murray, Alexander. 1978. *Reason and Society in the Middle Ages.* Oxford: At the Clarendon Press.

Nakamura, Hajime. 1964. *Ways of Thinking of Eastern Peoples.* Rev. trans. by Philip P. Wiener. Honolulu: East West Center.

Nakayama, Shigeru, and Nathan Sivin, eds. 1973. *Chinese Science: Explorations of an Ancient Tradition.* Cambridge, Mass.: MIT Press.

Nasr, S. H. 1968. *Science and Civilization in Islam.* New York: New American Library.

1971. "Qutb al-Dîn al-Shîrâzî." In *Dictionary of Scientific Biography,* vol. 11, 247–53.

Needham, Joseph. 1954–. *Science and Civilisation in China.* 7 vols., in progress. New York: Cambridge University Press.

1969. *The Grand Titration.* London: Allen and Unwin.

1970. *Clerks and Craftsmen in China and the West.* Cambridge: Cambridge University Press.

1976. "The Evolution of Oecumenical Science: The Roles of Europe and China." *Interdisciplinary Science Reviews* 1, no. 3, 202–14.

Nelson, Benjamin. 1968. "Casuistry." In *Encyclopedia Britannica,* vol. 5, 51–2.

1969. *The Idea of Usury: From Tribal Brotherhood to Universal Otherhood.* 2d, enlarged ed. Chicago: University of Chicago Press.

1981. *On the Roads to Modernity: Conscience, Science, and Civilizations. Selected Papers by Benjamin Nelson,* edited by Toby E. Huff. Totowa, N.J.: Rowman and Littlefield.

Oman, G. 1989. "Matba'a [Printing]." In *Encyclopedia of Islam,* 2d ed., vol. 6, 794–9. Leiden: E. J. Brill.

Onar, S. 1955. "The Majalla." In *Law in the Middle East,* edited by Majid Khadduri and Herbert Liebesny, pp. 292–308. Washington, D.C.: The Middle East Institute.

Parsons, Talcott. 1951. *Toward a General Theory of Action.* New York: Harper.

Pedersen, Johannes. 1984. *The Arabic Book*. Translated by Geoffrey French and edited with an introduction by Robert Hillenbrand. Princeton, N.J.: Princeton University Press.

Pedersen, Johannes, and G. Makdisi. 1985. "Madrasa." In *Encyclopedia of Islam*, 2d ed., vol. 5, 1123–34. Leiden: E. J. Brill.

Pedersen, Olaf. 1978. "Astronomy." In *Science in the Middle Ages*, edited by David C. Lindberg, pp. 303–37. Chicago: University of Chicago Press.

1984. "Galileo and the Council of Trent: The Galileo Affair Revisited." *Journal for the History of Astronomy* 14:1–29.

Peters, F. E. 1968. *Aristotle and the Arabs*. New York: New York University Press.

1973. *Allah's Commonwealth*. New York: Simon and Schuster.

Petry, Carl F. 1981. *The Civilian Elite of Cairo in the Later Middle Ages*. Princeton, N.J.: Princeton University Press.

Pierce, C. A. 1955. *Conscience in the New Testament. A Study of Syneidesis in the New Testament*. London: SCM Press.

Pines, Shlomo. 1963a. "Introduction" to Maimonides, *The Guide of the Perplexed*, pp. xi–lvi. Chicago: University of Chicago Press.

1963b. "What Was Original in Arabic Science?" In *Scientific Change*, edited by A. C. Crombie, pp. 181–205. New York: Basic Books.

1970. "Philosophy." In *The Cambridge History of Islam*, vol. 2, edited by P. M. Holt, pp. 780–823. New York: Cambridge University Press.

Pipes, Daniel. 1990. *The Rushdie Affair: The Novel, the Ayatollah, and the West*. New York: Birch Lane Press.

Plessner, Martin. 1974. "Science." In *The Legacy of Islam*, 2d ed., edited by Joseph Schacht and C. E. Bosworth, pp. 425–60. Oxford: The Clarendon Press.

Porkert, Manfred. 1974. *The Theoretical Foundations of Chinese Medicine: Systems of Correspondence*. Cambridge, Mass.: MIT Press.

Post, Gaines, 1964. *Studies in Medieval Legal Thought: Public Law and the State, 1100–1322*. Princeton, N.J.: Princeton University Press.

1973. "Ancient Roman Idea of Laws." *Dictionary of the History of Ideas* 2:685–90. New York: Scribners.

Price, Derek J. de Solla. 1959. "Contra-Copernicus: A Critical Re-estimation of the Mathematical Planetary Theory of Ptolemy, Copernicus, and Kepler." In *Critical Problems in the History of Science*, edited by Marshall Clagett, pp. 197–218. Madison, Wis.: University of Wisconsin Press.

Qian, Wen-yan. 1985. *The Great Inertia: Scientific Stagnation in Traditional China*. London: Croom Helm.

Rahman, Fazlur. 1960. "'Aql." In *Encyclopedia of Islam*, 2d ed., vol. 1, 341–2. Leiden: E. J. Brill.

1968. *Islam*. New York: Doubleday.

1979. *Prophecy in Islam*. Reprint. Chicago: Midway Reprints, University of Chicago Press.

Rashdall, Hastings. 1936. *The Universities of Europe in the Middle Ages*. 3 vols.

New ed., edited by A. B. Emden and F. M. Powicke. Oxford: Oxford University Press.

Rashed, Roshdi. 1973. "Kamâl al-Dîn al-Fârisî." In *Dictionary of Scientific Biography*, vol. 7, 212–19.

Reif, Patricia. 1969. "The Textbook Tradition in Natural Philosophy, 1600–1650." *Journal of the History of Ideas* 30:17–32.

Renan, Ernest. 1919. "Islam et La Science." In *Discours et Conferences*, 6th ed., pp. 375–402. Paris: Calmann-Levy.

Restivo, Sal P. 1979. "Joseph Needham and the Comparative Sociology of Chinese and Modern Science." In *Research in Sociology of Knowledge, Science, and Art*, vol. 2, edited by Robert A. Jones, pp. 25–51. Greenwich, Conn.: JAI Press.

Ropp, Paul S., ed. 1990. *Heritage of China: Contemporary Perspectives on Chinese Civilization*. Berkeley and Los Angeles: University of California Press.

Rosen, Edward, ed. and trans. 1971. *Three Copernican Treatises*. 3d ed. New York: Octagon Books.

1985. *Three Imperial Mathematicians: Kepler Between Tycho Brahe and Ursus*. New York: Abaris Books.

Rosen, Lawrence. 1980–1. "Equity and Discretion in a Modern Islamic Legal System." *Law and Society Review* 15:215–45.

Rosenthal, E. J. 1970. *Knowledge Triumphant*. Leiden: E. J. Brill.

Rosenthal, Franz. 1969. "The Defense of Medicine in the Medieval Islamic World." *Bulletin of the History of Medicine* 43:519–32.

1975. *The Classical Heritage in Islam*. Berkeley and Los Angeles: University of California Press.

1978. "The Physician in Medieval Muslim Society." *Bulletin of the History of Medicine* 52, no. 4, 475–91.

Ross, Sydney, 1962. "Scientist: The Story of a Word." *Annals of Science* 18, no. 2:65–85.

Sabra, A. I. 1967. *Theories of Light from Descartes to Newton*. London: Oldbourne.

1971a. "'Ilm al-Hisab." In *Encyclopedia of Islam*, 2d ed., vol. 3, 1138–41. Leiden: E. J. Brill.

1971b. "The Astronomical Origins of Ibn al-Haytham's Concept of Experiment." *Actes du XIIe congrès internationale d'histoire des sciences* (Paris), Tome IIIa:133–36.

1972. "Ibn al-Haytham." In *Dictionary of Scientific Biography*, vol. 5, 189–210.

1976. "The Scientific Enterprise." In *Islam and the Arab World*, edited by Bernard Lewis, pp. 181–92. New York: Knopf.

1984. "The Andalusian Revolt Against Ptolemaic Astronomy." In *Transformation and Tradition in the Sciences: Essays in Honor of I. Bernard Cohen*, edited by Everett Mendelsohn, pp. 133–53. New York: Cambridge University Press.

ed. and trans. 1989. *The Optics of Ibn al-Haytham: Books I–III on Direct Vision*. 2 vols. London: The Warburg Institute, University of London.

Saliba, George. 1982. "The Development of Astronomy in Medieval Islamic Society." *Arab Studies Quarterly* 4, no. 3, 211–25.

1984. "Arabic Astronomy and Copernicus." *Zeitschrift für Geschichte der Arabish-Islamischen Wissenschaften* Band 1, 73–87.

1987. "The Role of Marâgha in the Development of Islamic Astronomy: A Scientific Revolution Before the Renaissance." *Revue de Synthese* 4, no. 3–4, 361–73.

1987a. "Theory and Observation in Islamic Astronomy: The Work of Ibn al-Shatir." *Journal for the History of Astronomy* 18:35–43.

Sarton, George. 1927–48. *Introduction to the History of Science.* 3 vols. in 5 parts. Baltimore, Md.: Williams and Wilkins.

Saunders, J. J. 1963. "The Problem of Islamic Decadence." *Journal of World History* 7:701–20.

Savage-Smith, Emilie. 1988. "Gleanings from an Arabist's Workshop: Current Trends in the Study of Medieval Islamic Science and Medicine." *Isis* 79:246–72.

Sayili, Aydin. 1960. *The Observatory in Islam.* Ankara: Turkish Historical Society Series 7, no. 38.

1980. "The Emergence of the Proto-Type of the Modern Hospital in Medieval Islam." *Studies in the History of Medicine* 4:112–18.

Schacht, Joseph. 1950. *Origins of Muhammadan Jurisprudence.* Oxford: Oxford University Press.

1964. *Introduction to Islamic Law.* Oxford: Oxford University Press.

1974. "Islamic Religious Law." In *The Legacy of Islam*, 2d ed., edited by Joseph Schacht and C. E. Bosworth, pp. 392–403. New York: Oxford University Press.

Schacht, Joseph, and Max Meyerhof. 1937. *The Medico-Philosophical Controversy Between Ibn Butlan and Ibn Ridwan.* Cairo: The Egyptian University Faculty of Arts, Publication no. 13.

Schmitt, Charles. 1972. "The Faculty of the Arts at Pisa at the Time of Galileo." *Physis* 15, no. 3, 243–72.

1973. "Toward a Reassessment of Renaissance Aristotelianism." *History of Science* 11:159–93.

1975a. "Philosophy and Science in Sixteenth-Century Universities: Some Preliminary Comments." In *The Cultural Context of Medieval Learning*, edited by John Murdoch and Edith Sylla, pp. 485–537. Boston: Reidel.

1975b. "Science and the Italian Universities in the Sixteenth and Seventeenth Centuries." In *The Emergence of Modern Science*, edited by M. Crosland, pp. 35–56. London: Macmillan.

1983. *Aristotle and the Renaissance.* Cambridge, Mass.: Harvard University Press.

1988. "The Rise of the Philosophical Textbook." In *The Cambridge History of Renaissance Philosophy*, edited by Charles B. Schmitt and Quentin Skinner, pp. 792–804. New York: Cambridge University Press.

Schwartz, Benjamin. 1968. "On Attitudes Toward the Law in China." In *The Criminal Process in the People's Republic of China*, edited by Jerome A. Cohen, pp. 62–70. Cambridge, Mass.: Harvard University Press.

1985. *The World of Thought in Ancient China.* Cambridge, Mass.: Harvard University Press.

Shapiro, Martin. 1980. "Islam and Appeal." *California Law Review* 68:350–81.

Shaw, Stanford. 1976. *History of the Ottoman Empire and Modern Turkey.* vol. 1. New York: Cambridge University Press.

Siraisi, Nancy. 1973. *Arts and Sciences at Padua.* Toronto: Pontifical Institute.

    1987. *Avicenna in Renaissance Italy: The Canon and Medical Training in Italian Universities after 1500.* Princeton, N.J.: Princeton University Press.

    1990. *Medieval and Early Renaissance Medicine. An Introduction to Knowledge and Practice.* Chicago: University of Chicago Press.

Sivin, Nathan. 1973. "Copernicus in China." *Studia Copernicana* 6:63–122.

    1975. "Shen Kua." In *Dictionary of Scientific Biography,* vol. 12, 369–93.

    1976. "Wang Hsi-Shan." In *Dictionary of Scientific Biography,* vol. 14, 159–68.

    1984. "Why the Scientific Revolution Did Not Take Place in China – or Didn't It?" In *Transformation and Tradition in the Sciences,* edited by Everett Mendelsohn, pp. 531–54. New York: Cambridge University Press.

    1985. "Max Weber, Joseph Needham, Benjamin Nelson: The Question of Chinese Science." In *Civilizations East and West: A Memorial Volume for Benjamin Nelson,* edited by E. V. Walters et al., pp. 37–50. Atlantic Highlands, N.J.: Humanities Press.

    1988. "Science and Medicine in Imperial China – The State of the Field." *The Journal of Asia Studies* 47, no. 1, 41–90.

    1990. "Science and Medicine in Chinese History." In *Heritage of China,* edited by Paul S. Ropp, pp. 164–96. Berkeley and Los Angeles: University of California Press.

Skinner, G. William, ed. 1977. *The City in Late Imperial China.* Stanford, Calif.: Stanford University Press.

Sorokin, Pitirim, and R. K. Merton. 1935. "The Course of Arabian Intellectual Development, 700–1300 A.D.: A Study in Method." *Isis* 22 (Feb.):516–24.

Spence, Jonathan. 1978. *The Death of Woman Wang.* New York: Viking Press.

    1990. *In Search of Modern China.* New York: W. W. Norton.

van der Sprenkel, Sybille. 1966. *Legal Institutions in Manchu China: A Sociological Analysis.* London: Athlone Press.

Staunton, George T., ed. and trans. 1810. *Ta Tsing Lü Li: Being the Fundamental Law of the Penal Code of China.* London: Cadell and Davies.

Stern, S. M. 1970. "The Constitution of the Islamic City." In *The Islamic City,* edited by A. H. Hourani and S. M. Stern, pp. 25–50. Philadelphia: University of Pennsylvania Press.

Stiefel, Tina. 1976. "Science, Reason, and Faith in the Twelfth Century: The Cosmologists' Attack on Tradition." *Journal of European Studies* 6:1–16.

    1977. "The Heresy of Science: A Twelfth-Century Conceptual Revolution." *Isis* 68, no. 243, 347–62.

    1985a. *The Intellectual Revolution in Twelfth-Century Europe.* New York: St. Martin's Press.

    1985b. "'Impious Men': Twelfth-Century Attempts to Apply Dialectic to the World of Nature." In *Science and Technology in Medieval Society,* edited by

Pamel O. Long, pp. 187–203. New York: New York Academy of Sciences, vol. 441.

Strauss, Leo. 1973. *Persecution and the Art of Writing*. Reprint. Westport, Conn.: Greenwood Press.

Strayer, Joseph. 1970. *On the Medieval Origins of the Modern State*. Princeton, N.J.: Princeton University Press.

Swerdlow, Noel. 1973. "The Derivation and First Draft of Copernicus's Planetary Theory." *Proceedings of the American Philosophical Society* 117:423–512.

Swerdlow, Noel, and Otto Neugebauer, 1984. *Mathematical Astronomy in Copernicus's "De revolutionibus."* New York: Springer Verlag.

Talbot, Charles. 1978. "Medicine." In *Science in the Middle Ages,* edited by David C. Lindberg, pp. 391–428. Chicago: University of Chicago Press.

Thoren, Victor. 1990. *The Lord of Uraniborg: A Biography of Tycho Brahe.* New York: Cambridge University Press.

Tierney, Brian. 1982. *Religion, Law, and the Growth of Constitutional Thought, 1150–1650.* New York: Cambridge University Press.

Tillich, Paul. 1957. "The Transmoral Conscience." In *The Protestant Era,* pp. 136–49. Chicago: Phoenix Books.

Tu Wei-ming. 1990. "The Confucian Tradition in Chinese History." In *Heritage of China,* edited by Paul S. Ropp, pp. 112–37. Berkeley and Los Angeles: University of California Press.

Tyan, Emile. 1955. "Judicial Organization." In *Law in the Middle East,* edited by Majid Khadduri and Herbert Liebesny, pp. 236–78. Washington, D.C.: The Middle East Institute.

Ullmann, Manfred. 1978. *Islamic Medicine.* Edinburgh: Edinburgh University Press.

Unschuld, Paul. 1985. *Medicine in China: A History of Ideas.* Berkeley and Los Angeles: University of California Press.

    1986. *Medicine in China: A History of Pharmaceutics.* Berkeley and Los Angeles: University of California Press.

Vajda, George. 1971. "Idjaza." In *Encyclopedia of Islam,* 2d ed., vol. 3, 1021. Leiden: E. J. Brill.

Wallace, William O. 1981. *Prelude to Galileo: Essays on Medieval and Sixteenth-Century Sources of Galileo's Thought.* Dordrecht: Reidel.

Walzer, Richard. 1962. *Greek into Arabic.* Columbia, S.C.: University of South Carolina Press.

Watt, John R. 1972. *The District Magistrate in Late Imperial China.* New York: Columbia University Press.

Watt, William M. 1965. *A History of Islamic Spain.* Edinburgh: Edinburgh University Press.

    1973. *The Formative Period of Islamic Thought.* Edinburgh: Edinburgh University Press.

    1985. *Islamic Philosophy and Theology: An Extended Survey.* Edinburgh: Edinburgh University Press.

ed. and trans. 1953. *The Faith and Practice of al-Ghazali*. London: Allen and Unwin.

Weber, Max. 1951. *The Religion of China*. Translated by Hans Gerth. New York: Free Press.

1954. *Max Weber on Law in Economy and Society*. Edited and annotated by Max Rheinstein; translated by M. Rheinstein and Edward Shils. Cambridge, Mass.: Harvard University Press.

1958a. *The Protestant Ethic and the Spirit of Capitalism*. Translated by Talcott Parsons. New York: Scribners.

1958b. *The City*. Edited and translated by Don Martindale and Gertrud Neuwirth. New York: Free Press.

1969. *The Rational and Social Foundations of Music*. Reprint. Translated and edited by Don Martindale, Johannes Riedel, and Gertrud Neuwirth. Carbondale, Illinois: South Illinois University Press.

1978. *Economy and Society*. 2 vols. Edited by Guenther Roth and Claus Wittich. Berkeley and Los Angeles: University of California Press.

Weisheipl, James. 1964. "The Curriculum of the Faculty of Arts at Oxford in the Early Fourteenth Century." *Medieval Studies* 26:143–85.

Westman, Robert, S. 1975a. "Three Responses to the Copernican Theory: Johannes Praetorius, Tycho Brahe, and Michael Maestlin." In *The Copernican Achievement*, edited by Robert S. Westman, pp. 285–345.

1975b. "The Melanchthon Circle, Rheticus, and the Wittenberg Interpretation of the Copernican Theory." *Isis* 66 (June):165–93.

1980. "The Astronomer's Role in the Sixteenth Century." *History of Science*, 18:105–47.

1986. "The Copernicans and the Churches." In *God and Nature: Historical Essays on the Encounter Between Christianity and Science*, edited by David C. Lindberg and Ronald L. Numbers, pp. 76–113. Berkeley and Los Angeles: University of California Press.

1990. "Proof, Poetics, and Patronage: Copernicus's Preface to *De revolutionibus*." In *Reappraisals of the Scientific Revolution*, edited by David C. Lindberg and Robert S. Westman, pp. 167–205. New York: Cambridge University Press.

Westman, Robert, S. ed. 1975. *The Copernican Achievement*. Berkeley and Los Angeles: University of California Press.

White, Lynn, Jr. 1962. *Medieval Technology and Social Change*. Oxford: Oxford University Press.

1968. *Machina Ex Deo*. Cambridge, Mass.: MIT Press.

Wolfson, Harry. 1964. "The Controversy over Causality in the Kalam." In *Melanges Alexandre Koyré*, vol. 2, pp. 602–18. Paris: Hermann.

1976. *The Philosophy of Kalam*. Cambridge, Mass.: Harvard University Press.

Yan, Lî, and Dù Shíràn. 1987. *Chinese Mathematics: A Concise History*. Translated by John N. Crossley and Anthony W.-C. Lun. Oxford: The Clarendon Press.

Ziadeh, F. 1968. *Lawyers: The Rule of Law and Liberalism in Egypt*. Stanford, Calif.: The Hoover Institution.

Ziadeh, Nicola. 1964. *Damascus under the Mamluks.* Norman, Okla.: University of Oklahoma Press.

    1970. *Urban Life in Syria under the Early Mamluks.* Westport, Conn.: Greenwood Press.

Zilsel, Edgar. 1942a. "The Sociological Roots of Science." *American Journal of Sociology* 47:544–62.

    1942b. "The Genesis of the Concept of Physical Law." *The Philosophical Review* 51:245–79.

Zuckerman, Harriet. 1988. "The Sociology of Science." In *The Handbook of Sociology*, edited by Neil J. Smelser, pp. 511–74. Beverly Hills, Calif.: Sage Publications.

Index

For indexing purposes only, the term *Islam* is used to denote Arabic-Islamic civilization.

litigation trickster, 269, 272
Little, Lester K., 71n81
Liu, Adam, 281n63
Livingston, John, 53n30
Lloyd, G. E., 241
Lo, Winston W., 254n66, 255n69, 260n87
Loewe, Michael, 241n13
logic, 70, 106, 131, 155, 157, 158, 160, 171, 173, 182, 188, 229, 288, 302, 336, 354, 358, 360, 361; as a balance, 84, 86, 115; in China, 50, 338; as first of the arts, 229, 336; Islamic opposition to, 68, 77, 153, 159; as universal method, 195; *see also* dialectic, disputation
logic of discovery, 238
logics of decision, 203, 249, 302, 361; breakthrough in, 129, 133, 249
Long, Pamela, 231n93, 337n51
Lopez, Robert, 141n90, 315
Lui, Adam Y., 285n
lunar models, 54, 208
Luscombe, D. E., 108n63, 118
Luther, Martin, 110 and n76, 356, 357

MacDonald, D. B., 143n94
machina mundi, 337; *see also* world machine
Mackensen, Ruth S., 74n96, 224n70, 320n131
madhhab, see Islamic law, schools of
madrasas, see colleges, Islamic
Maestlin, Michael, 348, 357
maghrib (North Africa), 205
magic squares, 298
magnetism, 239n6, 358
Mahdi, Muhsin, 67n64, 143n94, 293n24
Mahoney, Michael, 49n8, 50n10, 287n3
Maimonides, 12n11, 57, 70n79, 206, 207, 212, 222
Maine, Henry Sumner, 44
Maitland, F. W., 119
majalla, 83
majlis (discussion room), 152, 172, 318
Makdisi, George, 68nn69,70, 71, 73n89, 75n98, 76n103, 77n104, 78n108, 79n111, 80n114, 81, 82n121, 98n30, 111n79, 124n18, 116n109, 151n4, 152n9, 153, 154n19–21, 155n23, 156, 157n29, 159n37, 160, 161n49, 163nn51–53, 164n55, 165n61, 166, 167n67, 168, 181n134, 206n11, 214n319, 215n40, 221, 224n71
Makkari, 234n103
Maliki school of law, 82, 214
Malinowski, Bronislaw, 65n
Malaysians, 263

Malpighi, Marcello, 177
malpractice legislation, 275
malpractice, medical, 77, 154, 173–4
Malta, 225
Mamluks, 87, 182, 185, 208
Mamura, Michael, 115n103
man: anthropologies of, 93, 141; as co-creator, 107; images of, 4, 38, 39, 90, 91; as possessor of reason, 1, 100, 104, 105, 107, 117, 129, 142, 145, 191, 194, 362; *see also* rationality; reason; reasonableness
man of knowledge, Chinese image of, 361
mandate of heaven, 240n11, 261, 272
Mangu, Mongol ruler, 241
Mansuri hospital, 172
manuals, for medical examinations, 175–6, 275n139
Marâgha observatory, 51, 54, 57, 87, 88, 179, 182, 204, 207, 208, 237, 241, 289, 294, 322, 340
marginality problem, in Arabic science, 84–7
Margoliouth, D. S., 293
market inspector (*muhtasib*), 140, 157, 175, 176, 186, 257; compared with Chinese censorate, 256; and medicine, 199, 222
market police, *see* market inspector
Mars, 323, 324, 325
Marsilius of Padua, 120
Martin, B. R., 362n5
Martindale, Don, 79n111
Marv, library in, 73
Mashler, C., 347n81
masjid, 81, 206n11
masjid-khan, 74, 151
masses, 222; distrust of, 224
master of arts, 227
Masterman, Margaret, 26n66
materia medica, 247
mathematical vs. empirical sciences, 197; fusion of, 32
mathematics, 84, 153, 203, 250, 276, 279, 350; and demonstration, 210; fractions, 287; and notation, 282; simultaneous equations, 287; university chair in, 334; written for mathematicians, 350; *see also* Arabic science; Chinese science; geometry; trigonometry
Mattingly, Garrett, 125n24
Mawardi, al-, 216
mazalim courts, *see* courts of complaints
McCarthy, Richard, 68n68
McConville, Sean, 96n24, 114n94, 182n137, 215n43
McIlwain, Charles, 147n106

law: demonstrates structure of
    rationality in a culture
science // modernity in evolution

Why didn't Arabic science "grow"?

book is repetitive, wordy, & sloppy

———————

why - Arabs
Arab law / religion too constricting
∅ corporate structure
no great concept of law, of universal rights
no real disputation   wrong
me - christian world was bigger —
    could escape —   more diverse
West & Islam share atomism (?!! ~
        250)